THE ETHICAL ENGINEER

THE ETHICAL ENGINEER

CONTEMPORARY CONCEPTS AND CASES

ROBERT MCGINN

PRINCETON UNIVERSITY PRESS

PRINCETON AND OXFORD

Copyright © 2018 by Princeton University Press
Published by Princeton University Press,
41 William Street, Princeton, New Jersey 08540
In the United Kingdom: Princeton University Press,
6 Oxford Street, Woodstock, Oxfordshire OX20 1TR

press.princeton.edu

ISBN 978-0-691-17769-4
ISBN (pbk.) 978-0-691-17770-0

Library of Congress Control Number: 2017958781

British Library Cataloging-in-Publication Data is available

This book has been composed in Sabon LT Std

Printed on acid-free paper. ∞

Printed in the United States of America

1 3 5 7 9 10 8 6 4 2

For Carol and Dick, Jan and Howard,
Kris and Steve, and Wanda and Joe,
in gratitude for abiding friendship

And for Birgit

CONTENTS

PREFACE

It is time for study of ethical issues in engineering to become an integral part of engineering education.

Reflecting that conviction, the main goal of this book is to help engineering students and practicing engineers enhance their understanding of ethical issues that arise in engineering practice. It is hoped that through reading it, they will become better able to recognize a wide range of ethical issues in engineering work and to think about them in a critical, comprehensive, and context-sensitive way. A secondary goal of this book is to raise awareness of technical, social, and personal characteristics of engineering situations that can induce or press engineers to engage in misconduct.

Some books on engineering ethics devote considerable space to classical ethical theories. The real-life cases they contain are often described and discussed in cursory fashion. Other engineering ethics books are anthologies of actual cases, each analyzed by a different author with her or his own approach. Still others are multiauthor hybrids, combining case studies of real-life episodes with philosophical essays on concepts or principles relevant to engineering ethics. Few show a practical, foundational approach to exploring ethical issues in engineering being brought to bear on a wide range of real-life cases. The bulk of this book is devoted to case studies of ethical issues in engineering work. Fundamental ethical responsibilities of engineers are applied to real-life engineering situations to shed light on engineers' context-specific ethical responsibilities.

Engineering students and practicing engineers who read this book with care will acquire intellectual resources useful for coming to grips with whatever ethical issues confront them in their future professional careers.

ACKNOWLEDGMENTS

Many individuals have supported my work on engineering ethics over the past two decades. School of Engineering colleagues who helped me in various ways include Brad Osgood, Steve Barley, Sheri Sheppard, Eric

Roberts, Walter Vincenti, James Adams, Peter Glynn, Nick Bambos, and Tom Kenny. Engineering students in my "Ethical Issues in Engineering" course offered valuable feedback on a number of ideas and issues explored in this book.

In 2003, James Plummer, then dean of the Stanford School of Engineering, put me in touch with staff engineers and scientists at the Stanford Nanofabrication Facility (SNF). The ensuing collaborations enriched my approach to engineering ethics. Sandip Tiwari of Cornell University, former director of the National Nanotechnology Infrastructure Network (NNIN); Yoshio Nishi, former SNF faculty director; and Roger Howe, SNF faculty director and former director of NNIN, supported my research on nanotechnology-related ethical issues. SNF associate director Mary Tang improved and facilitated my surveys of nanotechnology researchers' views about ethical issues related to their work.

Stephen Unger of Columbia University kindly read an early draft version of the text and offered valuable criticism and feedback. He also generously shared an explanation, a suggestion, and an example that have been incorporated into the text. My research collaborator and virtual colleague, Rafael Pardo Avellaneda, of the Spanish National Research Council (Consejo Superior de Investigaciones Científicas, CSIC) in Madrid, made valuable suggestions on survey design and data analysis. Howard and Janice Oringer, Mathieu Desbrun, Richard Herman, Alan Petersen, Roya Maboudian, and various engineering lab directors afforded me opportunities to present early versions of parts of this work at Cal Tech, the University of Illinois at Urbana-Champaign, Spectra-Physics, the Exploratorium in San Francisco, and the University of California, Berkeley. Finally, I would like to thank Eric Henney of Princeton University Press for his professionalism and wise counsel.

Palo Alto, California
October 2017

THE ETHICAL ENGINEER

CHAPTER 1

The Ethics Gap in Contemporary Engineering

TWO VIGNETTES

During the night of December 2–3, 1984, one of the worst industrial disasters in history occurred at Union Carbide's plant in Bhopal, Madhya Pradesh, India. Methyl isocyanate (MIC) liquid, an intermediate used in making Sevin, Union Carbide's name for the pesticide carbaryl, came into contact with water, boiled violently, and turned into MIC gas. Unchecked by various safety systems, tons of highly toxic MIC gas escaped from storage tank E610.[1] A cloud of MIC gas descended upon crowded shantytowns just outside the plant, as well as on Bhopal city. Estimates of the death toll from exposure to the gas, immediately or in the first few days afterward, range from 2,000 to 10,000.[2]

In February 1992, I attended a conference on professional ethics at the University of Florida, Gainesville. On the shuttle bus to the conference hotel, the only other passenger turned out to be a chemical engineer. I asked him whether there was any consensus in the chemical engineering community about what had caused the Bhopal disaster. His response was immediate and succinct: "Sabotage." Union Carbide has given the same explanation for three decades and continues to do so on its website.[3]

---

[1] Tank E610 contained 42 metric tons of MIC. See Chouhan (2005), p. 205. Estimates of how many tons of MIC gas escaped into the air range from "approximately 27 tons" (Cullinan, 2004) to "some 40 tons" (Peterson, 2009a).

[2] Edwards (2002), Broughton (2005), and Shetty (2014). If one counts those who died prematurely, months or years later, from effects of MIC exposure, the estimated death toll is much higher.

[3] See http://www.unioncarbide.com/history. On the company's historical timeline, the item for "1984" reads, "In December, a gas leak at a plant in Bhopal, India, caused by an act of sabotage, results in tragic loss of life." See also http://www.bhopal.com/Cause-of-Bhopal -Tragedy. Under "Frequently Asked Questions About the Cause of the Bhopal Gas Tragedy," the second question posed is "Who could have sabotaged plant operations and caused the leak?" The answer given reads, "Investigations suggest that only an employee with the appropriate skills and knowledge of the site could have tampered with the tank. An independent investigation by the engineering consulting firm Arthur D. Little, Inc., determined

On January 28, 1986, about 14 months after the Bhopal disaster, the U.S. space shuttle *Challenger* exploded and disintegrated 73 seconds after launch from Kennedy Space Center in Florida. The entire crew perished: six astronauts and Christa McAuliffe, the first "Teacher in Space."[4]

President Ronald Reagan appointed the late Arthur Walker Jr., at the time a faculty member at Stanford University, to serve on the Presidential Commission on the Space Shuttle *Challenger* Accident. Reagan charged the commissioners with determining the cause of the accident. In late 1987, after the commission had submitted its final report, I ran into Professor Walker on the Stanford campus and invited him to give a talk about his commission experience to a faculty seminar on technology in society. After his talk, I asked Walker what was the single most important lesson to be learned from the *Challenger* disaster. He replied, "Hire smarter engineers."

A GAP BETWEEN EDUCATION AND EXPERIENCE

The responses quoted in these vignettes are simplistic. The engineering outcomes involved cannot be explained as simply as those succinct replies suggest. The proffered explanations probably reflect the narrow educational backgrounds of those who offered them. Few intending engineers (or scientists) ever take ethics or social science classes that focus on engineering (or science) projects or practices. They are therefore predisposed to attribute the outcomes of destructive engineering episodes to technical failures or clear-cut, nontechnical factors. The latter include individual cognitive shortcomings, such as mediocre intellectual capability on the part of project engineers, and individual political motives, such as vengeful sabotage by a disgruntled employee.

Part of the appeal of such explanations is that they point up problems that can be readily "solved" by making specific changes, for example, hiring smarter engineers, and screening potential employees more rigorously. Engineers who never took ethics or social science classes closely related

that the water could only have been introduced into the tank deliberately, since process safety systems—in place and operational—would have prevented water from entering the tank by accident." On Union Carbide's sabotage theory, see Weisman and Hazarika (1987) and Peterson (2009b), pp. 9–11.

[4]Besides the loss of human life, the harm caused by this accident also had a financial component. "The space shuttle *Endeavor*, the orbiter built to replace the space shuttle *Challenger*, cost approximately $1.7 billion." See http://www.nasa.gov/centers/kennedy/about/information/shuttle_faq.html#1.

to engineering endeavor rarely consider the possibility that some harmful engineering episodes may be partly attributable to ethically problematic conduct on the part of engineer-participants. They also rarely consider the possibility that social or technical features of the often-complex contexts involved can help set the stage for and elicit such conduct.

Not only does contemporary engineering practice pose many ethical challenges to engineers, engineers are rarely adequately prepared to grapple with them in a thoughtful manner. There is an ethics gap in contemporary engineering, that is, *a mismatch or disconnect between the ethics education of contemporary engineering students and professionals, and the ethics realities of contemporary engineering practice*. One purpose of this book is to help narrow that gap.

EVIDENCE

Is there *evidence* of a gap between engineering ethics education for engineering students and the ethics realities of contemporary engineering practice? If there is, does it suggest that the ethics gap is substantial? Consider the following.

Between 1997 and 2001, the author conducted an informal survey of Stanford undergraduate engineering students and the practicing engineers they contacted about two topics: the study of engineering-related ethical issues in undergraduate engineering education, and the presence of ethical issues in engineering practice.[5]

Of the 516 undergraduate engineering majors who responded and ventured an opinion,[6] about 17 of every 20 (86.1%) indicated they expected to face ethical issues or conflicts in their engineering careers.[7] But how well did respondents believe their education had prepared them to deal "thoughtfully and effectively with such ethical challenges as they might encounter"? About a seventh (14.2%) responded "a good deal" or "a great deal," whereas more than half (54.3%) responded "a little bit" or "not at all."[8]

The undergraduates' responses did yield some encouraging findings. About three-fifths (62.2%) indicated that during their engineering

---

[5] McGinn (2003).
[6] One hundred forty-seven engineering majors did not respond because they did not plan to become practicing engineers; 28 others indicated they had no opinion.
[7] Ibid., p. 521.
[8] Ibid., p. 523.

education they had received the message that "there's more to being a good engineering professional in today's society than being a state-of-the-art technical expert."[9] However, that finding was offset by the sobering fact that only 14.9% of the respondents indicated they had learned "anything specific" from their engineering instructors "about what's involved in being an ethically and socially responsible engineering professional in contemporary society."[10]

Thus, while a healthy majority of the respondents had gotten a message that there's more to being a good engineering professional in contemporary society than being technically competent, the message often lacked specifics. Most students learned nothing concrete about the ethical responsibilities of engineers from their engineering instructors. As they left their classrooms and headed for workplaces where most expected to encounter ethical issues, few engineering students took with them specific knowledge of the ethical responsibilities of engineers.

But how likely is it that engineers will actually confront ethical issues in professional practice? Of the 285 practicing engineers who responded and expressed an opinion, 84.2%[11] agreed that current engineering students are "likely to encounter significant ethical issues in their future engineering practice."[12] Indeed, almost two-thirds (65.4%) of the responding engineers indicated they had already been personally "faced with an ethical issue in the course of [their] professional practice." Almost the same percentage (64.3%) stated they knew or knew of one or more other engineers "who have been faced with an ethical issue in their professional practice."[13] Not surprisingly, a remarkable 92.8% of the practicing engineer respondents who ventured an opinion agreed that engineering students "should be exposed during their formal engineering education to ethical issues of the sort that they may later encounter in their professional practice."[14]

Unless these two groups of respondents are atypical of engineering students and practicing engineers in general,[15] these findings suggest a

[9]Ibid., p. 524.
[10]Ibid., p. 525.
[11]Nine of the 294 practicing engineer respondents did not express an opinion on the matter. Ibid., p. 527.
[12]Ibid. Interestingly, this percentage is close to the percentage of surveyed engineering students who *expect* to encounter ethical issues in their future engineering careers.
[13]Ibid.
[14]Ibid.
[15]This possibility cannot be ruled out. The 691 Stanford undergraduate engineering students and the 294 practicing engineers who completed the relevant parts of the survey

serious disconnect: between the levels of engineering-student expectation and practicing-engineer experience of being confronted by ethical issues in engineering work, and the amount of effective engineering-related ethics education provided to U.S. undergraduate engineering students.

IMPORTANCE

I shall proceed on the assumption that this disconnect persists[16] and is substantial. Why is it important to bridge or at least narrow the gap between engineering-related ethics education and the ethics realities of contemporary engineering practice?

First, as the case studies in Chapter 4 make clear, misconduct by engineers sometimes contributes to causing significant harm to society. Making engineering students aware of ethical challenges in engineering practice and illustrating the serious social costs attributable to engineering misconduct could help prevent or lessen some of those societal harms.

Second, it makes sense for engineering students to learn upstream, for example, during their undergraduate studies, about material pertinent to challenges they are likely to face downstream, such as being faced with ethical issues during their engineering careers. For many years there was a disconnect between engineers' need for good technical writing and other communications skills, and the scarcity of training dedicated to cultivating such skills in undergraduate engineering education. Happily, in recent years technical communication classes and programs for undergraduates have emerged in a number of U.S. engineering schools, to the benefit of those able to access them. The same attention should be given to cultivating engineering-related ethics awareness and skills as it eventually was to technical communications skills. Failure to nurture the former does as much a disservice to engineering students as did failure to develop the latter. It sends them out into engineering workplaces ill-equipped to recognize and effectively grapple with another important type of professional challenge they are likely to face.

Third, acquiring intellectual resources useful for making thoughtful ethical judgments about engineering conduct can help empower engineers

---

questionnaire were not probabilistically random samples of the populations of U.S. undergraduate engineering students and U.S. practicing engineers, respectively.

[16] This disconnect might have decreased if a widespread increase in meaningful engineering-related ethics education had occurred since 2001. However, to the best of the author's knowledge, this has not happened.

to make up their own minds about the ethical acceptability of prevailing workplace culture and practices. Engineers who lack the skills to make thoughtful ethical judgments about questionable features of workplace culture or suspect work practices are more likely to yield to pressure to go along with prevailing attitudes and practices.

Fourth, equipped with an understanding of responsible engineering decision-making and practices, young engineers in the job market can better assess how committed the firms recruiting them are to supporting ethically responsible engineering work. It would be useful for would-be ethically responsible engineering students and practicing engineers in the job market to know to what degree the firms they are considering joining expect and exert pressure on their new engineer-employees to follow or-ders uncritically, even when the engineers have concerns about the ethical acceptability of some of the tasks they are assigned.

Fifth, the ability to recognize and comprehend the ethical issues in an engineering situation should make *inadvertent* irresponsible behavior by engineers less frequent. That recognition and understanding will dimin-ish appeals to the classic excuse "I didn't realize there were ethical issues involved in that situation." Presumably, some engineers who are able to recognize ethical issues in professional practice will choose to avoid con-duct they deem ethically irresponsible.

Sixth, a quite different kind of reason for the importance of bridging the ethics gap in contemporary engineering is that in recent years, pres-sure to provide engineering students with opportunities to study ethi-cal issues in engineering has grown. This pressure stems from multiple sources:

- In a 2003 request for proposals, the U.S. National Science Foundation (NSF) stipulated that each group of universities submitting a proposal for funding to establish a network of nanotechnology research laboratories had to indicate how it was going to "explore the social and ethical implica-tions of nanotechnology" as part of its mission.[17]
- In 2004, the U.K. Royal Academy of Engineering recommended that "con-sideration of ethical and social implications of advanced technologies . . . should form part of the formal training of all research students and staff working in these areas."[18]

[17] http://www.nsf.gov/pubs/2003/nsf03519/nsf03519.pdf.
[18] The Royal Society and the Royal Academy of Engineering (2004), Recommendation 17, p. 87.

- In 2006, a survey of 1,037 nanotechnology researchers at 13 U.S. universities posed this question: "How much do you believe that study of ethical issues related to science and engineering should become a standard part of the education of future engineers and scientists?" About three-tenths (30.1%) of the respondents replied "quite a bit," while another third (33%) replied "very much."[19] This suggests that significant interest in relevant ethics education exists among engineering students and young engineers themselves, not just on the part of accrediting agencies, professional societies, and engineering-related funding organizations.
- In 2009, NSF took a step toward requiring ethics education for engineering students. In implementing the America COMPETES Act of 2007, NSF stipulated that, as of January 2010, when an institution submits a funding proposal to NSF it must certify that it has "a plan to provide appropriate training and oversight in the responsible and ethical conduct of research to undergraduates, graduate students, and postdoctoral researchers who will be supported by NSF to conduct research."[20]
- The U.S. Accreditation Board for Engineering and Technology (ABET) currently requires that engineering programs seeking initial or renewed accreditation of their bachelor's degrees "document" that most graduates of the programs in question have realized 11 "student outcomes." Among them are "an ability to design a system, component, or process to meet desired needs within realistic constraints, such as economic, environmental, social, political, ethical, health and safety, manufacturability, and sustainability [constraints]"; and "an understanding of professional and ethical responsibility."[21]

In short, there are individual, organizational, and societal reasons why providing engineering students with meaningful engineering-related ethics education makes excellent sense.

UNFRUITFUL APPROACHES TO BRIDGING THE GAP

It is hoped that the reader is now persuaded that, all things considered, it would be worthwhile to expose engineering students to study of engineering-related ethical issues in their formal education. But even

---

[19] McGinn (2008), p. 117.
[20] https://www.nsf.gov/bfa/dias/policy/rcr.jsp.
[21] http://www.abet.org/DisplayTemplates/DocsHandbook.aspx?id=3149.

if that is so, the question remains: *what kind of approach to providing engineering students with education about engineering-related ethical issues is likely to be fruitful?*

I shall first describe two general approaches to engineering-related ethics education I believe are unlikely to be fruitful and then shall identify and briefly characterize one approach I regard as more promising. The two unfruitful approaches are (1) requiring engineering students to enroll in a traditional philosophy-department ethics course and (2) incorporating engineering-related ethics education into technical engineering classes.

Requiring a Typical Philosophy-Department Ethics Class

Requiring engineering students to enroll in a traditional philosophy-department ethics course is unlikely to be fruitful. Few such courses in the U.S. pay any attention to ethical issues in engineering. They tend to be concerned with ethical concepts and theories, the nature of ethical reasoning, and the status and justification of ethical judgments. With rare exceptions, the examples explored in such courses rarely involve professional contexts.[22]

It is not surprising that engineering-related examples and cases are typically absent from such courses. Few philosophy-department faculty members in U.S. research universities or liberal arts colleges have substantial knowledge of or interest in engineering (as distinguished from science). The same is true of the kinds of concrete situations in which engineers can find themselves that may give rise to ethical issues. In more than four decades of teaching at Stanford University, to the best of my knowledge no ethics course offered by the Department of Philosophy has paid any attention to ethical issues in engineering. I suspect that the same is true of philosophy-department ethics courses at virtually all U.S. universities and colleges.[23] Consequently, requiring engineering students to take a traditional philosophy-department ethics course with the hope

[22] The most common exception is that some such courses include exploration of phenomena that arise in medical professional contexts, for example, abortion, assistive reproductive technology, and organ transplantation.

[23] Engineering ethics courses are most often taught by instructors in academic units with names like General Engineering; Technology in Society; Engineering and Society; and Science, Technology, and Society, almost always at institutes of technology or universities with large engineering schools. Occasionally, an engineering ethics course is taught in an engineering department, such as computer science and civil engineering.

they will learn something useful about ethical issues in engineering would leave it completely up to the student to work out how the ideas and theories explored in such courses apply to engineering situations. It would therefore not be surprising if most engineering students perceived such courses as irrelevant to their future careers.

Integrating Ethics Study into Technical Engineering Classes

A second option is to attempt to cover engineering-related ethical issues in technical engineering classes. This could be done by a nonengineer guest instructor with expertise in engineering ethics, or by the primary engineer-instructor of the course.

If a nonengineer guest instructor with expertise in engineering ethics provides the engineering-related ethics education, it is likely to be limited to one or two lectures. Unfortunately, class members will almost inevitably perceive the (limited) material covered in such sessions as peripheral to the course. Moreover, the material covered will probably not be well integrated (by the main instructor) into discussion of the technical material encountered elsewhere in the course.

If the course's main engineer-instructor provides the coverage of ethical issues in engineering, then the consideration of ethical issues is likely to be intuitive and not grounded in ethics fundamentals. Having an engineer-instructor cover ethical issues in engineering is an excellent idea in principle; however, in practice it faces two problems: one pedagogical, the other temporal.

First, effectively integrating ethics into a technical engineering class is likely to be more pedagogically demanding for the engineer-instructor than getting back up to speed on a body of technical subject matter with which she or he was once familiar but has forgotten over time. Doing that integration *well* requires a grasp of key ethical concepts and principles, familiarity with a range of ethical issues in engineering, detailed knowledge of various cases, and the ability to apply key ethical concepts and principles to concrete cases in an illuminating way. It is difficult for an engineer (or anyone else) without formal ethics education and teaching experience to acquire such knowledge and ability in short order.

Second, required technical engineering classes are already tightly packed with technical subject matter. Engineer-instructors of such courses often complain that, in their classes as they now stand, they do not have enough time to cover even all the important *technical* subject matter that students need to know. But the more time that is devoted in such a class to studying engineering-related ethics issues, in hopes of making coverage of that topic nonsuperficial, the less time will remain for important technical engineering material. Hence, study of the latter would have to be diluted or truncated. That is extremely unlikely to happen.

Thus, what may sound ideal in principle—having instructors who are engineers provide education about ethical issues in engineering in technical engineering classes—faces serious practical barriers in the real curricular world of undergraduate engineering education.[24]

PREFERRED APPROACH

I favor a third kind of pedagogical approach to teaching engineering students about engineering-related ethical issues. In this approach, engineering students explore ethical issues in engineering in a separate course dedicated to such study. They read and discuss at length real-life cases in which engineering-related ethical issues arose, and make presentations on original cases of ethical issues in engineering they have researched and developed. The instructor has expertise and experience in teaching engineering ethics, has an abiding interest in engineering education, and is familiar with the realities of engineering practice. He or she is a full-time engineering school faculty member who believes analysis of ethical issues in engineering and evaluation of engineers' conduct from an ethics viewpoint are important tasks. Further, she or he believes such analysis and evaluation must be carried out with due attention to the specific contexts in which those issues arise and the related actions unfold.

* * *

[24]To "learn how to incorporate ethics into engineering science classes," one mechanical engineering professor attended an Ethics Across the Curriculum Workshop given by Illinois Institute of Technology's Center for the Study of Ethics in the Professions. Shortly thereafter, he added an ethics component to his Automatic Control Systems course. It included exploration of two "Ethics Cases" inspired by actual events. Students were asked to generate a list of possible courses of action open to the engineer(s) who faced an "ethical dilemma" about what to do. The instructor "asked students to vote on their preferred choice" of action in each case. Encouragingly, a survey revealed that most students believed that the course had "increased their awareness of ethics issues." However, given the limited time available in the course for discussion of ethical issues, the "mini-ethics lessons" do not appear to have tried to impart to students any ethics fundamentals that they could draw upon in making thoughtful ethical judgments about engineering conduct in the future. See Meckl (2003).

Chapters 2 and 3 present background and foundational materials intended to help engineering students and engineering professionals develop the ability to make thoughtful judgments about ethical issues in engineering and related engineering conduct. Then, making use of those materials, Chapter 4 explores a wide range of cases from different fields of engineering and analyzes various ethical issues raised therein. Almost all the cases are real-life ones, and some include engineers speaking in their own voice as they wrestle with the ethical issues involved.

Subsequent chapters discuss noteworthy ideas and lessons distilled from the case studies (Chapter 5), identify resources and options that might be useful to those who care about ethically responsible engineering practice (Chapter 6), and discuss the author's general approach to exploring ethical issues in engineering in somewhat greater detail (Chapter 7).

By reading and reflecting on the wide range of cases presented, and by grasping the intellectual resources used in exploring them, engineering students and practicing engineers should become more aware of and better able to come to grips with the ethical dimension of engineering practice. More specifically, such exposure should also help them develop sensitive antennae with which to detect ethical issues present in concrete engineering situations, and improve their ability to unpack and think clearly, critically, and contextually about such issues. With careful study, engineering students and practicing engineers will acquire concepts and principles that can be added to their personal ethics tool kits and used to come to grips in a thoughtful way with ethical challenges in their professional careers.

CHAPTER 2

Sociological and Ethical Preliminaries

Familiarity with background materials of two sorts—sociological and ethical—is useful for thinking about ethical issues in contemporary engineering practice. The sociological materials shed light on why the work situations of engineers in contemporary Western societies make it quite likely that they will face ethical issues in their professional practice. The ethical materials focus on a resource often cited and occasionally used by engineers to make judgments about the ethical acceptability of engineering actions, decisions, and practices.

The purpose of exploring these background materials early in this book is to refute two mistaken beliefs. The first such belief is that there is nothing qualitatively or quantitatively new about the presence of ethical issues in contemporary engineering practice. The second is that the question of how engineers should make ethical judgments about engineering conduct has been resolved and involves using the codes of ethics of the professional engineering societies.

I begin with the first kind of background materials: the sociological.

SOCIOLOGY OF ENGINEERING

Since the late nineteenth century, several noteworthy sociological changes have occurred in the engineering profession in the United States. These changes have made contemporary engineers more likely to face ethical issues in their work than previously.[1]

Over the last 125–150 years, the dominant locus of engineering employment has changed. Most engineers have gone from being independent engineer-consultants, owner-operators of machine shops or mines,

[1] The discussion of the first two changes is due to Reynolds (1991).

or employees in small firms to being employees in considerably larger organizations, whether private for-profit, private nonprofit, educational, or governmental. In the words of Terry Reynolds,

> early in the 20th century, organizational hierarchies (usually corporate) became the typical place of employment for American engineers. By 1925, for example, only around 25% of all American engineers were proprietors or consultants—the ideals of the previous century; 75% were hired employees of corporate or government organizations. By 1965 only around 5% of American engineers were self-employed.[2]

From the point of view of ethics, this was an important development, because it meant that since the early twentieth century, the autonomy of more and more engineers has been more tightly restricted than it was in the nineteenth century. As increasing numbers of engineers found employment in large-scale firms, they became subject to ongoing pressures to make decisions that gave top priority to their firms' interests. A declining percentage of engineers retained the freedom of the independent engineer-consultant and the engineer who owned her or his own machine shop or mining operation to determine his or her own projects, priorities, practices, and policies.

Engineers employed in private for-profit firms are always at risk of facing conflicts of interest. That is, they often find themselves in situations in which they are torn between the desire to protect the public interest and/or remain faithful to their professional and personal commitments to do excellent engineering work, and the need to serve the sometimes opposed private economic interests of their employers and/or their own private economic interests. The possibility of being faced with conflicts of interest in professional practice is a persistent fact of life for many if not most engineers employed in private for-profit corporations in contemporary societies.[3]

The incorporation of most engineers into large private firms engendered another new sociological trend: in the twentieth century, the typical career path of the engineer changed. The typical career trajectory of the engineer employed in a private for-profit firm increasingly followed a pattern, evolving from being a practicing engineer whose workday comprised largely or entirely technical engineering tasks, to being a corporate

---

[2] Ibid., p. 173.

[3] This is not to say that engineers who are not employees in private for-profit firms do not encounter conflicts of interest.

manager whose workday was devoted exclusively or primarily to non-technical managerial tasks. This development also provided fertile ground for new conflicts of interest. Engineers who become corporate managers are strongly expected to prioritize the profit and organizational interests of their firms. When that happens, the interest in doing or supporting excellent and responsible engineering work is sometimes relegated to a subordinate status. This tug of war can be ethically problematic.

Starting in the late nineteenth century, another new and important trend in engineering emerged, one that accelerated in the twentieth century: fundamental research took on unprecedented importance in engineering work. This development can be traced to the birth and development of large-scale sociotechnical systems of communication, transportation, and lighting in the nineteenth century. These systems were made possible (and given impetus) by the invention and diffusion of the telegraph, telephone, railroad, and incandescent lightbulb in a national market economy.[4] The enormous capital investments required to construct the large-scale systems such innovations enabled, something the prevailing market economy encouraged, made it imperative that the engineering work involved be well grounded.

AT&T could not afford to build a system like the nationwide telephone network on a trial-and-error basis. Fundamental research-based understandings of pertinent areas of physics and chemistry were required so that the huge capital investment needed to construct that large-scale system would not be at risk of being squandered. The bearing of this development on ethical issues in engineering is this: sometimes time and money pressures to push an engineering project forward are at odds with the need for fastidious, time-consuming research to achieve a better fundamental understanding of key aspects of the project situation. This tension can tempt engineer-managers to compress, curtail, or even not conduct the relevant research in order to meet the project-completion schedule and assure on-time delivery of promised goods and services. Negligence is an important form of professional malpractice, and failure to conduct or complete expensive and/or time-consuming research inquiries, and failure to do such research carefully, are noteworthy forms of negligence that sometimes taint engineering work.

---

[4] Another factor that made these large-scale sociotechnical systems possible was a social innovation in business: the emergence of the multiunit, professionally managed, vertically integrated, hierarchically organized modern business firm. See Chandler (1977).

Another significant sociological change in the engineering profession is that contemporary engineering endeavors undertaken by private engineering firms are often of unprecedented scale and scope. They are therefore extremely expensive to complete, but have enormous profit potential. This is relevant to ethics because, given the high stakes involved, the pressure to win or retain lucrative engineering contracts and to meet budgetary, profit, and market-share goals can be so great that engineer-managers and the engineers who report to them may resort to ordering, engaging in, or tolerating the use of ethically problematic engineering practices. Financial stakes are often so high that engineers can be tempted to deliberately overestimate engineering performance and reliability and/or underestimate project cost and risk.

Finally, several sociologically significant developments in the engineering profession have emerged in the U.S. since 1970. These include rapid growth in the employment of hardware and software engineers in the information technology (IT) industrial sector, in both established firms and start-ups; significant increases in the percentages of women and minorities among computer and engineering workers;[5] the unusually high employment job mobility enjoyed by engineers in Silicon Valley and other areas of California's cutting-edge computer industry;[6] the increasing importance of computers, the Internet, and email in engineering work; and the institutionalization in developed societies of increasingly rigorous regulation of the environmental impacts of mass-produced engineering products. These trends have given rise to new concerns and heightened others, such as the protection and theft of intellectual property; interactions between engineer-entrepreneurs and venture capitalists; recruitment and retention of engineering talent; human–computer interface design; work cultures of IT-industry firms; the privacy interests of engineer-employees; and trade-offs between enhanced performance, cost, and environmental friendliness of many engineering products.

These concerns have in turn engendered a range of specific engineering-related ethical issues, such as the following:

---

[5] While the percentages of women and minorities in engineering have not increased monotonically, compared with the situation that existed in 1970 the current percentages represent substantial increases. The increase in the percentage of Asians employed in computer and engineering occupations in the United States has been especially dramatic. See Landivar (2013), especially Figures 4, 10, and 11.

[6] Postrel (2005).

- whether it is ethically acceptable for an engineer to create software that covertly tracks the websites visited by civilian computer users;
- whether software engineers have an ethical responsibility to ensure that human–computer interfaces are readily usable by the lay public;
- whether a knowledgeable software engineer working for a firm has an ethical responsibility to confirm publicly the existence of a microprocessor flaw or software bug that has been publicly denied by her or his firm;
- whether an engineering department head in academia or industry has an ethical responsibility to try proactively to ensure that the work environment in his or her unit is not subtly biased against female or minority engineering students or engineers; and
- whether a software engineer working on a certain model of car has an ethical responsibility to refuse to design and to not acquiesce in the use of a "defeat device" to cheat on government pollution emissions tests.[7]

In short, because of a number of transformative sociotechnical developments in the engineering profession in the United States over the last 125–150 years, engineers today are more likely than ever to find themselves faced with challenging ethical issues at work. Consequently, it is all the more important that they be equipped to address such issues in an ethically responsible manner.

PROFESSIONAL ENGINEERING SOCIETY CODES OF ETHICS

A second kind of useful preliminary background material is the ethical. The focus of attention here will be on codes of engineering ethics.

Several engineering fields, for example, chemical, civil, electrical, and mechanical engineering, have had codes of ethics since the early twentieth century. In principle, engineers in any of these fields are expected to be familiar with and to act in accordance with the provisions and precepts of their professional engineering society's code of ethics. It looks impressive that engineering fields have their own codes of ethics.[8] Moreover, the latter are easily accessible on professional engineering society websites.

---

[7] For an account of how one kind of automotive "defeat device" worked, see Gates et al. (2017).

[8] The codes of ethics of various field-specific engineering professions, as well as of the multidisciplinary National Society of Professional Engineers (NSPE), can be found at http://www.onlineethics.org/Resources/ethcodes/EnglishCodes.aspx.

However, to the best of the author's knowledge, these codes do not play important roles in the socialization of young engineers by their respective professional societies. They seem to be invoked mostly when engineering societies are considering whether to discipline or expel members involved in controversial engineering episodes.

The Art Dealers Association of America (ADAA) has a "Code of Ethics and Professional Practices."[9] Like the codes of ethics of professional engineering societies, its provisions spell out a set of practices and norms that ADAA members are to "observe in their relations with clients, artists, other dealers, and auctions."[10] Interestingly, ADAA members must "acknowledge in writing their acceptance of, and compliance with," the provisions of the ADAA code.[11] To the best of the author's knowledge, professional engineering societies do not require such written acknowledgment by their members. When the members of a professional engineering society are not required to put in writing their acceptance of and promise of future compliance with the provisions of their society's code of ethics, doubt may arise about whether that society attaches serious operational importance, not just symbolic value, to its members' abiding by its code's provisions and precepts.

Professional engineering society codes of ethics have changed significantly over time. Today, many share a preeminent canon: namely, that engineers, in the performance of their professional duties, shall "hold paramount the safety, health and welfare of the public."[12] That was not always the leading provision of such codes. In March 1912, the American Institute of Electrical Engineers (AIEE) adopted a "Code of Principles of Professional Conduct." Section B.3 of that code reads as follows:

The engineer should consider the protection of a client's or employer's interests his first professional obligation, and therefore should avoid every act contrary to this duty.[13]

[9] "ADAA Code of Ethics and Professional Practices," http://www.artdealers.org/about/code-of-ethics-and-professional-practices.
[10] Ibid.
[11] Ibid.
[12] See, e.g., the codes of ethics of the National Society of Professional Engineers (NSPE) (2007), the American Society of Civil Engineers (ASCE) (2009), the American Society of Mechanical Engineers (ASME) (2012), and the American Institute of Chemical Engineers (AIChE) (2015). Other engineering society codes espouse essentially the same idea in different words.
[13] Leugenbiehl and Davis (1992), p. 41.

In September 1914, the American Society of Civil Engineers (ASCE) adopted its own Code of Ethics. It began as follows:

> It shall be considered unprofessional and inconsistent with honorable and dignified bearing for any member of the American Society of Civil Engineers: 1. To act for his clients in professional matters otherwise than as a faithful agent or trustee, or to accept any remuneration other than his stated charges for services rendered his clients.[14]

However, after World War II, the leading concern of many, if not most, professional society codes of engineering ethics in the United States shifted dramatically, from engineers' being loyal to their employer or client to holding paramount the health, safety, and welfare of the public in carrying out their professional duties.

What accounts for this major change is unclear. It may have been partly a result of expanding awareness by professional society engineers of several serious engineering accidents and disasters in North America in the late nineteenth and early twentieth centuries. Perhaps the vivid demonstration in World War II of the major impact of contemporary engineering on public welfare, in the form of protecting national security, contributed to bringing about this change. Perhaps it was partly due to spreading recognition of the key role engineering played in creating the physical infrastructure of twentieth-century American society that so enhanced public health and well-being. Whatever factors were at work, it is probable that the potency, impact, and pervasiveness of engineering in twentieth-century U.S. life helped foster the realization that it was critical that engineers keep the health, safety, and welfare of the public uppermost in mind in their professional practice. Recently, some professional society codes of engineering ethics have begun to come to terms with more contemporary societal concerns, such as sustainable development and privacy.[15] The key point here is that professional society codes of engineering ethics are historical artifacts that have evolved over time and continue to develop, albeit slowly. They tend to lag the changing foci of societal concern about engineering activity.

Beyond evolving content and shifting priorities, professional society codes of engineering ethics vary considerably in level of detail and clarity.

[14]Ibid., p. 45.
[15]On privacy, see Association for Computing Machinery (ACM) (1992), sec. 1.7. On sustainable development, see the ASCE Code of Ethics, Canon 1, parts c, e, and f; the ASME Code of Ethics of Engineers, Fundamental Canon 8; and the NSPE Code of Ethics, sec. III.3.d.

The "Code of Ethics" of the IEEE,[16] "the world's largest technical professional society,"[17] consists of one preliminary sentence and 10 succinct propositions.[18] In contrast, the "Code of Ethics and Professional Conduct" of the Association of Computing Machinery (ACM) is considerably more detailed. It contains 24 "imperatives" or responsibilities, some quite general, others more specific. Most apply to all computing professionals, while some apply only to those in organizational leadership positions. Each imperative is followed by comments on what it means and why it is important.[19]

Engineering society codes of ethics are sometimes quite vague. For example, according to the "Fundamental Principles" of the "Code of Ethics of Engineers" of the ASME, engineers are to "uphold and advance the integrity, honor, and dignity of the engineering profession" by, among other things, using their knowledge and skill for "the enhancement of human welfare," and by "striving to increase the competence and prestige of the engineering profession."[20] However, a specific engineering endeavor could enhance human welfare in some respects and reduce or dilute it in others. Is such a course of action consistent or inconsistent with the "enhancement of human welfare" that is ostensibly required of ASME members? Is the "enhancement" in question to be understood as *net* enhancement? How does the engineer seeking to adhere to this code provision determine that? What exactly is meant by "human welfare"? Beyond being a technically competent engineer, what reasonably counts as "striv[ing] to increase the competence and prestige of the engineering profession"?

The fourth Fundamental Canon of the codes of ethics of the ASME and the NSPE states that "in professional matters" engineers shall act "for each employer or client as faithful agents or trustees."[21] But what exactly does it mean for engineers to be "faithful agents or trustees" of their employers or clients, and what does being a loyal or faithful employee of an employer or client require of an engineer?[22] Such vagueness

---

[16] The IEEE, which had over 395,000 members in 2010, was formed in 1963 by merging the American Institute of Electrical Engineers, founded in 1884, and the Institute of Radio Engineers, founded in 1912. See http://www.ieee.org/about/ieee_history.html.

[17] Ibid.

[18] http://www.ieee.org/about/corporate/governance/p7–8.html.

[19] http://www.acm.org/about/code-of-ethics?searchterm=Code+of+Ethics.

[20] https://www.asme.org/getmedia/9EB36017-FA98–477E-8A73–77B04B36D410/P157_Ethics.aspx.

[21] Ibid. See also http://www.nspe.org/resources/ethics/code-ethics, sec. II.4.

[22] See the discussion of FERE4 in Chapter 3.

and uncertainty diminish the usability and practical value of professional engineering society codes of ethics. Vague precepts in a code of ethics may open the door for engineers to interpret the code in ways that prioritize their firm's or their own economic interests.

Besides playing a minor role in the socialization of young engineers, lagging current societal concerns about engineering, often being short on detail, containing key expressions that are vague, and being open to self-serving interpretations, codes of engineering ethics can be ethically problematic when relied upon as the basis for deciding whether engineering conduct is ethically acceptable. The engineer who consults a professional society code of ethics to determine the ethical acceptability of a certain course of action may be tempted to think that if it is not explicitly prohibited by the code being consulted, then acting thus is ethically acceptable. But deciding whether a course of engineering action is ethically acceptable is not like determining whether an expense incurred in doing business is legally tax deductible by scrutinizing the provisions of the relevant tax code. A course of action not explicitly prohibited by the letter of a professional society code of engineering ethics might nevertheless still violate its spirit. Developing the engineering student's and young engineer's ability to make an *independent* thoughtful ethical judgment after considering all relevant aspects of an engineering situation is arguably much more important than being able to efficiently scan the precepts of a code of engineering ethics to see if any explicitly addresses the course of action under consideration.

In fact, relying heavily on a professional engineering society's code of ethics arguably hampers engineers from developing the ability to make their own independent, carefully considered judgments about ethical acceptability and responsibility. That ability is especially important when it comes to making ethical judgments in novel or complex kinds of engineering situation not explicitly referenced in one's field's code of ethics. Education about ethical issues in engineering provided to engineering students in the United States has arguably been too dependent on finding rules, canons, or precepts in engineering society codes of ethics that appear to apply to the case at hand or that can be stretched to do so.

Even conscientious adherence to the precepts of the code of ethics for one's field of specialization is not the last word on being an ethically responsible engineer. Society needs engineers with sensitive "ethics antennae," not just ones who adhere to the letter of an engineering society code of ethics. Such individuals take into consideration the interests of all stakeholders who stand to be affected by the engineering conduct, decisions, and practices in question, whether they be individuals or the public at large. Engineers must also be attentive to subtle harms (and significant risks of same), not just to obvious physical and financial harms, whether directly or indirectly caused, in whole or in part, by engineering activity, and whether they manifest themselves immediately, in the short term, or in the foreseeable future. Developing habits of mind in engineers that reliably take such considerations into account is an important goal, one that use of the codes of ethics of most professional engineering societies does not foster.

In short, while the various professional engineering society codes of ethics can be useful to some engineers for some purposes,[23] they are not especially helpful for grappling with ethical issues that arise in many different kinds of engineering situations. In the next chapter, a different kind of foundational ethics resource will be presented and discussed, one that enables engineers who master it to make thoughtful, well-grounded judgments about the ethical merits of engineering conduct, decisions, and practices in a wide range of engineering situations.

---

[23] One way in which a professional engineering society code of ethics can be useful to a member-engineer is by serving as a basis for her or his refusal to act as directed by an employer or client when the directed action would violate a provision of the code. Some employers and clients may respect that refusal, but others may be unmoved and insist that the engineer-employee act as directed.

CHAPTER 3

The Fundamental Ethical Responsibilities of Engineers

AN ETHICAL RESPONSIBILITIES APPROACH

In this book, the approach taken to analyzing ethical issues in engineering work and to making ethical judgments about the conduct of engineers is termed "an ethical responsibilities approach."[1] "Taking an ethical responsibilities approach" means that in evaluating an engineer's conduct in an engineering situation from an ethics perspective, one begins by looking long and hard, widely and deeply, spatially and temporally, for all ethical responsibilities incumbent on the engineer in question.

But how does one go about identifying those ethical responsibilities? There are three general steps for doing so. First, the evaluator must have clearly in mind the **fundamental ethical responsibilities of engineers.**[2] Second, by examining the engineering activity in the situation being studied, including the activity's technical and social characteristics, the features of the work product, and characteristics of the contexts of production and use, the evaluator determines which of those fundamental responsibilities apply to the situation at hand. Third, the evaluator brings the applicable fundamental ethical responsibilities of engineers to bear on that situation to determine what *specific* ethical responsibilities are incumbent on the engineer(s) involved. I shall often refer to the latter as **derivative ethical responsibilities** of the engineer in that situation. "Derivative" connotes that these specific responsibilities can be derived by applying one or more of the fundamental ethical responsibilities of engineers to the specific

---

[1] In the book's final chapter, I argue that this approach might be better termed "a foundational-contextual ethical responsibilities approach." See Chapter 7.

Taking this approach is not meant to deny that engineers have ethical rights as well as responsibilities. See McGinn (1995) for discussion of several ethical rights of engineers. Nor does taking an ethical responsibilities approach exclude the possibility that, occasionally, considerations of harm unrelated to ethical responsibilities may outweigh those that do, thus calling for an ethical judgment about engineering conduct different from what taking an ethical responsibilities approach would suggest.

[2] These are discussed in detail in this chapter.

features of the engineering activity, work product, and/or social situation in question.

Before discussing the fundamental ethical responsibilities of engineers, items critical for exploring the ethical issues raised in the cases discussed in Chapter 4, I shall first make clear what I mean by a key expression used in this book: "an ethical issue."

ETHICAL ISSUES AND HARM

It is important to understand clearly what is meant by "an ethical issue." Such an understanding helps one recognize when one is faced with such an issue, rather than remaining oblivious to its presence. Norman Augustine, a former chairman of the U.S. National Academy of Engineering and chair of its committee on Engineering Ethics and Society, noted a potential problem that can plague someone unable to recognize when he or she is faced with an ethical issue:

> [M]ost of the engineers whom I have seen get into trouble on ethical matters did so not because they were not decent people, but because they failed to recognize that they were confronting an ethical issue.[3]

One impediment to attaining a clear understanding of what "an ethical issue" is stems from the fluid use of that expression in everyday English. When someone says that an action or practice raises "an ethical issue," that individual may mean any of several things.

She or he may simply be claiming that if the action occurs or the practice persists, some (possibly nonobvious) *harmful consequences* may result, ones that, in the speaker's view, merit scrutiny from an ethics perspective.

Alternatively, the speaker may be doing something other than calling attention to harmful consequences the person believes merit scrutiny from an ethics perspective. Suppose a woman or man uses a potent perfume or cologne that induces nausea in a significant fraction of those exposed to it who cannot readily escape its smell. Saying that wearing such a perfume or cologne in that kind of situation raises an ethical issue might be the way the speaker expresses a personal belief that doing so under such circumstances is *ethically questionable.* That is, she or he might be implying that there's a question about the propriety of the agent's action in that

---

[3] Augustine (2002), p. 5.

kind of situation that needs addressing: namely, whether wearing such a substance under such circumstances is ethically acceptable.

However, it is also possible that a person referring to "an ethical issue" may be implying that supporters and opponents of the action or practice in question disagree about its likely effects on the well-being of parties affected or likely to be affected by it, and/or that they disagree whether the action or practice is intrinsically good/right/proper or intrinsically bad/wrong/improper.

In what follows, when reference is made to some engineering action, decision, or practice as raising an ethical issue, it should be clear from context which of these three kinds of claims is being made: that *nonobvious harmful consequences may be involved that merit scrutiny from an ethics perspective*, that *some action or practice is ethically questionable and needs to be addressed*, or that *interested parties disagree about the action's well-being-related consequences and/or intrinsic ethical acceptability*.

That said, most of the ethical issues that arise in engineering work involve disagreements over whether some engineering action, practice, or policy is likely to cause harm, create an unreasonable risk of harm, or yield an unjust distribution of harm (or risk of harm) among parties who are or might be affected by it. To my knowledge, few if any engineering-related ethical issues involve disagreements about the *intrinsic* unacceptability of some engineering action, practice, process, or policy.[4]

Therefore, in what follows, "an ethical issue in engineering" refers to

a matter of engineering action or practice about which there is disagreement over what the engineer(s) involved *should* do, where the disagreement stems from differing beliefs about how the action or practice will affect the well-being of affected parties or, rarely, about its intrinsic acceptability or unacceptability.[5]

[4]If there are ethical issues in engineering that involve disagreements about the alleged intrinsic unacceptability of some engineering actions or practices, the fields in which such ethical issues are most likely to arise are probably bioengineering and environmental engineering.

[5]This claim reflects the author's general position regarding what "ethics" (as a cultural institution) is about. At bottom, ethics and the judgments about people's conduct made under its auspices have to do with the relationships between agents' actions and practices and the well-being of parties affected by them. Thus, for example, judgments that certain actions or practices are ethically right, wrong, or permissible depend at bottom on beliefs about their harm- and well-being-related consequences. Ethical judgments to be discussed about engineering actions and practices reflect beliefs about the extent to which their

It is also important to indicate how "harm" is to be understood here. In what follows, harm consists of the (incremental) damage done to a party when its well-being is significantly violated, undermined, or set back by some action, practice, product, or policy. The incremental damage done may, of course, be physical or financial. But harm can also take social, cognitive, and psychical forms. Harm occurs when consensually important interests of humans (and other sentient beings) are violated or significantly set back. At this stage of human development, there is consensus in most societies that humans have interests worthy of protection in the nonviolation of certain physical, social, and psychical states and conditions, such as continuation of their lives, preservation of their bodily integrity, retention of and control over their properly acquired property, preservation of their good reputation, protection of their privacy, and preservation of their freedoms of thought, action, and expression (as long as the exercise of those freedoms does not unjustifiably harm or pose an unreasonable risk of unjustifiably harming others).

Besides such "private" harms that result when such consensually important individual interests are violated, some actions or practices can also cause public or societal harms. Such harms could include outcomes such as a national border's being insecure; a public health epidemic; a weakened banking or legal system; a zoning regime or a currency's integrity being undermined; an important societal resource such as air, water, or electricity being degraded; or an important element of societal infrastructure, such as its transportation or communication system, being rendered inoperative.

More generally, it is critical that engineering students and practicing engineers acquire and keep in mind *comprehensive notions of harm*, ones that include but go beyond violations of the important interests in the protection of the individual's life, physical integrity, and legitimately acquired property. Only with a comprehensive notion of harm can an engineer recognize when engineering activities or practices are or may

---

consequences enhance, preserve, jeopardize, set back, or undermine the well-being of the various parties they affect. Of course, sometimes a single action or practice may both harm some parties and enhance the well-being of others, in which case one must look closely at whose well-being is enhanced, whose is harmed, and to what extents. For thoughts on such situations, see the discussion later in this chapter of Rawls's Difference Principle in relation to utilitarian decision-making.

be causing nonobvious harms (or creating unreasonable risks of such harms), and choose to act accordingly.

Put differently, just like medical students and doctors, engineering students and practicing engineers must grow *sensitive ethics antennae*. They must be able to detect when nonobvious ethical issues are present in engineering situations, for example, by being sensitive to inconspicuous harms that might result from engineering conduct or practices in those situations. Besides the changed sociological situation of the contemporary engineer, one of the major reasons the field of engineering has become a fertile area of ethical issues in contemporary times is that engineering activity has gradually come to be recognized as bearing on human well-being in ways that are sometimes more subtle, indirect, and/or intangible than when humans lose their lives, are injured, have their health degraded, or are deprived of their legitimate property.

THE FUNDAMENTAL ETHICAL RESPONSIBILITIES OF ENGINEERS

Many people, including more than a few engineering students and practicing engineers, believe that making ethical judgments is like expressing ice cream flavor preferences, that is, a matter of subjective taste that cannot be rationally debated. To the contrary, I submit that it is important for engineering students and engineers to realize that with thoughtful use of certain intellectual resources, it is possible to make ethical appraisals of engineers' conduct and practices that are *not* reducible to mere subjective expressions of individual taste. The intellectual resources I have in mind are *the four fundamental ethical responsibilities of engineers*.

Along with the injunction to preserve life, the most widely recognized foundational ethical responsibility of medical doctors is to "do no harm" to their patients through their work, whether by acts of commission or omission.[6] Engineers (and scientists) have a similar but not as widely recognized foundational ethical responsibility: to do no harm to *their* "patients," that is, to their fellow workers, employers, clients, users of

---

[6]This precept is also referred to as "the principle of nonmaleficence." See http://missinglink.ucsf.edu/lm/ethics/content%20pages/fast_fact_bene_nonmal.htm.

their products and systems, and, most generally, to all those affected or likely to be affected by their work and its products.[7]

Some contend that because the relationship between physicians and their patients is typically more direct than that between engineers (and scientists) and those affected by their work, engineers (and scientists) do not have the same fundamental ethical responsibility as doctors, that is, to do no harm to their patients. However, the fact that the relationship between engineers (and scientists) actions and the parties affected by their work products is often more *indirect* than in the case of physicians and their patients does not by itself exempt engineers (or scientists) from the fundamental ethical responsibility to do no harm to their patients (in the broader sense specified above). After all, *harm indirectly caused is still harm caused.*

Instead of limiting engineers' ethical responsibility to "do no harm" to their "patients," I contend that engineers have the broader ethical responsibility to **combat harm** that might be caused to others by their professional work (and/or by the work of others in which they are involved or by work about which they are technically knowledgeable).

Why is the more inclusive notion of "combating harm" preferable to that of "not doing harm"? First, it would ring hollow if someone, while not directly causing harm to another party, just observed, stood idly by, and did not even attempt to prevent harm that he or she was well positioned to keep from occurring to that party, and then attempted to justify that passive posture by noting that he or she had not *caused* the harm in question. Second, it would ring equally hollow for someone, knowing that a party was in some unpreventable harm's way, opted to remain silent about it and did nothing to let the party at risk of being harmed know what was coming, and attempted to justify that posture by stating that, after all, she or he had not actually *done* any harm to the party in question and could not have *prevented* it from happening. In short, "to combat harm" better captures the ethical responsibility that humans have in relation to harm than does "to do no harm." There is more to that relationship than simply not doing or causing (unjustified) harm to others.

What "combating harm" involves must, of course, be made explicit. The overarching ethical responsibility to "combat harm" can be

---

[7]This responsibility is essentially a special case of the bedrock ethical principle that it is wrong and impermissible to unjustifiably harm another human being (or, more broadly, another sentient being) through one's actions.

unpacked into three **Fundamental Ethical Responsibilities of Engineers** (henceforth: FEREs). Engineers, when functioning as engineers, have fundamental ethical responsibilities

> *to not cause harm*[8] *or create an unreasonable risk of harm* to others (or to public welfare or the public interest) through their engineering work. (**FERE1**)
>
> *to try to prevent harm and any unreasonable risk of harm* to others (and to public welfare and the public interest) that is caused by their engineering work, or by the engineering work of others in which they are involved, or about which they are technically knowledgeable. (**FERE2**)
>
> *to try to alert and inform about the risk of harm* those individuals and segments of the public at unreasonable risk of being harmed by their engineering work, by the engineering work of others in which they are involved, or about which they are technically knowledgeable. (**FERE3**)

Engineers employed by an organization or engaged by a client have a fourth fundamental ethical responsibility:

> *to work to the best of their ability to serve the legitimate interests of their employer or client.*[9] (**FERE4**)

Let us discuss in greater detail and indicate with more clarity and precision what each of these FEREs means and implies.

FERE1

FERE1 is the basic ethical responsibility of engineers to not *cause* harm or *create* an unreasonable risk of harm to others or to public welfare or the public interest through their work. More precisely, engineers have an ethical responsibility to not do anything in their work that will *cause or*

---

[8] In the statements of FERE1, FERE2, and FERE3, "harm" should be read as meaning "unjustifiable harm."

[9] This is a more defensible version of the categorical employee loyalty–to-employer canons found in a number of codes of engineering ethics. For example, as noted, Fundamental Canon I.4 of the *NSPE Code of Ethics for Engineers* states, "Engineers, in the fulfillment of their professional duties, shall act for each employer or client as faithful agents or trustees."

contribute to causing harm, or that will *create or contribute to creating* an unreasonable risk of harm, to parties affected or likely to be affected by that work.

It is essential to realize that FERE1 applies not only to acts of *commission*—acts deliberately undertaken and carried out—that cause harm or create an unreasonable risk of harm to others. It also applies to acts of *omission*—failures to do things or failures to do them with the level of care normally and reasonably expected of someone in the engineer's position—that cause or contribute to causing harm (or that create or contribute to creating an unreasonable risk of harm). An individual whose failure to do something, or failure to do it with the care normally and reasonably expected of someone in that individual's position, causes or contributes to causing harm or an unreasonable risk of harm to another and is guilty of the form of ethical irresponsibility called *negligence*. Negligence, including carelessness and recklessness, on the part of an engineer is no less incompatible with FERE1 than is an act done deliberately that the engineer realizes will unjustifiably harm another or put her or him at unreasonable risk of harm.

A parent has an ethical responsibility to not harm a small child deliberately, for example, by locking her or him into the family car parked in the sun and leaving her or him there for a long time on a very hot day, because, say, the child had been crying a lot that day. But suppose that a parent leaves her or his child in the car on a very hot day, although with no *intention* of harming it. Suppose further that, having run into an old friend in the store with whom the parent fell into extended conversation, the parent forgets that she or he left the child in the car, and when she or he finally returns finds that the child is injured, unconscious, or dead. Such a parental deed, although unintentional, would count as negligence and violate the parent's ethical responsibility not to cause harm or create an unreasonable risk of harm to a child. Similarly, failure by an engineer to check a vital engineering system for which he or she is responsible, or failure to do so in a diligent manner or with appropriate frequency, although acts of omission, would count as negligence and violate the FERE1-based derivative ethical responsibility to carry out the system checks with due care. Either omission would be ethically irresponsible and help cause whatever harm or unreasonable risk of harm results from it.

Some engineering students are skeptical that engineers would ever behave in an ethically irresponsible manner in their professional practice. Some who feel this way are uninterested in or even opposed to studying ethical issues in engineering work. Granted, engineers who *deliberately* harm someone through their work, or deliberately do something in their work that foreseeably puts others at unreasonable risk of incurring harm, are probably few and far between. However, the frequency of negligence in engineering is probably considerably greater than the frequency of actions deliberately undertaken even though the engineer-actor realizes they will cause harm or create an unreasonable risk of harm. Becoming aware of the diverse, sometimes subtle forms of negligence in engineering work is one reason the study of ethical issues in engineering is worthwhile. Negligence on the part of an engineer shows that engineering conduct need not be intentional to be ethically irresponsible.

It is worth noting that engineers have an ethical responsibility to not cause harm (or unreasonable risks of harm) not only to *individual parties* affected by their work but also to *public welfare* or *the public interest*. For example, FERE1 implies that the engineer has a basic ethical responsibility not to do anything that would pose (or contribute to posing) a significant risk to national security or to the safe, effective, and healthy state or smooth functioning of important resources on which the public depends, such as clean air and water, safe roads, and reliable power, transport, and communication systems. Thus, incompetently, malevolently, or negligently designing or mismanaging a public electrical grid, water supply, sanitation, fishery, or traffic control system to a point at which its safety, reliability, quality, or sustainability is impaired, diminished, or jeopardized would be incompatible with FERE1.

Engineering work subject to FERE1 is not limited to work *initiated* by the engineer in question. It can include work launched and directed by another but in which the engineer is a participant. This is so because the activities of an engineer participating in engineering work initiated by another can also be a factor that contributes to causing harm to parties affected by that work, or to the creation of an unreasonable risk of harm from that work.

Sometimes, actions or practices that cause harm or create an unreasonable risk of harm to some parties also have beneficial consequences to the same and/or other parties. Indeed, an engineer's activity or practice that causes or creates an unreasonable risk of harm to some might yield much more aggregate benefit than harm or risk of harm to others. This real possibility raises a difficult question: *does the fact that an engineering action or practice likely to result in harm to some parties but nevertheless likely to produce* positive net benefit *imply that the engineers involved have no ethical responsibility to refrain from performing that action or practice, and/or that, on the contrary, it would be ethically right for them to order or to carry out that action or practice?*

Some ethics thinkers take a strictly utilitarian view on this question and believe that the only consideration relevant to determining whether it is ethically proper to carry out some action, project, or practice is whether its projected net benefit is positive or, put differently, whether the projected benefits outweigh the projected harms.[10] This author has a different view. Sometimes an engineer can have an ethical obligation to *refrain* from carrying out a certain act or from engaging in a certain practice *even though* its projected beneficial consequences seem likely to substantially outweigh its projected harmful consequences. That an engineer's action seems likely to have more beneficial than harmful consequences does not necessarily imply that she or he has no ethical obligation to refrain from carrying it out and, on the contrary, should support it. Let us explore this idea in more detail.

Many decision-makers in contemporary U.S. society, in various professions, seem to operate in accordance with the following decision-making rule (**R1**): as long as the projected benefits of a proposed course of action exceed or outweigh its projected harms, or, put differently, as long as the ratio of the project's projected benefits to its projected harms is greater than 1, then it is permissible or right to carry out that action, practice, or policy, and there is no ethical responsibility to refrain from that action because of those projected accompanying harms.[11] Let us call R1 "The 'If Positive Net Benefit, Then Go Ahead' Rule." This thinking often colors deliberations about engineering projects like the building of dams, the construction of high-rise buildings and freeways, and the creation of nuclear power plants and waste-disposal facilities.

What rarely seems to enter into assessments by engineers (and other professionals) about such undertakings is whether any considerations apply in a particular situation that should *trump* calculations that yield positive net benefit. For example, suppose one social group is likely to bear a significantly disproportionate share of the harm, the risk of harm, or, more generally, the costs and risks of the action, practice, or project. If so, perhaps that fact should trump a benefit-harm-risk analysis that finds

---

[10] Some hold that for it to be ethically right or obligatory (as opposed to being ethically permissible or acceptable) to carry out that action, its net benefit must be greater than, or at least as great as, any feasible alternative course of action.

[11] Some go further and hold that when projected benefit exceeds projected harm, *failure* to act is ethically *wrong*.

positive net benefit for the action under consideration and block the use of R1 to justify proceeding with that action.

That there might be considerations that sometimes reasonably trump a standard benefit-harm-risk analysis that yields a positive net benefit, is suggested by John Rawls's Difference Principle. As elaborated in his *A Theory of Justice*,[12] Rawls's Difference Principle states that for the *unequal distribution* of a benefit or a burden among members of a group to be distributively just, two conditions must be satisfied. In the case of a benefit, (i) everyone in the group must partake of the benefit to some positive degree under the distribution **and** (ii) *those currently worst off must benefit the most* under the proposed or actual distribution. If what is being distributed is a burden, then an unequal allocation of the burden is distributively just only if (i) everyone in the group bears some part of the burden **and** (ii) *those currently worst off are burdened or disadvantaged the least* under the proposed or actual distribution. In short, for an unequal distribution of a good/benefit or a bad/burden among members of a group to be distributively just, it must make the greatest positive or least negative difference to the currently worst off.[13]

---

[12] Rawls (1999).

[13] This principle helps explain why outrage is often expressed when, in a difficult economic environment, cutbacks in salary, benefits, or resource allocations fall disproportionately more on the shoulders of workers or the least powerful citizens, and disproportionately less on those of the executive, top management, or the politically powerful class.

Rawls's ingenious justification for this principle is that it would be adopted as a rule for running society by a committee of disembodied rational beings gathered at a pre-societal conference and tasked with making the rules that will govern society when it comes into being. More precisely, Rawls holds that this principle would be adopted if these decision makers deliberated "behind a veil of ignorance," i.e., in circumstances in which they had no knowledge of their eventual human characteristics (e.g., race and gender) and their eventual societal characteristics (e.g., social class and economic resources). Rawls contends that these legislators would vote to adopt a policy under which the greatest benefit (smallest burden) of an unequal distribution of a social good (bad) would go to those currently worst off, since the legislators themselves could easily wind up being amongst the worst off when they were born, assumed a gender, acquired an ethnicity, had a nationality, and became members of a particular or economic social class. Rawls's Difference Principle is a rational insurance policy adopted to protect against the possibility of winding up among the worst off in society. It also reflects the moral intuition that the currently worst off deserve the greatest break when a social benefit or burden is being unequally distributed, whether it be tax breaks or water resource allocations. This principle is useful to deploy as a check against the possibility that one's view of the ethical propriety of a distributive action or policy is colored by the favorable position of the viewer. Rawls's Difference Principle invites one to become aware of the elements of one's current privileged situation that are due to good fortune and to adopt a policy that is not shaped by that privilege, but, rather, driven by recipients' degree of genuine need or other pertinent distributive criteria.

Is Rawls's Difference Principle applicable to situations that might be faced by practicing engineers? If Chinese civil engineers involved in the Three Gorges Dam project (1994–2012) were challenged about the effects of that project on the public, some of them might argue that although it did cause harm to some people—to make way for the dam, about 1.3 million rural people were involuntarily uprooted from their communities and moved to newly built settlements[14]—the benefit to Chinese society as a whole dwarfed the harm incurred by the forcibly displaced. To that argument, one could respond, in Rawlsian spirit, "Perhaps the project benefit did exceed the harm caused, but the way that project-caused harms and risks of harm were distributed is unjust. Rural people who were living along certain parts of the Yangtze River, now the Yangtze River valley, were among the economically worst off, yet they were forced to bear the bulk of the burden of harm so that the economically better off urban dwellers downstream could realize the lion's share of the benefits." Under Rawls's Difference Principle such distributions of the harm and risk of harm burden would be distributively unjust.

For adherents of Rawls's Difference Principle, the attempt to invoke rule R1 to justify proceeding with the Three Gorges Dam, or to justify its construction after the fact, should be rebuffed, because the harm-risk burden was unjustly distributed. Such individuals might view failure to satisfy Rawls's Difference Principle as a *trumping factor*, that is, as a factor that, when it applies, deserves to take precedence over any attempt to ethically justify a decision on a project by invoking rule R1, which requires only that there be positive net benefit, that is, more benefit than harm (or a benefit-harm ratio greater than 1).

Even if the benefit-harm ratio for a proposed course of action is greater than 1, or the net benefit is positive, it could well be that, rather than the harm/risk being inequitably *distributed* in the Rawlsian sense, the *magnitude* of the actual or risked harm might make it ethically appropriate to decline to act in order to realize the greater benefit. If the projected harm *exceeds some significant threshold of acceptability*, it might be ethically right, all things considered, to decline to carry out the action in order to avoid incurring the accompanying smaller but threshold-exceeding harm. Let us call this second decision-making rule of benefit-harm-risk analysis, "The 'Thanks, But No Thanks' Rule."

The upshot of this rule is that an engineer can have an ethical responsibility to not carry out a certain action or project, or to not engage in a

---

[14]Eckholm (1999).

certain engineering practice, *even if* the projected benefit of doing so exceeds in magnitude the projected harm or risked harm. But this will be so *only if* at least one trumping factor makes it reasonable to set aside rule R1 and decline to realize the greater benefit in order to avoid incurring the harm or distribution of harm that would accompany it. Thus far, two possible trumping factors have been identified that might induce an engineer to decline to act in accordance with rule R1: (i) the expected harm or significant risk of harm is unjustly distributed, and (ii) the expected harm, although less than the expected benefit, exceeds in magnitude the burden deemed the maximum acceptable one. Each engineer must reflect and decide for herself or himself what items are on her or his *personal list of factors that, if they apply, trump the invocation of R1 in a particular engineering situation.*

FERE2

FERE2 is engineers' basic ethical responsibility to *try to prevent harm* (and any unreasonable risk of harm) to others, to public welfare, or to the public interest—harm that is or would be caused by their own engineering work, by engineering work of others in which they are involved, or engineering work about which they are technically knowledgeable. Even if an engineer is not doing anything in his or her work that will deliberately or negligently *cause* harm or create an unreasonable risk of harm, he or she has an ethical responsibility to try to *prevent* any impending harm or unreasonable risk of harm that may have been set in motion by the work in question. As used here, "prevent" encompasses various modalities of prevention, such as "lessen the severity of," "diminish the scope of," and "prevent the repetition of" the harm or risk of harm in question.

Standing by and simply observing the harm that is about to occur or is taking place is no more an ethically responsible option for the engineer than it would be for someone to go to a neighbor's home to borrow some flour, discover an unaccompanied infant in a bathtub rapidly filling up with water, choose not to intervene, and simply watch the child drown in the tub. While the person who goes to the neighbor's house normally has the mental and physical ability to prevent the harm by removing the endangered child from the bathtub, an engineer *may* have sufficient technical credibility and/or insider knowledge about some work/project or product such that her or his public warning that the former will cause harm if it goes forward or the latter is produced might help prevent or lessen the harm.

Two additional clarifications are in order here. First, FERE2 is not an ethical responsibility to *prevent* harm but to *try to prevent* harm. There is no ethical responsibility to prevent harm when it is impossible for an engineer to do so. One can have an ethical responsibility to do x only if doing x is practically possible.

Second, suppose the very attempt to prevent harm or the creation of an unreasonable risk of harm might itself cause harm or create a risk of harm to the would-be preventer and/or to others. In that case, the strength or weight of the ethical responsibility to try to prevent the original harm on the would-be harm preventer is directly proportional to the magnitude of the original harm that he or she might prevent, but inversely proportional to both the magnitude of the (new) harm that the would-be harm preventer might *cause* and the likelihood that he or she would cause it by trying to prevent the original harm.

FERE3

FERE3 is the basic ethical responsibility of engineers to *try to alert and inform* individuals and segments of the public put at significant risk of harm by their engineering work, work with which they are involved, or work of others about which they are technically knowledgeable. Even if an engineer is not *causing* harm or creating an unreasonable risk of harm, and even if some harm or unreasonable risk of harm from pertinent engineering work cannot realistically be *prevented*, the engineer may still have an ethical responsibility to *try to alert and inform* parties at risk of incurring harm that they are vulnerable. Being thus alerted and informed might at least allow them to prepare for and take steps to minimize the impact of the harm the parties are at risk of incurring.

Here too, several clarifying comments are needed. First, as with FERE2, note the qualifying phrase "to try." An engineer who took all reasonable steps to try to alert and inform vulnerable parties about the risk of harm would fulfill FERE3 even if circumstances kept her or him from succeeding in doing so. Second, the harm involved in an FERE3 situation could be any of a range of kinds of harm, from a physical injury to a major property value loss to a serious violation of privacy. Third, the engineer who has this ethical responsibility can owe it to various kinds of "parties," for example, another individual, a social group, a valuable cultural institution, or society as a whole.

FERE4

FERE4 is the basic ethical responsibility of engineers working for an employer or a client to work to the best of their ability to serve the legitimate interests of that employer or client. Several clarifications and qualifications are also needed here.

First, FERE4 is a *conditional* responsibility. It is binding on the engineer only as long as her or his employer or client treats her or him reasonably as regards compensation and working conditions (including safety, health, and career development opportunities). It ceases to apply to and is no longer binding upon an engineer if her or his employer or client treats her or him poorly, unfairly, or unreasonably in more than a fleeting way.

Second, it is critical to note the qualifier "legitimate" in "legitimate interests." Engineers do not have a fundamental ethical responsibility to do the best they can to serve or promote *any and every* interest of their employer or client. Rather, engineers have that responsibility only with respect to employer or client interests that are "legitimate," such as the employer interest in having its engineer-employees do high-quality, cost-effective work in a timely way. FERE4 does not apply to and is not incumbent upon an engineer if an employer or client interest is illegitimate. Thus, even if her or his employer or client instructs an engineer to do so, she or he does not thereby have an ethical responsibility to do her or his best to (i) improperly obtain or use a competitor's intellectual property, (ii) bring a new engineered product to market without adequately testing its safety, or (iii) manipulate a government automotive emissions test by covertly installing engine software that keeps noxious vehicle emission levels below regulatory limits when the vehicle is tested statically but permits them to significantly exceed regulatory limits when it is operated on the road.

Third, consider the word "client" in FERE4. In contemporary societies, public resources enable or facilitate, directly or indirectly, wholly or in part, the work of many if not most engineers. "Public resources" include government grants, contracts, fellowships, scholarships, loans, and various sorts of publicly initiated or supported infrastructure, such as the Internet, libraries, national research laboratories, the transportation network, the communication network, and the electrical power grid.

Therefore, whether he or she works for a private or governmental employer or client, it makes sense for the contemporary engineer to view *society at large* as her/his **omnipresent background client.**

Under FERE4, engineers (and scientists) must *always* work to serve the legitimate interests of this client—the public—to the best of their ability. Moreover, *when the private interests of the individual engineer, or the interests of her or his private sector or governmental employer or client, conflict with the legitimate interests of society at large—the "omnipresent background client"—the interests of the latter must take precedence over the former.*

Societal resources are the sine qua non of much if not most contemporary engineering activity. If engineering activity repeatedly harmed public welfare or violated the public interest as a by-product of serving private or bureaucratic interests, the engineering profession would risk losing not only significant enabling resources provided to it by the public but also the authority that society has granted the profession to set its own standards for education and admission to practice.[15]

Fourth, FERE4 is binding on the engineer only if a second condition is satisfied. Suppose an engineer, Smith, is an employee of the U.S. National Security Agency (NSA). Protecting U.S. national security is clearly a legitimate NSA interest. Suppose further that, pursuant to that interest, NSA instructs engineer Smith to design a new software program that, when applied, would covertly gather personal medical and financial data about a large number of U.S. Internet and email users. Use of this work product would arguably cause significant harm by violating the privacy of those whose records were covertly captured. Put differently, Smith's design activity, while in accord with FERE4, would conflict with FERE1.[16]

The point is that even if a legitimate employer interest is asserted, FERE4 is binding on an engineer-employee only if her or his action, undertaken to serve that interest, does not violate FERE1 by causing significant unjustified harm or creating an unjustified unreasonable risk of harm. If the endeavor does or is likely to do so, FERE4 *by itself* does not warrant or justify such an action. In a specific situation when FERE4 and FERE1 conflict, the engineer should bring to that situation a *presumption* that FERE1 takes precedence over FERE4. However, he or she should also be open to the possibility that in that specific situation a compelling

---

[15] Society has granted that authority to the engineering profession in return the profession's pledge to protect public welfare and the public interest in engineering work.

[16] I am indebted to Samuel Chiu for posing a question that prompted me to ponder this qualification on the applicability of FERE4.

case might be able to be made that overcomes that presumption, thereby allowing FERE4 to take precedence over FERE1 and the engineer to act accordingly.[17]

* * *

It is important for engineering students and practicing engineers to become adept at applying the FEREs to the kinds of work situations they are likely to face. If an engineer has internalized the FEREs and learned to determine which of them apply to a particular work situation; if she or he has discerned the key technical and social features of the work situation in which she or he finds her or himself; if she or he is aware of the key features of the contexts in which the conduct is occurring; and if she or he makes a serious effort to estimate the likely effects of the engineering conduct in question on the well-being of those affected by it, then she or he will be able to ascertain the more *specific* ethical responsibilities that are incumbent on her or him in that work situation. This grounded approach is profoundly different from resorting to intuition, religious beliefs, or the precepts of a static engineering society code of ethics as the basis for making ethical judgments about engineering conduct.

The importance of the FEREs notwithstanding, they and whatever derivative ethical responsibilities are incumbent on an engineer in a concrete situation do not always provide the last word on what the engineer should do in that situation. A particular derivative ethical responsibility, for example, the responsibility of an engineer-employee to blow the whistle publicly on her or his employer for engaging in what she or he believes to be harmful misconduct, could be outweighed by considerations that justify acting otherwise, such as concern that blowing the whistle under the prevailing circumstances might imperil national security. That caveat recognized, identifying the applicable FEREs is *always* a relevant and

---

[17] In the example of engineer Smith and the NSA, FERE1 and FERE4 appear to conflict. Smith should approach her or his consideration of the situation with the presumption that not violating FERE1 takes precedence over not violating FERE4. However, Smith should also be open to the possibility that a compelling case might be able to be made that would justify overriding that presumption and justify his or her designing the requested software. Whether such a case could be made would depend on *the specifics of the situation*, such as the magnitude and scope of the harm likely to be done to civilians through use of the software, the weight of the NSA's national security interest that prompted the assignment to develop the software, the magnitude and risk of the harm to the country that could result from not developing and using that software, and the possible existence of alternative courses of action that would further NSA's legitimate national security interest as much as would developing and using the software at issue but would cause significantly less harm to civilian informational privacy interests.

important aspect of evaluating an engineering situation from an ethics perspective. Engineers' situation-specific derivative ethical responsibilities are *always important initial guides* to what they should do in that situation and should *always* be given serious consideration. The guidance the FEREs provide should be viewed as provisionally binding, with those who believe the engineer should act otherwise bearing the burden of making a compelling case to that effect.

Case Studies of Ethical Issues in Engineering

This chapter contains eighteen case studies of ethical issues in engineering. Reflecting on them should impart to the reader a more comprehensive notion of harm. Such a notion is important for ethically responsible engineering practice. If engineers recognize the various kinds of harm that engineering activity can produce, they can shape their practice to avoid causing them. However, a narrow notion of harm, for example, one with only physical and financial components, may leave an engineer unaware that his or her professional activities can have subtle harmful effects.

In the author's experience, the cases of ethical issues in engineering that work best with engineering students are ones anchored in real-life episodes. Almost all the cases that follow are rooted in actual events. They involve various fields of engineering: from mechanical, civil, and chemical, to electrical, computer science and systems, and biological engineering. Some focus on hardware, others on software. They have diverse outcomes: some positive and encouraging, others tragic and troubling. Some of the cases have the useful feature of incorporating the words and perspectives of engineers involved, rather than relying solely on the views of nonengineers. About half of the cases have previously been explored in the scholarly literature on engineering ethics, though not by using the FEREs. The others are partly or wholly new contributions.

Each case discussion begins with some *background* on the situation in question. This is followed by *ethical analysis* of the case, with attention to, among other things, how the FEREs apply to the actions carried out by engineers in the specific case situation. Discussion then follows of some of the more important *lessons* the case suggests, sometimes coupled with explicit statement of the key ethical issues that arose and were explored. The exploration of each case concludes with several *discussion questions* for the reader to ponder.

Before turning to the cases, a critical intellectual issue merits brief discussion. The ethical judgments about engineering conduct made in a number of the case discussions take as their point of departure the fact that the conduct of some engineer(s) contributed to bringing about harm or creating a risk of harm for affected parties. For this reason, it is important to clarify the pivotal (and notoriously slippery) concept of "cause," as in "*x* caused *y*."

When a speaker says that someone or something, *x*, "caused" some outcome or result *y*, the term "caused" can have any of several senses. What the speaker might mean is that *x precipitated, triggered,* or *initiated* the occurrence of *y*. However, the speaker could also mean something quite different and less obvious: that *x* was *conducive to* the occurrence of *y*, that *x facilitated* the occurrence of *y*, that *x* was the *stimulus* that elicited *y* as a *response*, or that *x enabled y* to occur.

For example, suppose someone lights a firecracker fuse with a match, and a few seconds later, the firecracker explodes. The lighting of the fuse would be the precipitating or initiating cause of the explosion. But it is also true that the explosive powder's being dry was an enabling causal factor in the explosion. While this factor did not directly cause the explosion to occur, it made it possible, in that without it, the explosion would not have occurred. Depending on the outcome phenomenon under scrutiny, an adequate account of what "caused" it might well need to include both kinds of factors: precipitating/triggering and enabling. Indeed, in some cases, an adequate causal explanation of an outcome might also need to include various "conducive," "facilitative," and "stimulative" factors.

In short, depending on the nature of the outcome being examined, multiple factors of different sorts may have contributed to bringing it about. I shall call them **contributory causal factors**. Unfortunately, observers (and policy makers) are often tempted to focus exclusively on one kind of contributory causal factor: **foreground precipitating factors**, like an engineer's breakthrough invention or decision to take some action. They may overlook other kinds of contributory causal factors, such as **background enabling** or **facilitating factors**. Factors of such sorts often set the stage for some foreground factor to initiate or trigger the outcome.

When making an ethical judgment about an engineer's conduct in a particular engineering-in-society situation in relation to the outcome that ensued, *it is essential to not limit one's attention just to foreground precipitating or initiating causal factors.* One must also be alert to and on the lookout for possible background causal factors of various sorts, each of which may have helped set the stage, in different ways, for the precipitating factor that triggered the outcome of interest. Indeed, the foreground precipitating or initiating factor may have exercised its causal influence on the outcome only because the various background stage-setting factors configured the engineering-in-society situation in a particular way. With these thoughts about causation in mind, we now turn to the case studies.

Table 1: Case Relevance by Field of Engineering

Case	Aeronautical	Bio-	Chemical	Civil/Environmental	Computer Science	Electrical	Management Science & Engineering	Materials Science & Engineering	Mechanical
1					X	X	X		X
2	X				X	X			
3						X		X	X
4					X	X	X		
5					X				
6				X					
7			X	X					X
8	X						X		X
9								X	X
10			X		X	X		X	X
11									X
12				X					X
13	X				X	X			
14				X					
15				X					
16							X	X	X
17					X				
18		X							

CASE 1: THE CADILLAC DEVILLE/SEVILLE ENGINE-CONTROL CHIP

Background

In 1990, General Motors (GM) received reports that Cadillac DeVille and Seville passenger cars were exhibiting a tendency to stall when the system responsible for heating and air conditioning (HAC) was functioning.[1] To solve this problem, GM engineers designed a new engine-control computer chip (ECC). When a sensor detected that the climate-control system was turned on, the chip sent signals that resulted in the injection of more fuel into the car's engine, thereby enriching the fuel/air mixture beyond what it was when the climate-control system was turned off.[2] The 1991-model-year DeVilles and Sevilles were equipped with the new chip, and it was used in these Cadillac models through the 1995 model year.[3]

The new ECC solved the stalling problem per se.[4] However, combustion of the increased amounts of fuel injected into the engines to prevent stalling produced exhaust that overwhelmed the cars' catalytic converter and sent additional carbon monoxide (CO) out the tailpipe. Tailpipe emissions reached levels of up to 10 grams of CO per mile per car, compared with the 3.4 grams per mile permitted under the Clean Air Act.[5] The Environmental Protection Agency (EPA) discovered the excess pollution during routine testing in 1993.[6] According to the EPA, the roughly 470,000 cars equipped with the new chip sent about 100 million extra tons of poisonous CO emissions into the environment.[7] The Department of Justice sued GM for exceeding the Clean Air Act's limits for CO.

In 1995, GM reached a settlement with the federal government, paying $11 million in fines, at the time the second largest civil fine ever imposed under the Clean Air Act and "the largest involving pollution from

---

[1] Cushman (1995). The HAC system is also called the "climate-control system."

[2] Ibid.

[3] Ibid.

[4] The new ECC also enabled the cars to meet government emissions standards. Since the car's climate-control system was turned off while the vehicle was undergoing stationary emissions testing, the ECC did not send a signal to enrich the fuel/air mixture. By sending signals to enrich the fuel/air mixture only when the car's climate-control system was turned on, typically when the vehicle was being used in normal driving, the new ECC had the effect of concealing the fact that more pollution was being emitted during normal driving than during stationary testing. The reader may note similarities (and differences) between this case and the Volkswagen TDI diesel emissions scandal that came to light in 2014–15.

[5] http://usdoj.gov/opa/pr/Pre_96/November95/596.txt.html.

[6] Ibid.

[7] Ibid.

motor vehicles."[8] GM also agreed to recall and fix the 470,000 Cadillacs at an estimated cost of $25–$30 million, and to pay "up to $8.75 million on projects to offset emissions from these vehicles."[9]

EPA's concern with the effects of the increased CO emissions on the environment stemmed from carbon monoxide air pollution's long-term harmful effects on human well-being. As the then attorney general Janet Reno observed, "Carbon monoxide can cause cardiopulmonary problems and can lead to headaches, impaired vision, and a reduced ability to work and learn."[10]

This narrative raises two questions. First, what engineering-related ethical issues did the engineers involved face in this case? Second, did how the engineers involved in this episode chose to act in relation to those issues fulfill the specific ethical responsibilities derivable by applying the FEREs to the particulars of this situation?

Ethical Analysis

A striking feature of this case is the public anonymity of the engineers involved. To the best of the author's knowledge, to date, outside of GM, nothing is known about the identities or specific deeds of the engineers involved in this case, other than that they designed a new ECC. None of them has pulled back the veil of corporate secrecy, stepped forward and self-identified, and testified about what went on behind the scenes. Nor, to the author's knowledge, is it known publicly whether management specifically *instructed* the engineers tasked with solving the stalling problem to do so by designing a new ECC that altered the fuel/air mixture injected into the cars' engine. It is also possible that management gave the engineers latitude to solve the problem by whatever means or approach they thought best. Given these gaps in knowledge, let us consider the ethics ramifications of several possible scenarios.

Suppose management instructed the engineers to take the approach of designing a new ECC, and literally or effectively ordered them not to consider any other approach. Under FERE1, the engineers in question should at least have taken all reasonable steps to assure that the "cure" they devised did not also create a serious new "illness" (i.e., a new harm

[8]Ibid.
[9]Ibid.
[10]Ibid.

or significant risk of harm). That is, they should have taken appropriate steps to ensure that the new ECC would not only solve the stalling problem but that in doing so it would not cause significant, comparable, or greater harm (or risk of harm) than the harm (or risk of harm) it prevented. If the engineers focused only on quickly devising a technically viable and cost-effective way of stopping the stalling and did *not* check to see whether the new chip would significantly negatively affect the natural environment by indirectly producing substantially increased levels of noxious CO tailpipe emissions, then they would arguably have been negligent, something incompatible with FERE1, the engineer's ethical responsibility not to cause harm or an unreasonable risk of harm, by commission or omission.[11]

Against this line of reasoning, it might be argued that the engineers in question were probably trained as electrical engineers, mechanical engineers, or computer systems engineers. Therefore, it would be unreasonable to expect them to have seriously considered the environmental consequences of the new ECC that solved the stalling problem. But even if recognition of the importance of not harming the environment is deemed something that would be unreasonable to expect of such engineers—a claim with which, given the era in which we live, the author disagrees—heeding environmental regulations is critically important to automaker profitability. This alone makes it extremely hard to believe that none of the engineers tasked with solving the stalling problem thought to inquire whether the engines into which a richer fuel/air mixture was being injected to stop the stalling would produce incremental exhaust, and, if so, how much more and to what effect.

It is even harder to believe that none of the design engineers involved was aware that catalytic converters do not have unlimited conversion capabilities. That they don't implies they can be overwhelmed by sufficiently large amounts of engine exhaust. When that happens, some toxic CO from the engine is not oxidized to produce carbon dioxide (CO_2) but, instead, is emitted unchanged through the tailpipe and into the environment. It would have been prudent and responsible engineering practice

---

[11]Solving the stalling problem with the new ECC prevented some harm by saving the lives of certain drivers and passengers, namely, those whose cars, equipped with the old chip, would have stalled in traffic and been at serious risk of a deadly collision or of leaving the roadway when the car stalled. However, the greatly increased CO pollution emitted by the 470,000 DeVille and Seville cars in which the new chip was installed probably adversely affected the health of many vulnerable people exposed to the pollution. Those negative effects cannot be neglected under FERE1 because they did not instantly kill or injure those exposed to them, as a stall-caused collision or sudden out-of-control road departure could have.

for the engineers to check whether the catalytic converters in the DeVilles and Sevilles slated to be equipped with the new ECC would be able to handle the increased CO exhaust generated by combustion of the richer fuel/air mixture. Indeed, under FERE1, it would have been ethically irresponsible for the engineers not to explore this matter.

If, however, the engineers *did* check out the emissions consequences of the redesigned ECC, either personally or via commissioning an inquiry by qualified specialists, and knew or suspected that the resultant emissions would be significantly incrementally harmful to the environment, the question arises of whether they fulfilled FERE2. This they could have done by, at a minimum, speaking up in house, letting their supervisors or managers know clearly and explicitly about the significant incremental pollution they believed or suspected would be produced if the stalling problem was solved in this way. Doing so would arguably have fulfilled the engineers' derivative ethical responsibility to try to prevent the incremental harm or unreasonable risk of such harm that would result from use of the new ECC.

Suppose the engineers involved tried but, because they were rebuffed, failed to prevent the incremental harm that would foreseeably result from installing the new chip. Then, under FERE3, another derivative ethical responsibility would remain: to try to alert and inform appropriate parties at risk of incurring harm, especially children and elderly people—more practically, those charged with protecting the health interests of those groups—that they were at risk of harm from the environmental effects of running the engine with the newly programmed chip. The author knows of no evidence that the engineers tried to prevent the increased harm by making the effects of the new ECC known to their immediate supervisors or other superiors in hopes they would be directed to pursue another approach to solving the problem, or that they did anything to alert representatives of the groups at greatest risk of being medically or cognitively harmed. However, without pertinent evidence, the possibility that the engineers actually did the former in house, and the latter in a discreet way, cannot be ruled out.

Now consider the second scenario. Suppose the engineers tasked with solving the stalling problem were actually given the latitude to pursue *whatever* affordable and effective technological approach they thought best. Then either they *knew* of another such approach that would solve the stalling problem and was less harmful to the environment than the redesigned ECC, or *didn't know but should have known* that another such approach existed, or *reasonably did not know* of any such approach.

Suppose they *did* know of another such approach but set it aside and recommended what they knew was the more-pollution-producing new ECC. Then, under FERE1, they arguably would have violated the derivative ethical responsibility to not act so as to foreseeably produce more net harm than needed to solve the stalling problem.[12]

The same would be the case if they did *not* know of any such approach but *should* have known of one. In this situation, the general ethical responsibility to not cause harm or an unreasonable risk of harm (FERE1) implies a derivative ethical responsibility to do due diligence before acting, here by checking to see if there were other affordable, technically viable, less-pollution-producing approaches. Failing to do so would arguably be negligent and inconsistent with FERE1.

Even if the engineers *reasonably* did not know of any other less noxiously polluting alternative way of resolving the stalling problem,[13] under FERE2, they would still have had a derivative ethical responsibility to at least inform management that the only solution they knew of would yield significant incremental CO emissions. Moreover, if the company publicly denied or remained silent about the incremental CO emissions, then under FERE3, the engineers might have had a derivative ethical responsibility to try to alert and inform representatives of those groups whose health would be especially negatively affected as a result of using the new ECC.[14]

Lessons

Arguably the most important ethics-related lesson suggested by this case is that choosing too narrow a system boundary or context for thinking about and carrying out design work on an embedded device can lead to unintentional violation of an engineer's fundamental ethical responsibilities. Situating one's engineering design work within an unduly narrow *technical* system context, for example, the {*engine* + *ECC*} system, or the less narrow {*engine* + *ECC* + *HAC equipment* + *sensor* + *fuel injectors*

---

[12] One possible exception: if the more-harm-producing solution seemed likely to yield a more equitable distribution of the harms borne than the less-harm-producing one.

[13] Of course, launching a major redesign of the car's fuel-electrical-emissions-control system might have been an option, but it was probably ruled out on grounds of time and cost.

[14] See Case 11 for discussion of the conditions under which an engineer arguably has an ethical responsibility to publicly "blow the whistle" in an attempt to prevent harm or an unreasonable risk of harm.

+ *emissions-control equipment*} system, increases the chances the engineer will overlook important, possibly harmful consequences of using the device being designed. In this case, under FERE1 the engineers had a derivative ethical responsibility to situate their design work in a broad *socio-technical-environmental* system context, for example, the {*driver + engine + ECC + HAC + sensor + fuel injectors + emissions control equipment + tailpipe + natural environment + humans living in that natural environment*} system. That was the broad system context within which the new chips were going to function and within which they could (and did) have noteworthy harmful effects.

In short, an important general ethics-related lesson of this case is that *choosing too narrow a system boundary or context within which to situate one's engineering design work on an embedded technological device can lead to cognitive myopia.* This can result in failure to consider some possible downstream harms, an inadvertent but blameworthy violation of the FERE1 responsibility to not cause harm or contribute to creating an unreasonable risk of harm, by commission or omission.

This case also suggests a second noteworthy ethics-related lesson. The environmental harm at the center of this case was made up of individually small components. Put differently, the harm to the environment was aggregative in nature. I call such harms **public harms of aggregation**. The extra CO pollution emitted by one or a few individual cars with the new ECC was arguably negligible, since it did not by itself significantly increase the risk of harm to those breathing the more polluted air. But the aggregated harm to the public from almost a half million cars over five model years was far from innocuous. It posed an unreasonable risk of significant harm and was clearly ethically unacceptable.

To see that the outcome was ethically unacceptable, imagine if the aggregate CO emitted from the almost half a million cars equipped with the new chip had been produced by just one or a small number of cars in one or a small number of episodes. The resultant environmental degradation would clearly and properly have been perceived as ethically unacceptable. Such an outcome would have constituted a blameworthy harm that would have justified limiting the liberty of the party or few parties that caused it.

The negative effects of one or a few specimens of a defective technological design may not be large enough to capture our attention, induce us to pass critical ethical judgment, and act accordingly. However, when that design is mass-produced, the aggregate effect of a large number of individually small negative effects can be extremely detrimental. *Would-be ethically responsible engineers must therefore be alert not only to whether their products are individually harmful but also to whether they yield or are likely to yield any public harms of aggregation.* They must acquire a disposition to reject or, at the very least, give serious consideration to rejecting, as ethically irresponsible, otherwise-effective engineering designs that in practice lead to significant aggregative harms as by-products.

* * *

Little is known about the conduct of the specific engineers involved in this case. Given that fact, my analysis of the different possible scenarios did not permit me to make categorical ethical judgments about the actions of the individual engineers involved, whatever those respective actions actually were. I was obliged to limit myself to making hypothetical, scenario-specific judgments. But reaching a definitive or categorical ethical judgment about the engineers' actions was *not the key point of the discussion.* More important than reaching such ethical judgments about the engineers' conduct are (i) unpacking the ethical dimension of the case, (ii) identifying the most important and nonobvious ethical issues, and (iii) seeing under what sets of conditions it would be fair to say that the engineers involved had or had not fulfilled specific ethical responsibilities incumbent upon them because of how the FEREs applied to the circumstances of the case.

The reader should keep this in mind in exploring the remainder of the cases. Becoming familiar with the ethical concerns explored and the ethical issues targeted in the cases, and with the ideas and kinds of thinking brought to bear upon them, is more important than whether the reader agrees with the author's views about the specific ethical responsibilities of particular engineer-participants.

Discussion Questions

1. Was the choice of "system boundary," in relation to which the new ECC was to be designed, an ethically significant act for the design engineers involved? Why or why not?
2. Did the engineers tasked with designing the new ECC have a derivative ethical responsibility to situate their work in a suitably broad and inclusive context? Why or why not?
3. Did the engineers tasked with designing the new ECC have a derivative ethical responsibility to make their design reflect the best

estimate of the *aggregate* incremental CO pollution that would be emitted in normal driving by all cars to be equipped with the device? Why or why not?

CASE 2: SDI BATTLEFIELD MANAGEMENT SOFTWARE

Background

In March 1983, the then U.S. president Ronald Reagan launched the Strategic Defense Initiative (SDI). The professed goal of this R&D project was to render nuclear ballistic weapons "impotent and obsolete."[15] This was to be done by developing a sophisticated multisatellite-based computer-controlled intercontinental ballistic missile (ICBM) defense system. The system would sense and identify enemy long-range nuclear missiles in their booster phase, then destroy them with emissions from powerful lasers. These devices would be located on satellites and powered with energy drawn from onboard nuclear reactors or solar energy systems.

In contrast with Case 1, an engineer involved in an advisory capacity in this case has provided personal testimony about what transpired.[16] David L. Parnas, a U.S.-born software engineer who became a Canadian citizen in 1987, earned BS, MS, and PhD degrees in electrical engineering from Carnegie-Mellon University.[17] In May 1985, Parnas became a consultant-member of the Department of Defense Strategic Defense Initiative Organization's (SDIO) Panel on Computing in Support of Battle Management.[18] For serving on this panel, he received an honorarium of $1000 per day.[19] Considered an expert on "the organization of large-scale software systems,"[20] Parnas eventually reached the conclusion that building a trustworthy computer-controlled space-based anti-ICBM weapons

---

[15] http://www.fas.org/spp/starwars/offdocs/rrspch.htm.

[16] Parnas (1987).

[17] In 2007, Parnas was corecipient of the IEEE Computer Society's 60th Anniversary Award for "insights into making large-scale systems development manageable with the concepts of encapsulation and information hiding, and helping to establish software development as an engineering discipline firmly rooted in mathematics." See http://www.computer.org/portal/web/awards/60thaward-parnas.

[18] Parnas (1987, p. 46.

[19] Ibid.

[20] Ibid.

system was not realistic. He eventually resigned from the panel and campaigned publicly against SDI.

Ethical Analysis

At least two ethical issues in this case are worth examining. The first is what Parnas should have done when he reached the conclusion he did about the practical viability and reliability of SDI. The second revolves around the conflicts of interest that allegedly confronted various technical practitioners involved in this case.

I begin by exploring the issue of what Parnas should have done when he reached the conclusion he did about SDI.

Parnas was keenly aware that as a technical professional he had certain ethical responsibilities, ones closely related to several FEREs. He referred to "the individual responsibility of the professional" as going "beyond an obligation to satisfy the short-term demands of the immediate employer," a claim that evokes one of the limitations of FERE4 discussed in Chapter 3. He also recognized another ethical responsibility: "I cannot ignore ethical and moral issues. I must devote some of my energy to deciding whether the task I have been given is of benefit to society." Although he put it positively in this formulation, Parnas clearly believed he had an ethical responsibility to avoid devoting himself to endeavors that would be harmful to society. Further, as indicated by his claim that "blow[ing] the whistle" is "sometimes necessary," he also believed that the technical professional has a responsibility to try to prevent harm to "society"—in FERE2's terms, to public welfare or the public interest—and that under certain conditions that responsibility may require the technical professional to resort to public whistleblowing.

As a member of the SDIO Panel, Parnas came to the conclusion that a satellite-based computer-controlled anti-ICBM system could not be devised that could be known to be "trustworthy."[21] The main reason he reached that conclusion involved the software of the system, "the glue that holds such systems together."[22] Unless the software was tested under battlefield conditions, there was no way of knowing whether it had bugs that would disable the entire software-hardware system. "No experienced person trusts a software system before it has seen extensive

[21] Ibid., pp. 47, 49, and 52.
[22] Ibid., p. 48.

use under actual operating conditions."[23] "Realistic testing of the integrated hardware and software is impossible. Thorough testing would require 'practice' nuclear wars, including attacks that partially damage the satellites."[24] Moreover, there could be no way of knowing a priori that communication between satellites tracking the launched missile(s) would remain intact in the face of enemy attacks on any part of the integrated system. The trustworthiness of such a system was also undermined by the fact that "SDI software must be based on assumptions about target and decoy characteristics, characteristics that[,] since they are controlled by the attacker," are not knowable in advance by the defender.[25]

Because of such considerations, Parnas could not convince himself either "that it would be possible to build a system we could trust" or "that it would be useful to build a system we did not trust."[26] Because of his disbelief in the practical technical viability of the proposed system, Parnas was concerned about the immense cost of "the whole multibillion-dollar system."[27] NATO, he believed, was "wasting an immense amount of effort on a system it couldn't trust."[28]

Moreover, since the system was intrinsically unable to be known to be trustworthy, the United States would not give up its nuclear weapons. In turn, the Soviet Union, not sure that the U.S. system *wouldn't* work, would launch a major effort to come up with new offensive weapons that couldn't be defeated by the system. This in turn would prompt the United States to develop a new round of weapons of its own to defeat its enemy's new weapons. For this reason, Parnas believed SDI risked triggering a costly and dangerous new arms spiral.[29]

On June 28, 1985, Parnas *resigned from the SDIO advisory panel*. He did so because he believed that its support of SDI and approval of spending much more money on SDI software R&D was harmful to the country, and because he thought that the new arms race likely to be ignited by SDI would make international relations more risky: "SDI endangers the safety of the world."[30] In effect, Parnas believed that by continuing to participate in the panel's work, he was lending his name, prestige, and intellectual capital to actions that were harmful to the United States and

[23] Ibid.
[24] Ibid.
[25] Ibid.
[26] Ibid., p. 47.
[27] Ibid., p. 48.
[28] Ibid., p. 47.
[29] Ibid.
[30] Ibid., p. 52.

to the cause of international peace.[31] Parnas's resignation from the committee was in accord with FERE1.

Later he decided to *campaign publicly* against SDI. That decision had a different origin. When Parnas decided to resign from the SDIO panel, he believed he had a professional obligation to explain to SDIO why he was doing so. Therefore, with his letter of resignation to SDIO, he submitted eight short essays he had written while a member of the panel, "explaining why he believed the software required by the Strategic Defense Initiative would not be trustworthy."[32]

> When [SDIO] did not respond, I went up the executive structure, up to the President's Science Advisor. When that was clearly not going to be effective, I wrote to key Congressmen and Senators. It was only after those steps that the resignation went public.[33]

Parnas claims he never gave or sent a copy of his letter of resignation to the *New York Times*. Nevertheless, the reasons he gave SDIO for believing that the software required by SDI could never be trustworthy eventually became the focus of a *New York Times* article about Parnas's opposition to SDI.[34] SDIO's public response to this article, as Parnas perceived it, was not an attempt to rebut his arguments on their merits but "to mislead rather than inform the public."[35]

> When I observed that the SDIO was engaged in "damage control," rather than a serious consideration of my arguments, I felt I should inform the public and its representatives of my own view. I want the public to understand that no trustworthy shield will result from the SDIO-sponsored work. I want them to understand that technology offers no magic that will eliminate the fear of nuclear weapons. I consider this to be part of my personal professional responsibility as a scientist and an educator.[36]

This account has a distinct aura of FERE2 about it. Parnas wanted to make sure the public did not succumb to the illusion that SDI would render it safe from nuclear weapons, hence enabling and perhaps tempting

[31] Another reason Parnas resigned from the panel had to do with his concern about fraud: "[T]aking money allocated for developing a shield against nuclear missiles, while knowing such a shield was impossible, seemed like fraud to me. I did not want to participate, and submitted my resignation." Ibid., p. 49.

[32] Parnas (1985), p. 1326.

[33] David Parnas, email to the author, January 22, 2015.

[34] Mohr (1985).

[35] Parnas (1987), p. 49.

[36] Ibid.

the U.S. government to take actions, based on that false belief, that would prove harmful downstream to the country and the international community.

I now turn to the conflict-of-interest issue. In this connection, Parnas made relevant statements about four partially overlapping groups of technical practitioners. When he reached his conclusion that the envisioned SDI system could never be known to be trustworthy, Parnas "solicited comments from other scientists and found none that disagreed with my technical conclusions."[37] As for his fellow SDIO software panelists, they continued to support the development of the SDI system. While some defended the viability of the projected system, Parnas found their views vague and based on assumptions that he regarded as "false." Regarding researchers who were taking SDIO funds, he stated, "I have discussed my views with many who work on SDIO-funded projects. Few of them disagree with my technical conclusions."[38] Finally, he discussed his views "with scientists at Los Alamos and Sandia National Laboratories. Here, too, I found no substantive disagreement with my analysis."[39] Nevertheless, Parnas did not find significant support in any of these groups for his resignation from the SDIO advisory committee or for his subsequent public opposition to SDI. Why?

Regarding the other members of the SDIO advisory panel, "Everyone seemed to have a pet project of their own that they thought should be funded."[40] As for the "other scientists" whom he consulted, "they told me that the program should be continued, not because it would free us from the fear of nuclear weapons, but because the research money would advance the state of computer science!"[41] Those who whose work was being funded by SDIO "gave a variety of excuses. 'The money is going to be spent anyway; shouldn't we use it well?' 'We can use the money to solve other problems.' 'The money will be good for computer science.'"[42] Finally, as for the researchers at Los Alamos and Sandia National Laboratory, Parnas "was told that the [SDI] project offered lots of challenging problems for physicists."[43]

In short, from Parnas's viewpoint, various kinds of conflicts of interest—intellectual as well as financial—prevented even those who

[37] Ibid., p. 48.
[38] Ibid., p. 51.
[39] Ibid.
[40] Ibid., p. 47.
[41] Ibid., pp. 48–49.
[42] Ibid., p. 51.
[43] Ibid.

agreed with his technical analysis from opposing the project. Parnas even found *himself* with a conflict of interest. He had "a project within the U.S. Navy that could benefit from SDI funding" and thought this might disqualify him from serving on the panel. When he pointed this out to the organizer of the SDIO panel, the latter "assured me that if I did not have such a conflict, they would not want me on the panel."[44] Parnas resigned from the SDIO panel because, in his view, protecting the public interest (both by not "wasting taxpayers money"[45] and by fighting against a course of action that he believed would heighten the risk to national security) and serving his own private interest as a consultant paid an honorarium of $1000 per day for being on the SDIO panel and with easy access to SDIO funding for more research of his own, conflicted; those interests did not align as long as he continued to serve on the panel.

Some fellow panel members resolved their conflicts of interest by making a questionable assumption. They assumed that in the long run the public welfare of the United States would best be served by continuing to fund the development of an antimissile system that, according to Parnas, most acknowledged couldn't be known to be trustworthy.[46] Continued funding for SDI would provide ample new funding for the computer science and high-technology fields, something that would surely yield great benefit to civil society and possibly to researchers in those fields as individuals. That assumption could not be refuted in the short term, because one could always claim that as-yet-unknown long-term beneficial consequences would later confirm it. It was also convenient and self-justifying, in that it allowed those who made it to view the public interest and their own private interests as aligned rather than as in conflict, thereby resolving in their minds the vexing ethical issue of conflict of interest.

Lessons

One ethics-related lesson extractable from this case is that an engineering research expert who is invited to become a consultant-member of a committee of experts that provides advice to a firm or an organization

[44] Ibid., p. 47.

[45] Ibid., p. 52.

[46] After Parnas had made known his technically grounded conclusions about SDI, one panelist argued before a government committee that it would be useful to continue spending money on SDI R&D because "even if [SDIO] could not do everything that Reagan had asked, they could do something that would move us closer to that capability." Parnas's argument was that "it was dishonest not to state clearly that we could not do anything close to what Reagan had promised." (Parnas, email to the author, January 22, 2015).

with a questionable or controversial project it wants to push should be extremely careful before agreeing to enter into such a relationship, however lucrative or prestigious. Why? Because if she or he enters into such a relationship without having done due diligence on the integrity of the firm or organization, including the degree to which it bases its decisions on facts and solid arguments, she or he may find that her or his name, professional reputation, and/or ideas are being used to confer legitimacy and credibility on the goals, plans, and projects of the entity *independent of their actual technical viability*. Moreover, her or his views on the technical merits of the project in question may not be taken seriously unless they support the firm's or organization's entrenched views and plans.[47]

A second lesson of this case that is worth underscoring has to do with the fact that Parnas was able to engage in a form of before-the-fact whistleblowing without his career being damaged. The reason he was able to do so is that, unlike most twentieth-century engineers, he was not an employee of a large-scale private corporation. He was an academic software engineer who was not liable to be fired "at will" by his employer because of the positions and actions he took regarding SDI. Unlike most contemporary engineers, Parnas retained his autonomy and the freedom to speak his mind as he saw fit.[48]

Finally, Parnas's judgment about the conflict of interests he observed in this case was based on the key prioritization rule noted in Chapter 3: when a course of action would violate the public interest while serving one or more private interests, then protecting the public interest, that is, the interest of the omnipresent background client—society at large— must take priority over fulfilling the private interests of individual researchers seeking government funding.

In effect, Parnas fulfilled *a variant of FERE4 that applies to those working for government*. FERE4 asserts that engineers have an ethical responsibility to work to the best of their ability to serve the legitimate

---

[47]Parnas regards this lesson as "over-simplified," because, at least in his case, doing such due diligence was impossible for anyone *not* on the committee. He "assumed that [SDIO] knew what they were doing" and "knew that I would not know what they had in mind unless I joined the committee." However, once he joined the committee, "it became clear to me that they did not know what they were doing and, at least as important, they did not care." (Parnas, email to the author, January 22, 2015).

[48]However, Parnas did pay a price for his position and actions regarding SDI. He had been doing "a lot of lucrative consulting for companies that were DoD contractors bidding on SDI contracts. They were told that they could no longer use me as a consultant and I lost this." (Parnas, email to the author, January 22, 2015).

interests of their employer or client, with one implication being that the employee does *not* have an ethical responsibility to serve any *illegitimate* interests of the employer or client, for example, to steal a rival's intellectual property. What I shall call FERE4* states that "engineers have an ethical responsibility to do the best they can to serve the *legitimate* interests of the state or government of which they are employees or that they serve as consultants." By implication, the state employee or consultant to government does not have an ethical responsibility to serve any *illegitimate* interest of the state. This is important, since, as Parnas put it,

> it is not professional to accept employment doing "military" things that do not advance the legitimate defense interests of the country. If the project would not be effective, or if, in one's opinion, it goes beyond the legitimate defense needs of the country, a professional should not participate. Too many do not ask such questions. They ask only how they can get another contract.[49]

David Parnas believed that, for technical and other reasons, SDI could never be known in advance to be trustworthy. He also believed that pursuing SDI would likely spawn a new arms race and thereby pose an increased risk to international security. Given these beliefs, the actions David Parnas took in connection with his membership on the SDIO Panel on Computing in Support of Battle Management mark him as a researcher whose engineering work was fully in accord with the fundamental ethical responsibilities of engineers.

Discussion Questions

1. Did Parnas have an ethical responsibility to resign from the SDIO advisory committee once he concluded that an SDI system could never be known to be trustworthy? Why or why not?
2. How did Parnas's decision to speak out publicly against SDI relate to the FEREs?
3. Did any of the parties with actual or apparent conflicts of interest in this episode behave ethically irresponsibly? Why or why not?
4. Parnas asserted that it was not "professional" for a technical professional "to accept employment doing military things that do not

[49] Ibid., p. 52.

advance the legitimate defense interests of the country." Do you agree with that claim? Why or why not? If you agree with that claim, what is it about what being a "professional" requires of an engineer or scientist that could lead her or him to that conclusion?

CASE 3: COLLABORATIVE RESEARCH PRACTICES AT BELL LABORATORIES

Background

In 1997, Jan Hendrik Schön received a PhD from the University of Konstanz in Germany and was hired by Bell Laboratories. He soon started publishing prolifically in elite research journals. In 2001, along with various coauthors, Schön published an astonishing average of one paper every 8 days.[50] The papers he and his coauthors published reported advances in

> condensed-matter physics and solid-state devices, including organic field-effect transistors, organic single-molecule transistors, organic junction lasers, field-induced high-temperature superconductivity in several materials, plastic Josephson junctions, and tunable superconductivity weak links.[51]

In 2001, after noticing that two published articles of Schön's, ostensibly reporting the results of different experiments, contained identical noise graphs, two university researchers called some of the published papers into question. Lucent Technologies, of which Bell Laboratories was then a unit, formed an investigation committee in May 2002 to explore the possibility of research misconduct by Schön and his coauthors.

In September 2002, the committee issued its report.[52] It claimed to have confirmed 16 of the 24 allegations of research misconduct it had identified and investigated.[53] In addition, the committee discovered and disclosed other questionable research practices. Eventually, 28 papers by Schön and his coauthors were withdrawn.[54] Schön was fired from his position at Bell Labs, and his PhD was revoked. However, all coauthors of

---

[50] Beasley et al. (2002), p. 8.
[51] Ibid., p. 9.
[52] https://media-bell-labs-com.s3.amazonaws.com/pages/20170403_1709/misconduct-revew-report-lucent.pdf. Henceforth, this document will be cited as *Committee Report*.
[53] Ibid., p. 3.
[54] http://en.wikipedia.org/wiki/Sch%C3%B6n_scandal.

Schön on the works in question were "completely cleared"[55] of "research misconduct"[56] by the investigation committee.

Before we proceed, a seeming anomaly needs comment. Schön's PhD was in physics, not in a field of engineering. Why, then, has this case been included in a book on ethical issues in engineering? It is included for several reasons.

While one of Schön's specialties was condensed-matter physics, another was solid-state devices at the nanoscale level, a burgeoning field of engineering research. He worked on topics closely connected with important engineering applications, such as organic semiconductors. Moreover, one of Schön's coauthors at Bell Labs was arguably as much an engineer as a scientist. That person is a professor in an engineering school and works on nanotechnology.

The fact is that at the nanoscale it is often difficult to decide whether a particular activity should be viewed as "science," "engineering," and/or "engineering science." It can also be difficult to decide whether the practitioner carrying out the activity is functioning as an engineer, a scientist, or both, simultaneously.

Further evidence of blurred lines here lies in the fact that Dr. Lydia Sohn, among the first researchers[57] to claim that two of Schön's papers, apparently reporting on different experiments, had the same noise graphs, is a professor of mechanical engineering. Her main research interests are "micro-nano engineering" and "bioengineering."[58]

In addition, as discussed in Chapter 2, research is increasingly common in and characteristic of contemporary engineering activity, a trend clearly reflected in this case. So, apart from this case's intrinsic interest, its linkages with engineering applications and personnel, and the characteristic importance of research in contemporary engineering, are reasons that it seems justifiable and desirable to include the Schön case in a book on ethical issues in engineering. Moreover, some of the ethically challenging situations that Schön faced in a research setting are ones that

---

[55] *Committee Report*, p. 4.

[56] In 2000, the U.S. Office of Science and Technology Policy adopted and published a definition of "research misconduct" as "fabrication, falsification, or plagiarism" [FFP] in "proposing, performing, or reviewing research, or in reporting research results." See *Federal Register*, Vol. 65, No. 235 (December 6, 2000), pp. 76260–76264, DOCID:fr06de00–72. However, FFP does not encompass many other kinds of ethically problematic research practices, such as some present in this case.

[57] For details, see Reich (2009), pp. 214–220.

[58] http://www.me.berkeley.edu/faculty/sohn/.

engineers have faced in a development setting in the past and are increasingly likely to face in research settings in the future.[59]

Ethical Analysis

The Schön case illustrates several familiar kinds of research misconduct.[60] The Lucent Investigation Committee found that Schön had manipulated and misrepresented data in a number of instances. For example, he was found to have engaged in "substitution," that is, to have used the same "whole figures, single curves and partial curves in different [papers] or [in] the same paper to represent different materials, devices, or conditions"; to have engaged in "nontrivial alteration or removal of individual data points or groups of data points within a data file"; and to have utilized values of "fitted or assumed mathematical functions . . . where measured data would be expected."[61] Schön also was found to have deviated significantly from established research practices by not systematically maintaining "proper laboratory records" and by deleting "virtually all primary (raw) electronic data files."[62]

But the Schön case also encompasses phenomena that, while not falling under the prevailing U.S. government definition of research misconduct, exemplify what one researcher calls "'other deviations'" from "acceptable research practices,"[63] deviations many regard as unethical. In what

[59] For an interesting case of alleged data fabrication and falsification in macro-engineering product development, see Vandivier (1972).

It is telling that at Stanford University, the Nanoscale Sciences and Engineering Shared Facility Program Committee is cochaired by a professor of applied physics and physics, and a professor of electrical engineering. See http://snf.stanford.edu/pipermail/labmembers/2013-September/005550.html.

[60] According to the U.S. Department of Health and Human Services,' Office of Research Integrity, "Research misconduct is defined as fabrication, falsification, or plagiarism in proposing, performing, or reviewing research, or in reporting research results. . . . Fabrication is making up data or results and recording or reporting them. Falsification is manipulating research materials, equipment, or processes, or changing or omitting data or results such that the research is not accurately represented in the research record. . . . [The research record is the record of data or results that embody the facts resulting from scientific inquiry, and includes, but is not limited to, research proposals, laboratory records, both physical and electronic, progress reports, abstracts, theses, oral presentations, internal reports, and journal articles.] Plagiarism is the appropriation of another person's ideas, processes, results, or words without giving appropriate credit. Research misconduct does not include honest error or differences of opinion." See http://ori.hhs.gov/federal-research-misconduct-policy.

[61] *Committee Report*, pp. 3 and 11–12.

[62] Ibid., p. 3.

[63] Resnick (2014).

follows, I call them *ethically questionable research practices.* Discussion of three of them follows.

One such practice involves the attribution of coauthorship of scholarly articles to certain research collaborators. The *Committee Report* states:

> [Bell Labs researcher] Christian Kloc grew many single crystals that were absolutely essential to the research in question. This clearly qualifies him as a coauthor in the large number of papers [of Schön's] on which he is listed.[64]

What, if any, are the ethical responsibilities of a researcher X who produces key materials that enabled X's colleague Y to perform a certain experiment, in return for which Y lists X as a coauthor of the article Y published that reports the results of that experiment?

A second example from the Schön case involved a different research collaborator and a different research practice:

> Zhenan Bao participated in a relatively small number of the papers in question. In addition to synthesizing molecules for Hendrik Schön, she also prepared thin films of organic material on patterned electrodes. One of the papers combined quite reasonable electrical results that Zhenan Bao measured on her own apparatus, on structures she fabricated, with spectacular measurements by Hendrik Schön, on quite different structures he fabricated.[65]

This raises the question of what, if any, are the ethical responsibilities of a researcher X who carries out work that yields expected or reasonable results on structures that X produced, when those results are to appear in the same manuscript in which X's coauthor Y reports surprising or spectacular results on structures that Y produced?

A third collaborative research practice in the Schön case involved the senior researcher who headed the Bell Labs team studying electronic properties of organic crystals, of which Schön was a member. Bertram Batlogg was listed as a coauthor of more than half of the 28 papers of Schön's that were eventually withdrawn. Yet, according to the *Committee Report*, Batlogg never did any device preparation or measurement work for these papers.[66] This raises the question, what, if any, are the ethical responsibilities of the senior leader of a research group, when a junior

---

[64] *Committee Report*, p. 17.
[65] Ibid., p. 18.
[66] Ibid.

member of that group repeatedly obtains extraordinary results and writes them up in multiple papers that she or he submits to prestigious journals?

FERE1 is relevant to this case. It is clear that manipulation and misrepresentation of data are apt to cause harm to other researchers. This harm stems from the time, money, and energy that researchers expend trying futilely to replicate claimed findings in papers based on that data. But what about the above-mentioned collaborative research practices of Schön and some of his coauthors? Did they involve the violation of any FEREs? Let us discuss each of the three practices in question.

Awarding coauthorship to a research "collaborator" whose sole contribution to an inquiry was to produce the material on which the experiment was carried out arguably distorts the meaning and dilutes the significance of being an author of a paper. The *Committee Report* claims that the fact that Christian Kloc "grew many single crystals that were absolutely essential to the research in question . . . clearly qualifies him as a coauthor in the large number of papers on which he is listed."[67] While listing a person as a coauthor for producing an important material for an experiment done by another is apparently a familiar practice in contemporary research, the report does not specify any criterion in light of which the researcher who grew the crystals "clearly" deserves to be a coauthor. It is far from obvious that such a person's contribution deserves to be on a par, credit-wise, with those of individuals who did things like come up with the idea for an experiment, contribute significantly to designing the experiment, contribute significantly to obtaining or analyzing the experimental data, or contribute significantly to drawing the key conclusions from the findings. That a research practice is common does not by itself guarantee that it is intellectually and ethically justifiable. In my view, a researcher who produces and contributes an important material that makes it possible for another researcher to carry out an experiment is not thereby entitled to be listed as a coauthor of the resultant paper. Indeed, I submit that she or he has an ethical responsibility to *decline* to be listed as a coauthor of the paper that her or his material provision made possible if that status is offered to her or to him.

It is curious that the committee states that Kloc, who grew the "high-quality single crystals of organic materials" that made many of Schön's experiments possible but did not make any physical measurements on the devices involved, "clearly" deserved to be a coauthor, with Schön, of the resultant papers. The *Committee Report* also states that "co-authors, through their explicit association with the papers, by implication endorse the validity of the work."[68] Viewing providers of materials for experiments of others as legitimate coauthors increases the number of those who are perceived as implicitly endorsing the validity of the work, something likely to increase the confidence of others in the credibility of the reported results, and/or fuel repeated futile efforts to replicate claimed findings.

The International Committee of Medical Journal Editors (ICMJE) has done valuable work on the nature and ethics of authorship. One result of its work is a set of "Recommendations for the Conduct, Reporting, Editing, and Publication of Scholarly Work in Medical Journals." Authorship-related topics addressed include the nature of authorship, responsibilities of authors (including matters related to conflicts of interest), responsibilities of authors and referees in the submission and peer-review processes, responsibilities of journal owners, and the protection of research participants.[69]

One ICMJE authorship recommendation sheds useful light on the Kloc coauthorship question. The 2013 document recommends that a person be deemed an author of a work if and only if his or her contribution to it satisfies the following four criteria: (1) she or he made "substantial contributions to the conception or design of the work; or the acquisition, analysis, or interpretation of data for the work"; (2) she or he "draft[ed] the work or revis[ed] it critically for important intellectual content"; (3) she or he gave "final approval of the version to be published"; and (4) she or he agreed "to be accountable for all aspects of the work in ensuring that questions related to the accuracy or integrity of any part of the work are appropriately investigated and resolved."[70]

Does Kloc qualify as a coauthor of the relevant Schön papers under the ICMJE criteria? Consider the first of these criteria. Kloc's growing single crystals for Schön was clearly not a substantial contribution to the "conception or design" of Schön's work. But did not his growing crystals for Schön make a substantial contribution to "the acquisition, analysis, or interpretation of data for the work"? It might seem so. But the answer hinges critically on whether doing something that *enables* an experiment to be conducted by another reasonably counts as making a "substantial contribution" to "the acquisition . . . of data for the work." In my view,

[68] Ibid., p. 16.
[69] International Committee of Medical Journal Editors (2013).
[70] http://www.icmje.org/recommendations/browse/roles-and-responsibilities/defining-the-role-of-authors-and-contributors.html.

a contribution of materials does not so count. For Kloc himself did not acquire, analyze, or interpret the data involved. He *enabled* its acquisition by someone else; he did not participate in acquiring, analyzing, or interpreting it. Hence, at least under the ICMJE authorship criteria, Kloc should not have been listed as a coauthor of the papers in question. In my view, his role should have been duly recognized in the *acknowledgments* section of the appropriate papers for making the experiments possible.

Turning to a different coauthorship issue, one raised by the papers coauthored by Schön and Bao, the committee found that the latter made "substantial contributions" to acquiring the data and producing the findings reported in some of the papers in question. Hence, Bao clearly met the coauthorship criteria. The question the Schön-Bao coauthored papers raises is different: namely, what, if any, are the ethical responsibilities of the coauthors of a paper when each has done all the work on one part of the paper and none on the other parts? That a single paper can contain unsurprising results obtained by one coauthor and startling results obtained by the other underscores how important it is that each coauthor read and sign off on the paper as a whole, not just on his or her part of it.

If, however, each coauthor scrutinizes only the part which he or she has done and is an expert on, and signs off on the paper without scrutinizing it as a whole, then, all things being equal, that makes it more likely that erroneous or fraudulent claims in such papers will not be detected upstream. This in turn increases the chances of wastes of time, money, and energy downstream by other researchers who futilely attempt to replicate the paper's surprising or extraordinary claimed findings, as in fact happened when researchers in various laboratories were unable to replicate Schön's claimed findings. My view is that each coauthor has a derivative ethical responsibility to read and sign off on the paper as a whole, for not doing so further opens the door to harm (or an unreasonable risk of harm) from publishing erroneous or fraudulent findings (FERE1), and because doing so can be viewed as an attempt to prevent the harm that might well result from publishing such findings (FERE2).

Regarding the withdrawn Schön-Batlogg coauthored papers, if the "distinguished leader"[71] of a research group fails to scrutinize numerous yet-to-be-submitted manuscripts of one or more junior group members that contain extraordinary findings, but is nevertheless listed as a coauthor of those papers, this makes it more likely that papers will be published that eventually prove to be seriously flawed or the result of research misconduct. This in turn risks diminishing the credibility of other

---

[71] *Committee Report*, p. 18.

publications coming out of such groups, and possibly also that of the journals publishing such papers.

The *Committee Report* limited itself to asking a *prudential* question: "Should Batlogg have insisted on an *exceptional* degree of validation of the data in anticipation of the scrutiny that a senior scientist knows such *extraordinary* results would surely receive?"[72] The author's view is that because failure to do so arguably contributed to creating (indirectly) an unreasonable downstream risk of harm to others—in the form of potential significant wastes of time, money, and effort—Batlogg had a *derivative ethical responsibility* to carry out such validation. That view goes beyond the claim, implicit in the *Committee Report*'s question, that it would have been prudent of the senior scientist to carefully validate the data upstream to avoid possible downstream embarrassment.

By engendering credibility-undermining effects and risks of same, such authorship practices can indirectly harm public welfare, the interests of individual researchers, and even erode belief by researchers in the integrity of the research publication process in scholarly journals. Under FERE1, researchers have derivative ethical responsibilities to refrain from engaging in such practices because of the harms—public and private—and unreasonable risks of harms they can contribute to causing.

Moreover, under FERE2, senior leaders of research groups arguably have a derivative ethical responsibility to proactively create and implement policies aimed at preventing the harms (and unreasonable risks of harm) apt to result when papers by junior team members containing radically novel or exciting findings are not scrutinized before being quickly submitted for publication consideration, with the senior leader listed as a coauthor. The situation becomes even more ethically urgent if there is competition in the same area between groups of researchers at different institutions, each striving strenuously to stake a claim to having been the first to achieve the breakthrough in question.

Lessons

To date, the U.S. research community and the U.S. government have limited "research misconduct" to "fabrication, falsification, and plagiarism" (FFP) in the planning, carrying out, reporting, and reviewing phases of the research process. The Schön case demonstrates that what is still lacking is a clearly articulated set of ethically unacceptable research practices

[72]Ibid., p. 18.

beyond FFP, including but not limited to the problematic authorship practices exhibited in the Schön case.[73] Researchers have ethical responsibilities under FERE1 and FERE2 to avoid engaging in such practices and to take steps to try to prevent the harms that can result when such practices are followed in response to the publication and administrative pressures that weigh on many contemporary researchers.

There is evidence that researchers themselves would welcome guidance about acceptable and unacceptable research practices. In 2006, I surveyed more than a thousand nanotechnology researchers at 13 universities about ethical issues related to their work.[74] One survey question asked the nanotechnology researchers which of three given responses they agreed with the most regarding a proposal that researchers be given "clear ethical guidelines for the conduct of nanotechnology research." The three responses were that having such guidelines is (i) "necessary," (ii) "desirable but not necessary," and (iii) "neither desirable nor necessary."[75] Remarkably, almost three-fifths (58.7%) chose "necessary," about a third (34.7%) chose "desirable but not necessary," but only about a sixteenth (6.6%) chose "neither desirable nor necessary."[76] This suggests strongly that many contemporary researchers, at least in the burgeoning field of nanotechnology, believe that having guidelines for the ethical conduct of research in the laboratory would be very helpful, and that leaving such conduct up to the intuition of each researcher is insufficient.

The larger ethical significance of the Schön case stems from the fact that in contemporary society, research in engineering (and science) is increasingly a group, or at least a collaborative, process carried out by multiple individuals often working under pressure and under an intense division of labor, sometimes simultaneously at different sites. More often than not, multiple people are involved in each research project, either as members of the research team or as associates or collaborators of one or more of the primary researchers.[77] This trend has given rise to significant ethical issues about the fairness of new (and often convenient) authorship practices, the responsibilities of coauthors, and the responsibilities of those in charge of labs in which exciting research is being carried out by young lab members and rapidly published with the senior lab directors listed as coauthors.

[73] For one effort to do so, see Resnick (2014).
[74] McGinn (2008), pp. 101–131.
[75] Ibid., p. 128.
[76] Ibid., p. 110.
[77] Wuchty et al. (2007), pp. 1036–1039.

It is striking that the Lucent Investigation Committee completely cleared all of Schön's collaborators and coauthors of all charges of research misconduct but was noncommittal on the non-FFP ethically questionable research practices involved. This strongly suggests that no consensus yet exists in the research community about the ethical responsibilities of researchers to avoid forms of ethically problematic research conduct beyond FFP, including dubious authorship practices that could be just as harmful as cases of FFP.

Whether a set of authorship recommendations, comparable to those developed by the ICMJE but adapted to engineering, emerges and gains significant traction in the engineering research community remains to be seen. Until that happens, the ICMJE authorship recommendations deserve attention by would-be ethically responsible engineers as a way of doing more than simply following prevailing research practices or relying on intuition or quid pro quo convenience.

Discussion Questions

1. Are there ethically questionable research practices that do not involve coauthorship? If you believe there are, give examples of several of them.
2. Which coauthorship practices explored in this case discussion are ethically irresponsible? For each practice you list, justify your position.
3. Are there coauthorship practices not explored in this case discussion that you believe are ethically questionable or irresponsible? If so, give an example of one such practice and indicate why you regard it as ethically questionable or irresponsible.
4. Is it ethically appropriate for a researcher who simply provides another researcher with material that makes the latter's work possible to accept the latter's offer of coauthorship on the resultant research paper(s)? Why or why not?

CASE 4: THE APPLE NEWTON MESSAGEPAD

Background

John Sculley was appointed CEO of Apple Computer in April 1983. He had studied architectural design as an undergraduate at Brown, earned an MBA from the Wharton School at the University of Pennsylvania, and had a successful career in marketing with PepsiCo.[78] In 1990, Sculley named himself chief technical officer (CTO) of Apple.[79]

In 1987, Apple engineer Steve Sakoman came up with an idea for a tablet-like device radically different from the personal computer as then known. Jean-Louis Gassée, then head of Apple R&D, put Sakoman in charge of a skunkworks project to develop the idea.[80] As envisioned, the device would have the ability to "read handwritten notes jotted on its screen with a special pen, convert them to text, store them for future use, even transmit them wirelessly to other computers," and allow appointments to be entered remotely and automatically in the device's memory and onboard calendar.[81] Sculley eventually became enthusiastic about such a device. In January 1992, he dubbed it a "personal digital assistant" (PDA).[82]

In early 1990, Sculley gave the product team 2 years to move "from research project to marketable product."[83] A handheld device "that cost less than $1,500 would have to be ready by April 2, 1992."[84] At the January 1992 Consumer Electronics Show (CES) in Las Vegas, without mentioning the projected device by name, Sculley predicted that PDAs, with a range of remarkable communications abilities, would one day become widespread and "that the market for devices and content would reach $3 trillion by the end of the decade."[85]

The approaching deadline was at odds with the slow progress being made in developing the necessary, complicated software. According to one account of the development of the "Newton MessagePad," "[i]t was not unusual for the engineers to be working [on it] between fifteen and twenty hours every day."[86]

A specific software engineer was responsible for "writing the portion of the Newton software that controlled how text and graphics were displayed on the screen." His "incredibly long hours" allegedly left his wife "miserable, stuck all day and all night" in their house.[87] Three weeks before an announced public demonstration of the still-unfinished Newton at the CES in Las Vegas in January 1993, the engineer took his own

---

[78] http://en.wikipedia.org/wiki/Sculley.
[79] Carlton (1997), p. 136.
[80] Markoff (1993).
[81] Ibid.
[82] McCracken (2012).
[83] Hormby (2010).
[84] Ibid.
[85] Ibid.
[86] Ibid.
[87] Markoff (1993).

life. Within a week of this tragedy, "another Newton programmer had a breakdown. He attacked his roommate and ended up in jail on an assault charge."[88] While such extreme occurrences were rare and may have been partly attributable to other factors, it is notable that "[t]he company hired psychologists to make sure that everybody on the team was mentally healthy and even started a 'buddy system' for engineers who were on the verge of burning out."[89]

A less feature-rich model of the Newton MessagePad was finally launched in August 1993, sixteen months after the initial launch date Sculley had set.[90]

Ethical Analysis

This case suggests a significant engineering-related ethical issue that seems to have been neglected to date: namely, how should an engineer react and respond to the introduction for an extended time period of an arduous work schedule, one said or thought to be required if an ambitious/aggressive product deadline or ship date set by management is to be met? Let us explore this issue.

The idea that engineers have general ethical responsibilities to society and to end users of their products and systems has gained traction over the years among engineers and professional engineering societies. However, to my knowledge, the notion that engineers can also have ethical responsibilities to their significant others, spouses, and other family members has not been recognized by engineers or incorporated into engineering society codes of ethics. Nor has it been explored in the literature of engineering ethics. The responsibilities the author has in mind probably go unmet quite often and seem to be unrecognized by many practicing engineers.

Recall that the first three FEREs hold that, among other things, engineers have fundamental ethical responsibilities to not impose an unreasonable risk of harm on any party affected by their work, to try to prevent such risks as are being imposed on such parties, and to try to alert and inform upstream those at unreasonable risk of being harmed by or through their work (or work about which they are technically knowledgeable).

[88] Ibid.
[89] Hormby (2010).
[90] Markoff (1993) and Hormby (2010).

Besides the ethical responsibilities that all engineers have to be attentive to the effects of their work *products* on the well-being of others, the Newton MessagePad software engineers arguably should also have been attentive to possible effects of their work *practices and processes*, for example, their apparently extremely demanding work regimes, on the psychosocial well-being of those likely to be (indirectly) affected by them, namely, their significant others, spouses, and other family members.

While the evidence for the following claims is only anecdotal and sketchy, not systematic or extensive, I suspect that many engineers in the rapidly changing, highly competitive IT industry do not think carefully about the effects of adhering to their prescribed work regimes on the well-being of those close to them. Most probably acquiesce in rather than resist arduous and onerous work regimes engendered by sometimes overly ambitious management-prescribed ship dates and product development schedules. I also suspect that many engineers fail to alert and inform significant others, spouses, or other family members upstream about coming disruptive changes in their work regimes. Finally, I suspect that many engineers assume that their significant others, spouses, and children will and perhaps should *adapt* to whatever psychosocial effects result from their often highly pressurized and grueling work regimes. To the extent that these suspicions correspond to reality, important derivative ethical responsibilities that such engineers arguably have to their significant others, spouses, and other family members probably go unmet.

It might be argued that Apple management had a "legitimate interest" in imposing an ambitious Newton MessagePad ship date/product development schedule on the team. If it did, then, under FERE4, the engineers would have had an ethical responsibility to work to the best of their ability to serve it. It is difficult to say with certainty what the actual work situations were of the engineers who worked on the MessagePad project, from, say, April 1990 until August 1993. However, to the extent that an engineering work deadline or schedule is so ambitious, aggressive, or demanding that adhering to or meeting it over an extended period of time risks impairing engineers' health and/or the psychosocial well-being of some of those close to some of them, then that deadline/schedule should be deemed unreasonable. Further, no employer has a "legitimate interest" in imposing an unreasonable deadline or schedule on its engineers. Hence, engineer employees who are subjected to one do not have an FERE4-based derivative ethical responsibility to go all out to meet it.

Given the entrenched intensive work cultures of many Silicon Valley firms, upstream acquiescence by engineers on IT product-design teams who are subjected to overly ambitious or aggressive ship dates or product development schedules is apt to increase the chances that the well-being of some individuals close to some team members will be impaired downstream as a by-product.[91] Engineers, like all workers, are entitled to reasonable work schedules and regimes, ones that do not force them to choose between striving mightily to get their assigned tasks done on time and jeopardizing their mental or physical health and/or disrupting the lives of those close to them. Engineers should be wary of unreasonable time schedules imposed on them, especially (but not only) by nontechnical managers or executives with little appreciation for the difficulty of the R&D tasks that remain, and express their reservations or disagreement about such schedules as a group and as early and vigorously as possible.

The health of an engineer's significant other, spouse, or children can be harmed psychologically or socially through her or his work *practices or processes*, just as individuals, organizations, and societies can incur physical and financial harm from defective work *products*. The employer has an ethical responsibility to its engineers to give them reasonable schedules and adequate work resources, and engineers have an ethical right to a reasonable work schedule and regime. However, engineer-employees also have an FERE1-based derivative ethical responsibility to their significant others, spouses, and other family members to not create an unreasonable risk of harming them indirectly through adhering to extremely demanding work regimes—whether management-imposed or self-adopted.[92]

If such harms cannot be prevented (without jeopardizing one's job), then, under FERE3, engineers subjected to an unreasonable schedule have a derivative ethical responsibility to at least alert and inform their spouse/significant other/family member as far upstream as possible about changes in their work regime that may create an unreasonable risk of psychosocial harm to by disrupting core relationships. In my view, it is unfair of employers to impose work conditions on their engineer-employees that make burnout likely. However, it is also irresponsible of engineers to acquiesce in such work regimes upstream if they are reasonably or

---

[91] On the "Silicon Syndrome," see Hayes (1989), p. 116.

[92] An engineer's individual decision to adapt to such a demanding work regime might be shaped by the "can-do" culture that prevails in her or his work group or firm.

foreseeably likely to have detrimental effects downstream on unsuspecting significant others or families.

A claim (dubiously) attributed to Arthur Schopenhauer may apply here: "All truth passes through three stages. First it is ridiculed. Second, it is violently opposed. Third, it is accepted as being self-evident."[93] For the foreseeable future it is unlikely that many IT employers will recognize and respect the engineer-employee's ethical right to a work schedule and regime compatible with her or his mental and physical health, and undisrupted family life. Given that state of affairs, I submit that hard-driven IT engineers have a derivative ethical responsibility to avoid casual acquiescence in and to try to redress an onerous work regime because of its possible deleterious psychosocial effects on the well-being of those close to them. Should engineers be unable to ameliorate an extended arduous work regime, under FERE3 they have an ethical responsibility to at least alert and inform those close to them about its possible negative effects, and to not simply expect them to adapt to whatever demanding changes they face at work.[94]

Discussion Questions

1. Do you believe that engineers can have work-related ethical responsibilities to their spouses, families, or significant others? If you believe they can, justify your view and give an illustrative example. If you believe they cannot, justify your position.
2. Do engineers at risk of having a more arduous work regime imposed on them, one that they can foresee will have a disruptive

---

[93] See Jeffrey Shallit (2005), especially p. 5.

[94] An engineer and engineering-ethics expert who read this case in draft shared the following in correspondence: "I saw a good example of this many years ago while working at Bell Labs in the department designing the first ESS (electronic switching system), which was to replace the existing electromechanical systems, such as Number 1 Crossbar and Number 5 Crossbar. At one point there was a lot of pressure to get the first field trial to the point where they could show the FCC that they were far enough along to justify accelerated depreciation of the existing plant. Testing was being carried out on a 24-hour basis. One engineer was working night shifts, while his wife was in the late stages of pregnancy. Appeal to a higher-level manager resulted in his being given permission to switch to normal hours, but then, feeling that this would be letting the team down, he decided to continue on the night shift. One sub-department head suffered a heart attack during this period." (Stephen Unger, email to the author, September 22, 2014). The Bell Labs manager deserves credit for approving the appeal, but the engineer in question seems not to have recognized that he also had an ethical responsibility to his spouse, not just to his team, and that at some point in the pregnancy the former deserved to trump the latter.

effect on the lives of those closest to them, have an ethical responsi-bility to try to prevent or resist the imposition of such a regime? If so, why? What kinds of actions might such engineers take in trying to fulfill that responsibility?

3. Can engineers have work-related ethical responsibilities to their spouses, families or significant others that do not involve the psy-chosocial consequences of arduous work regimes? Explain your po-sition and, if you believe they can, give an illustrative example.

CASE 5: AN EMPLOYEE DATABASE MANAGEMENT SYSTEM

Background

Of the case studies discussed in this chapter, almost all are anchored in specific real-life episodes. This case is one of the few that are not. How-ever, software engineers could easily find themselves in situations similar to the following.[95]

"Three years ago Diane started her own consulting business. She has been so successful that now she has several people working for her, and many clients." Her firm's work includes "advising on how to network microcomputers, designing database management systems, and advising about security." Recently she was engaged to design a database manage-ment system (DBMS) for "the personnel office of a medium-sized com-pany." Its intended use was to maintain and analyze the contents of a database containing "extremely sensitive" employee information, such as performance appraisals, salary information, insurance-related medical records, and the like.

Diane "involved the client in the design process, informing the CEO, the director of computing, and the director of personnel about the prog-ress of the system." The client had the option to choose "the kind and level of security to be built into" the DBMS being designed for it. Regard-ing this choice, Diane described to her client various levels of security and corresponding costs and risks of unauthorized breaches. As it turned out, the DBMS was going to cost more than the client anticipated. Therefore, the client chose to save money by opting for "a less secure system," one

---

[95] See Anderson et al. (1993), "Case 2: Privacy," p. 100. Unless otherwise indicated, the quotations in Case 5 are taken from p. 100 of this article. The case description given in "Case 2: Privacy" was adapted from Johnson (1993).

that would offer only "weak security." Diane believes strongly "that the system should be much more secure." However, the client has made its choice of security level, and Diane must now respond to it.

Ethical Analysis

The key ethical issue in this case is how Diane should respond to the client's choice of a DBMS design that would offer only weak security, that is, one that would offer only a low level of informational privacy protection to the client's employees.

For Diane to accept the client's choice of security level as directions to be followed diligently and unquestioningly, and to design a DBMS that provides only weak security for the information involved, would be to create an unreasonable risk of harm to the client's employees. If she were to proceed as directed, sensitive personal information about the employees would be accessible without much difficulty to individuals, be they fellow employees or hackers, with no legitimate right to see, capture, or exploit it. Such individuals could easily use the information they were able to access in ways that could cause significant harm or create unreasonable risks of significant harm to the client's employees.

If Diane genuinely believes that the weak protection/less secure system for which the client opted would in fact put the client's employees at an unreasonable risk of harm, then she has a derivative ethical responsibility, under FERE1, not to accommodate her client's money-saving system-design choice.

Since she is being asked to design a DBMS that would engender unreasonable risks of harm to employees, Diane also has a derivative ethical responsibility to try to persuade the client to opt for a DBMS that provides stronger protection against unauthorized access to such sensitive information. But even if Diane were to decline to design such a DBMS for her client, the client could commission the same kind of low-security system from another software engineer. Making a serious effort to persuade the client to make the system more secure would be acting in accordance with FERE2; not doing so would not.

If Diane were to try but fail to persuade her client to opt for a higher level of security, FERE3 would imply that she should attempt to alert and inform the employees in question that they are at significant risk of being harmed by having their personal information accessed by unauthorized parties if the client gets another software engineer to design the DBMS with the low security level the client specified. If Diane were to simply acquiesce in the company's cost-saving security-level choice, then she would have effectively turned herself into a skilled "hired gun" indifferent to ethically responsible engineering practice. She would also have failed to act in accord with FERE1, FERE2, and FERE3.

Under FERE4 engineers have an ethical responsibility to do their best to serve the "legitimate interests" of their employers and clients. But does the client in this case have a *legitimate* interest in opting for weak security? If the client does *not* have a legitimate interest in specifying "weak security" as the DBMS protection level, then Diane does not have an FERE4-derived ethical responsibility to do her best to serve that client interest.

If Diane believes, on technical grounds, that designing the DBMS as requested by the client would make unauthorized access relatively easy, thereby engendering a significant or unreasonable risk of harm to the employees in question, then she does not have an ethical responsibility to do the best job she can to fulfill that client request. For a client to prescribe that its consultant design for it a DBMS, involving sensitive employee information, that will foreseeably put its employees at risk of incurring harm is no more a "legitimate" interest of a client than is an employer's interest in having a manufacturing process designed on the cheap that foreseeably puts line workers at a serious risk of physical injury or illness. Hence, because commissioning such a DBMS with weak security is not a legitimate interest of an employer or client, Diane does not have a derivative ethical responsibility to her client to do the best she can to serve it.

But might not Diane have an ethical responsibility to *her own* employees to design the DBMS as specified by the client? To turn down this job could have the unintended but foreseeable effect of causing her to lay off some employees or cut back on their working hours, which would clearly harm them economically and perhaps psychologically. Perhaps this consideration should trump any qualms Diane might have about agreeing to design the requested DBMS.

But this line of reasoning is not compelling. It is extremely unlikely that turning down the project of designing the DBMS as the client specified would force Diane to fire or cut back on the hours of her employees. After all, her business is "so successful" and has "many clients."[96]

---

[96] It is possible that turning down the consulting work because of the client's specification of weak privacy protection would tarnish Diane's reputation, and hence her ability to secure consulting jobs in the future. However, the fact that her firm is very successful indicates that Diane is widely regarded as an excellent software engineer. This could well

If, however, contrary to what is suggested by the case description, the situation was such that declining to design the DBMS as requested really would compel Diane to fire some employees or cut back on employee hours, then Diane would face an ethical dilemma. She would be forced to choose between two difficult alternatives: completing a design that she realizes would create a significant risk of harm to the client's employees if implemented, and having to fire some of her own employees or cut back on their hours. In such a case, Diane should carefully consider the projected magnitudes, scopes, likelihoods, durations, and degrees of reversibility of the harms in question; the extent to which her own at-risk employees have viable employment options if they are let go or have their hours cut back; the probability that the DBMS she could decline to design as requested would be completed as requested by another, less scrupulous engineer; and, per Rawls's Difference Principle, how comparatively well off are the two sets of employees—hers and her client's—who are at risk of incurring harm.

Given the case description, under FERE2 Diane should first try to persuade the client to opt for a significantly stronger level of protection for the sensitive employee information to be kept in the system. If unsuccessful, given FERE1 and given that her successful company currently has many clients, she should decline to implement the client's economically driven choice of weak security. Moreover, if she believes that another engineer-consultant is likely to acquiesce in and fulfill the client's design request, and that the client is unlikely to inform its employees that sensitive information about them will be kept in a company DBMS with weak security, under FERE3 Diane should try to alert and inform the client's employees that their privacy would be at risk of being violated if such a system is implemented. The economic success of her company affords Diane an opportunity to act in accordance with FERE1 by declining to complete the client's DBMS as requested.

Lessons

The lessons suggested by this case extend well beyond it. Consider a specific example. There is a prospective real-life system that in key respects is analogous to the DBMS discussed above: the so-called Intelligent Vehicle-Highway System (IVHS).[97]

---

offset whatever negative consequences might follow from her having declined to meet the client's request.

[97] The following account of the IVHS relies on Agre and Mailloux (1997).

Since the mid-1980s, a group of transport and computer professionals has been doing preliminary planning for an integrated system of computers, highways, vehicles, sensors, electronic vehicle tags, and ID numbers that would work together to, among other things, smooth traffic flow, monitor freight deliveries, and combat vehicular theft in the United States. As with the aforementioned DBMS for sensitive employee information, the informational privacy of individual drivers, for example, whether it can be determined from the system when and where particular drivers have gone in their vehicles, alone or with others, will depend on what technological design is selected and what design choices are made and incorporated into the system. The ethical value of (individual user) privacy could be given a high priority, relegated to a subordinate status, or eliminated altogether from the final system architecture in favor of a managerial system goal like "efficiency" or "cost-effectiveness."

Different companies are likely to propose different designs of the IVHS. Each will have a notion of what the system should look like, most likely one that leverages its own in-house technological strength and experience. The degree to which the competing designs build in protection for individual privacy will likely vary considerably. For example, one company envisioned using electronic tags (with built-in, randomly chosen ID numbers) and beacons, with only the driver knowing her or his personal ID, except in cases of stolen vehicle recovery efforts and toll collection, when the individual driver would voluntarily disclose the vehicle's ID.[98]

In contrast, rather than taking the approach of *minimizing the collection* of personal information in the operation of the IVHS, another company envisioned secure *central storage* of the personal information gathered.[99] Thus, the degree to which the ethical value of individual privacy will be embedded in the eventual final architecture of the system remains to be seen. If protecting the informational privacy of individual drivers is consensually deemed a component of public welfare—a public *good* of aggregation—then the engineers involved in designing the system have an ethical responsibility to be attentive to the extent to which the final system architecture design protects or undermines the informational privacy component of public welfare.

More generally, both the Diane-DBMS and IVHS cases raise the more general ethical issue of *whether and the extent to which important ethical values are embedded in specific engineering product and system designs*. The reader may have encountered the expression "design for

[98] Ibid., p. 300.
[99] Ibid., p. 301.

manufacturability" (DFM), which typically refers to the engineering goal of designing a product or system in such a way as to make it relatively easy and cost-effective to manufacture. The author submits that DFM has an ethics counterpart: **design for ethical compatibility** (DFEC). This refers to the goal or objective of designing a product or system that is compatible with and perhaps even conducive to—rather than antithetical to—core ethical values, such as justice, liberty, and privacy. DFEC applies not only to DBMSs and the envisioned IVHS, but also to many kinds of products and systems.

For example, consider the low-tech example of *el cortito*, meaning literally, "the short one." *El cortito* refers to the short-handled hoe that California farm owners and managers required their migrant farmworkers to use in the 1960s.[100] The hoe embodied and served the value of rigorous managerial control of farmworker labor. Management claimed that farmworkers could control the short-handled hoe better, hence making them less likely to damage crops. However, the short-handled hoe offered another benefit to management: when farm workers were required to use *el cortito*, supervisors could tell at a glance who was not working in the field, namely, anyone standing upright, because given the hoe's short length—its handle was about 18 inches long—farmworkers could use it for weeding and thinning work only when they were bent or stooped over. Thus, whether intended or not, the design of *el cortito* clearly gave supervisors tighter control over the labor process.

In the early 1970s, an organization called California Rural Legal Assistance (CRLA) gathered evidence from farmworkers about the negative health effects on them of using the short-handled hoe over a period of time. On September 20, 1972, CRLA petitioned the Industrial Safety Board (ISB) of California's Division of Industrial Safety (DIS) to prohibit the use of *el cortito* as a hazardous tool that farmworkers were compelled to use. After hearings, the ISB "denied the CRLA petition to ban the short-handled hoe on July 13, 1973."[101] However, in light of provisions of the California Labor Code, on January 13, 1975, the California Supreme Court ruled that the DIS had given the state's occupational safety and health standards "an unduly narrow interpretation supported neither by the language of the regulation nor by the authorizing statutory

[100] See Murray (1982).
[101] Ibid., p. 35.

Fig. 1. DILBERT © Scott Adams. Used by permission of Andrews McMeel Syndication. All rights reserved.

provisions."[102] Consequently, several months later, in April 1975, the chief of the DIS issued Administrative Interpretation Number 62, which stated that "the use of the short-handled hoe shall be deemed a violation of safety order Section 8 CAC 3316."[103] This effectively sounded the death knell for the short-handled hoe, although it was not definitively banned in California until 1985.[104]

In short, one can speak of the ethical responsibility of the design engineer working in the context of the following system: {**product, producer, user, environment of use**} and stress the importance of designs that are at least compatible with if not conducive to, rather than antithetical to or subversive of, important ethical values, such as justice, health, liberty, privacy, and sometimes even *usability*. The cartoon in Figure 1 indirectly makes this point regarding IT products with a human–computer interface. The more important being able to use a piece of software is to functioning effectively in everyday life in contemporary society, and the smaller the number of realistic alternatives members of society have to using it, the more its designers have a derivative ethical responsibility (under FERE1) to design software interfaces that are user-friendly.[105]

This ethical responsibility of software product designers parallels the ethical and legal responsibilities of urban planners, theater owners, building owners, and commercial shop designers under the Americans with Disabilities Act of 1990.[106] They are legally (and ethically) responsible for

[102] http://scocal.stanford.edu/opinion/carmona-v-division-industrial-safety-30281.

[103] Murray (1982), p. 36.

[104] Ferris and Sandoval (2004).

[105] For historical perspective on user-unfriendliness in personal computers, see Corn (2011), pp. 177–202.

[106] http://www.dol.gov/dol/topic/disability/ada.htm.

designing equitable, accessible, and user-friendly hardware interfaces that do not impede or prevent members of society from functioning effectively in everyday life, whether at banks, stores, movie theaters, ATMs, parks, or government facilities.

Discussion Questions

1. Can it be ethically responsible for an engineer to design and "sign off on" a product that she or he knows is conducive to behavior (or has use consequences) incompatible with or subversive of important ethical values? Why or why not?
2. Are there circumstances under which it would have been ethically responsible for Diane to unquestioningly accept and uncritically implement her client's specification that only a minimal level of security be built into the data base management system in question? If not, why not? If so, what are they?
3. Can you think of a product of engineering design that by virtue of its properties (a) fosters an important ethical value of society, as well as one that (b) hinders or violates such a value? Give an example of each.
4. Is designing for ethical compatibility a derivative ethical responsibility of all engineers under all circumstances? Why or why not?
5. Is it possible for an engineer to be an ethically responsible practitioner even if her or his design is incompatible with one or more important ethical values of society? Explain your position and give an illustrative example.

CASE 6: THE CITICORP CENTER TOWER

Background

In the early 1970s, Citibank decided to build its new world headquarters in New York City. A midtown Manhattan block on the east side seemed like an appealing possible location. Part of that block was owned by St. Peter's Evangelical Lutheran Church. The owner was willing to sell its property to Citibank, but only if a new church was built for it at the northwest corner of the block, such that it was not connected

Fig. 2. Citicorp Center. © Edward Jacoby.

to the proposed Citibank tower and no columns from the tower ran through the new church. To accommodate the church, the project's distinguished structural engineer, William LeMessurier, devised the idea of placing the planned 59-story, 915-foot-tall Citicorp tower (Figure 2) on top of 114-foot-high columns situated at the midpoint of each of its four sides, and of cantilevering the northwest corner of the building out over the new church. This was agreeable to all stakeholders, and the building, christened Citicorp Center (CC), was erected and put into service in 1977.

As it happened, the building's structural design was flawed. The first public account of how LeMessurier came to realize that his building's structural system was problematic appeared in *The New Yorker* in

1995.[107] According to this account, in 1978, with the building already oc-
cupied, LeMessurier "received a phone call at his headquarters, in Cambridge, Massachusetts, from an engineering student in New Jersey."[108]
The "young man" told him that his professor had "assigned him to write
a paper on the Citicorp tower" and that he believed that "LeMessurier
had put the columns in the wrong place."[109]

Years later, this dramatic account was challenged. In 2011, a woman
named Diane Hartley stated she was the "engineering student in New
Jersey" who had called LeMessurier's New York office in 1978.[110] At
the time, Hartley was a civil engineering student at Princeton University
writing her senior honors thesis under civil engineering Professor David
Billington. She was exploring the "scientific, social and symbolic implications" of a major office complex and had examined the Citicorp Center
building. Hartley claimed she spoke with Joel S. Weinstein, then a junior
engineer working at LeMessurier's New York office. "Weinstein sent her
the architectural plans for the Citicorp tower and many of his engineering calculations for the building."[111]

Although LeMessurier, presumably after being contacted by Weinstein,
originally dismissed the student's concerns, he later went back over his
calculations thinking that the Citicorp building would make a good example to use in his structural engineering class at Harvard. He found that
while he had taken into account the stresses created by winds perpendicular to the sides of the building, he had not done so for those exerted
simultaneously on two sides of the building by "quartering winds," that
is, diagonal winds that strike the building at its corners. He may not have
done so because calculation of stresses produced by quartering winds
was not required by the NYC building code at the time, one that had been
designed with conventional high-rise buildings with their columns at the
four corners in mind.[112]

[107] Morgenstern (1995).

[108] Ibid., p. 45.

[109] Ibid.

[110] Whitbeck (2011a).

[111] Ibid.

[112] It could be that Diane Hartley's concerned call to Weinstein about quartering winds
indirectly prompted LeMessurier to consider them when he revisited his calculations for
the building's structural support system. According to Hartley, there was no evidence on
the structural drawings or in the calculations that Weinstein sent her that stresses on the
building from quartering winds had previously been taken into account. Hartley "asked
Joel Weinstein for his calculations of the effects of quartering winds. He said he would send
them, but she claims she never received them. When she told Joel Weinstein of the increased

When he realized this, a concerned LeMessurier made inquiries about how the Citicorp building had been built in relation to his specifications. In particular, he was concerned about his specification that the structural members of the ingenious eight-tiered inverted-chevron-shaped bracing system he designed for the building be *butt-welded*, not bolted. When he inquired about the butt welds, LeMessurier claimed he was told that in a meeting of his New York office's engineering staff with representatives of Bethlehem Steel, the latter had argued that butt welds were stronger than necessary. LeMessurier's New York office had accepted Bethlehem's suggestion, and the decision had been made to go with bolted joints.[113]

The problem was that this change had made CC vulnerable to being toppled by a 55-year storm, that is, a storm of the strength that occurs on average only once every 55 years in New York City. Worse yet, if the structure's "tuned mass damper"—a heavy, roof-based system intended to reduce the lightweight building's sway under high wind—were to become inoperative, for example, through a loss of electricity, the building was vulnerable to being toppled by a 16-year storm. LeMessurier was horrified. What would he do? He considered committing suicide by crashing his car into a bridge abutment on the Maine Turnpike at 100 mph.[114] Eventually, he decided to sequester himself, go into emergency mode, and devise a plan to retrofit and strengthen the building before any possible summer hurricane might threaten to topple it.

Once he had devised his retrofitting plan, LeMessurier contacted the building's owners, informed them of the risk of the structure's being toppled, and persuaded them of the merit of his plan to retrofit the building and strengthen the connections with 2-inch-thick steel plates. LeMessurier and Citicorp then revealed the situation to New York City public safety officials and, with the cooperation of top Citibank executives, launched the emergency retrofit plan. In late August 1978, when the retrofitting work had reached the halfway point, a major storm, Hurricane Ella, approached the East Coast. Fortunately, before it reached New York City, Ella turned east and veered out to sea. The retrofitting

---

stresses that her calculations showed for quartering winds, he reassured her that the building was safe and its design was, indeed, 'more efficient.'" Ibid.

If her account is accurate, by deciding to contact LeMessurier's office and speaking with Weinstein, Hartley initiated a process that resulted in CC's being much safer than it had been before.

[113] Morgenstern (1995), p. 46.

[114] Ibid., p. 48.

work was completed in time. The tower—later renamed Citigroup Center, and currently known as 601 Lexington Avenue and owned by Boston Properties—is now allegedly one of the strongest and most stable high-rise buildings in the world.

Ethical Analysis

William LeMessurier, who died in 2007, is widely regarded as an ethical hero of engineering. He is listed as a "moral exemplar" on the website "Online Ethics Center for Engineering and Science."[115] As LeMessurier put it in the last paragraph of the Morgenstern article, as an engineer

> [y]ou have a social obligation. . . . In return for gaining a license and being regarded with respect, you're supposed to be self-sacrificing and look beyond the interests of yourself and your client to society as a whole. And the most wonderful part of my story is that when I did it nothing bad happened.[116]

This is tantamount to the rule brought up in Chapter 3, that when there is a conflict between private and public interests, the latter must take precedence over the former. In designing the tuned mass damper for CC, and in reacting as he did once he concluded that CC was vulnerable to being toppled by a 16-year storm, LeMessurier's conduct was impressive from an ethical responsibility perspective. That said, let us ask: was there anything ethically problematic about LeMessurier's actions that should also be taken into consideration in striving for a comprehensive and balanced ethical assessment of his engineering conduct in connection with the CC building?

For clarity, let us distinguish LeMessurier's conduct in the period beginning at the point when he first realized that CC was at risk of being toppled by a 16-year storm, from his conduct in the period prior to this point. Different ethical issues emerged in each period. Let us call the two periods the *pre-risk-recognition period* and the *post-risk-recognition period*.

Regarding the *post*-risk-recognition period, once LeMessurier determined that Citicorp Center was at risk of being toppled by a 16-year storm, he behaved in accord with FERE2. By devising a detailed plan to

---

[115] http://www.onlineethics.org/Topics/ProfPractice/Exemplars/BehavingWell.aspx.
[116] Morgenstern (1995), p. 53.

retrofit and strengthen the joints in the building's support structure, and by going to Citicorp management and New York City public safety officials to reveal the nature of the problem, he made a laudable effort to prevent the immense harm that would have resulted if the building had been toppled by a major hurricane in the summer of 1978. LeMessurier's retrofit program effectively eliminated the preexisting unreasonable risk of harm.

What to make of LeMessurier's conduct with respect to FERE3 is a more complex question. He did reveal the risk he had been led to recognize to top Citicorp executives, and he and representatives of the owner did inform New York City safety authorities about the existence and magnitude of the risk and about his retrofitting plan. This is all ethically responsible, indeed laudable, as far as it went. However, looking more closely at what transpired, Citicorp statements to the press about the retrofitting effort represented the engineer and firm as having decided, on the basis of new data, to have the building "wear suspenders as well as a belt."[117]

By issuing such a statement the company gave the misleading impression that the building was already safe and was just being made even safer at the suggestion of the structural engineer. LeMessurier went along with this deception. In addition, the existence and operation of the retrofitting activity was apparently kept hidden from the public living near the building, as well as from those who worked in or near the building. The retrofitting was carried out each evening and early morning, from 8 p.m. to 4 a.m., and those working in the building were kept in the dark about it, on the grounds that revealing the work through the news media might induce public panic. Given the tabloid tendencies of some of New York City's large-circulation daily newspapers of the day, that concern was not without merit. At one point, LeMessurier agreed to be interviewed by a *New York Times* reporter who suspected that there was something unusual going on at the building. However, just before the interview was to occur, a labor strike began at that newspaper, and LeMessurier avoided having to make a decision on the delicate ethical issue of whether to disclose the risk and the retrofitting program publicly, or to deny the risk and misrepresent the situation to the public.

In short, it is at least debatable whether LeMessurier acted fully in accordance with FERE3. Disclosing the risk of toppling to Citicorp and to New York City public safety officials might seem sufficient to justify

[117] Ibid., p. 51.

saying that he had fulfilled his FERE3-derived specific ethical responsibilities. But, had I been working in or near the Citicorp Center at the time, or been living at the time in the neighborhood of the building, I would have been livid to learn after the fact that I had been allowed to carry on with business as usual by being kept in the dark about the existing risk and the nocturnal retrofitting project.

Turning now to the *pre*-risk-recognition period, in my view two derivative ethical responsibilities merit exploration. One is organizational and the other involves cognition.

Organizationally speaking, by omission LeMessurier allowed a culture to exist in his firm under which structural changes in a high-rise building could be made without being approved directly by him. This is clear from the aforementioned incident by which LeMessurier came to learn that his specified butt-welded connections had been changed to bolted connections without input from him.

The Morgenstern article contains an important passage that, while not a direct quotation, clearly reflects LeMessurier's position and is intended to rebut the view that changes should not have been permitted without first being run by the firm's top engineering expert, namely, LeMessurier.

> The choice of bolted joints was technically sound and professionally correct. Even the failure of his associates to flag him on the design change was justifiable; *had every decision on the site in Manhattan waited for approval from [his firm's home office in] Cambridge, the building would never have been finished.*[118] (emphasis added)

The italicized would-be justification in the last sentence of this quote is a classic straw man argument. I do not believe or contend that "every decision on the site in Manhattan" should have "waited for approval from [LeMessurier in] Cambridge."[119] I do, however, submit that every *structural* change, or, put differently, *every change that would or might affect the structural integrity of the building*, should have been run past the person with the highest level of technical expertise, namely, William LeMessurier. The culture that existed at his firm did not mandate that that happen as standard operating procedure in the firm's everyday engineering practice. For that reason, I believe that LeMessurier was remiss in not establishing and maintaining a culture at his firm that would

---

[118]Ibid., p. 46.
[119]Ibid.

unambiguously prohibit such changes from being made without their being run by him for prior scrutiny and approval.

Cognitively speaking, in designing a radically innovative kind of high rise, one whose columns were at the midpoint of each of the four sides rather than at the four corners, LeMessurier used the conventional method of structural analysis to determine how strong the bracing system needed to be. To see precisely what is problematic about doing so, consider the following.

Historian of science Thomas Kuhn is credited with the influential notion of a disciplinary "paradigm."[120] He argued that at a given point in time in any *mature* field of science virtually all its practitioners share a certain set of fundamental assumptions, core beliefs, and epistemological commitments. Kuhn referred to the set of these shared elements that most practitioners in a mature field of science subscribe to at a given point in time as the field's or discipline's **dominant paradigm** at that juncture. Practitioners in the field are socialized into embracing this paradigm while graduate students, partly through using textbooks that reflect and apply it. Subscribers to the dominant paradigm or "disciplinary matrix"[121] share beliefs that certain entities exist and that certain forces operate, and feel impelled to employ certain consensually agreed upon methods in their work. They also follow certain consensually agreed upon rules of procedure and explore and explain only certain consensually agreed upon phenomena. In short, *the dominant paradigm of a mature field of inquiry exerts a subtle but powerful guiding or regulatory influence on* the what *and* the how—that is, *the substance and the manner—of the practice of science in that field.*

In recent years, scholars have adapted and applied Kuhn's ideas about the role and functioning of paradigms in the sciences to the study of technology and technological change.[122] These scholars contend that at a given point in time, the activity surrounding a particular mature technology proceeds under the auspices of a dominant paradigm, that is, a set of assumptions to which most if not all people working on that technology subscribe. At a given point in time, a mature technology's paradigm comprises consensual assumptions about the following:

- the *normal configuration* of the device or system in question;
- the *operational principle* that underlies its functioning;

---

[120] Kuhn (1970), pp. 23–34, and pp. 174–210.
[121] Ibid., p. 184.
[122] See, for example, Constant (1980).

- the *material(s)* out of which the technology is made;
- the *process(es) and method(s)* by which the technology is designed, made, and operated; and
- the standard *use(s)* to which the technology is put.[123]

Dominant paradigms exert subtle but powerful conservative influences on technologists and engineers who subscribe to them. Moreover, as can also happen with scientists, engineers can be not only *benefited* but also *betrayed* by their paradigm adherence. How can paradigm adherence be problematic?

Technology paradigm adherence can go wrong in two ways: (i) if paradigm adherents, for example, engineers who work on the technology in question, are unable to see or accept fruitful new uses of a technology; and (ii) if paradigm adherents wrongly assume that a design, material, or methodological approach appropriate for use with a conventional technological item of a particular type is also warranted for a technological item of that type that is deeply unconventional. In short, uncritical or rigid technology paradigm adherence can go wrong by fostering *undershooting* the proper domain of use or *overshooting* the domain of valid application of the design, material, or analytical methods in question.

LeMessurier succumbed to paradigm *overshooting*. He wrongly assumed that the approach to structural analysis taught in graduate structural engineering courses as the proper way to determine the wind-caused stresses on structural elements of high-rise buildings whose columns are *at the four corners* was also appropriate for the analysis of such forces in buildings whose columns are at *the midpoint of each of its four sides*. As LeMessurier acknowledged when I explained Kuhn's ideas to him,[124] he had unwittingly extended the paradigmatic method of structural analysis of a high-rise building beyond its legitimate domain of application and into a new domain where it did *not* apply, one that included his building. He had designed a brilliantly innovative building—one with its columns at the midpoint of each of the four sides, inverted-chevron

---

[123]The expression "normal configuration" is due to Walter Vincenti, while "operational principle" is due to Michael Polanyi. See Vincenti (1992), p. 20, and Polanyi (1962), p. 328. I added the additional paradigmatic elements listed. Using Vincenti's Kuhnian distinction, a technological design is an example of "radical" or "revolutionary" (as opposed to "normal" or "conventional") engineering design (at a given point in time) if it departs substantially from the reigning paradigm of that kind of item in one or more of its dimensions (configuration, operational principle, materials, methods, processes, uses).

[124]This took place during a multihour face-to-face conversation the author had with LeMessurier in San Francisco, California, on August 24, 1995.

structural-support system, and tuned mass damper—and then analyzed it in a conventional way (paying attention only to perpendicular winds)—as it would have been proper to do had its columns been at the four corners.

Thus, in light of the fact that radical engineering design is highly esteemed and handsomely rewarded in the United States and other contemporary societies, design engineers have a dual derivative ethical responsibility:

> to be alert to whether they are working on a conventional/normal engineering design, or on a radical/ revolutionary/paradigm-departing design;

and, if the latter,

> to be extra vigilant that they do not unwittingly extend the established paradigm of engineering analysis or practice for an object with a conventional design to an object whose design falls outside the paradigm's valid domain of application.

An engineer would do so by carefully checking and rechecking her or his assumptions, analyses, and procedures. LeMessurier inadvertently overshot the paradigm for high-rise structural design. Doing so was arguably imprudent and at odds with the aforementioned derivative ethical responsibility he had under FERE1 as applied to his innovative building.

Lessons

This case suggests a critically important general take-away lesson for engineers in all fields. Keeping FERE1 in mind, whenever an engineer is working on a product, system, or process that is radically innovative in one or more of its dimensions, for example, in its configuration, structure, operating principle, or material, she or he has an ethical responsibility *to not assume or take for granted that the usual ways of designing, fabricating, and operating conventional products or processes of the sort in question also apply to the innovative "outside-the-box" item in question, and to double-check whether they do in fact apply.*

This crucial derivative ethical responsibility helps compensate for the fact that there can be subtle psychological pressure on engineers—also on scientists and, for that matter, people working under the auspices of any entrenched disciplinary paradigm—to "normalize" revolutionary or radically innovative engineering designs or processes, leading them to implement and work with them "in the usual manner." Doing so can create a risk of serious harm that violates FERE1.

This danger is shown not only in the Citicorp Center case but in several other cases, some considered elsewhere in this book, for example, the cases of the space shuttle *Challenger*; the envisaged incredibly complex SDI computer-controlled weapons system; and the seemingly routine sign-offs by colleagues on Schön's astonishing organic electronics papers on which they were listed as coauthors.

The cartoon in Figure 3 is useful for making my central point about paradigm departure and ethical responsibility in engineering practice. Whenever an engineer is working on a product or system that rather than being an example of conventional or normal engineering design is better viewed as an example of radical engineering design—hence, is "outside the box"—then the message is *not* the one that the

Fig. 3. "Never, ever, think outside the box." Leo Cullum / *The New Yorker* Collection / The Cartoon Bank.

(conservatively dressed) gentleman is conveying to the cat—to "never, ever think outside the box"—but rather that *when* he or she is designing outside the box, the engineer, pursuant to FERE1, has a weighty derivative ethical responsibility to proceed in an extravigilant and precautionary manner.

Put differently, when an engineer is working on something outside the box, that is, is doing radical engineering design work involving a basically new configuration, operational or structural principle, constituent material, method of analysis, production process, or use, she or he has an ethical responsibility to make doubly sure that the approaches known to apply to conventional or normal engineering design work also apply in the domain of radical engineering design work. LeMessurier mistakenly assumed that the usual way of doing structural analysis on a conventional high-rise validly applied to his deeply innovative structure. Instead of first determining whether the usual way of calculating stresses on a conventional high-rise building still applied to his radically unconventional building, LeMessurier appears to have *assumed* that the domain within which the usual way of calculating stresses routinely applied included the radically unconventional building support structure he had conceived. When engaged in radically innovative or revolutionary engineering design, the engineer has *a derivative ethical responsibility to be extra vigilant and to take extra precautions to avoid unwittingly utilizing a technique of engineering analysis beyond its legitimate domain of application.* This derivative ethical responsibility results from applying FERE1—here, the engineer's fundamental ethical responsibility to not create an unreasonable risk of harm through his or her work by acts of omission—to the novel circumstances of radical engineering design work.

Discussion Questions

1. What is it about a particular engineering design that makes it an example of a revolutionary or paradigm-departing design, as opposed to being an example of normal engineering design?
2. Do engineers have a derivative ethical responsibility to take extra precautions and be extra vigilant when working on a paradigm-departing design? Why or why not? If they do, from which FERE can that responsibility be derived?

3. Is it fair to say that while LeMessurier acted in accord with FERE2 in going to Citicorp executives and New York City safety authorities with his plan for retrofitting the building, he was organizationally remiss or negligent regarding the culture at his firm, and cognitively remiss or negligent in determining the strength and margin of safety of his structural support system, since he unwittingly treated a paradigm-departing building design in a conventional way? Why or why not?

4. How would you assess LeMessurier's conduct vis-à-vis the public, both people working in the building and living in the neighborhood, as regards the FERE3 notion that the engineer has a fundamental ethical responsibility to try to alert and inform parties at risk of incurring harm from one's work that they are vulnerable to being harmed? Justify your position.

CASE 7: THE UNION CARBIDE PESTICIDE PLANT IN BHOPAL

Background

Methyl isocyanate (MIC) is a highly toxic chemical used as an intermediate, with alpha-naphthol and carbon tetrachloride, in the production of Sevin, Union Carbide Corporation's (UCC) trademarked name for the pesticide carbaryl. Until the early 1970s, UCC's Indian subsidiary, Union Carbide of India Limited (UCIL), had been importing finished MIC to make pesticide. However, to achieve agricultural self-sufficiency and to stem currency outflows, the Indian government encouraged UCC to build a plant to produce MIC from scratch at UCIL's facility in Bhopal, India. Production of MIC at the Bhopal plant commenced in 1980.

On December 2, 1984, around 11:30 p.m., the eyes of some shift workers at the MIC production facility in Bhopal began to tear. One worker noticed a drip of liquid about 50 feet off the ground, accompanied by some yellowish-white gas. Told of the discovery, the shift supervisor in charge, who later claimed he had been told only of a water leak,[125] postponed an investigation of the matter until after tea break. In fact, what was being released into the air, from holding tank E610, was toxic MIC gas. Instruments indicated that the pressure inside the tank was rising

---

[125] Diamond (1985b).

rapidly. Then, the thick concrete in which tank E610 was partly buried cracked. Soon thereafter, the tank emitted a blast of MIC gas that shot out a stack and formed a large white cloud. Part of the cloud hung over the factory, while another part descended upon the residential area near the plant.

The Bhopal disaster is widely regarded as one of the worst industrial disasters in history. Estimates of the number of deaths attributable to the MIC emission range from 3,000 to as high as 30,000, with a number of writers using the rough figure of 20,000.[126] Estimates of the number of injured range from 200,000 to about 560,000.[127] In 1989, the Indian Supreme Court upheld the legality of an agreement under which UCC paid $470 million to the Indian government to settle all claims arising from the disaster.[128] However, litigation spawned by the disaster continued for more than a quarter century. In 2010, eight former executives of UCIL, one already deceased, were convicted of negligence. Each was sentenced to two years in jail and fined $2,100.[129] Bhopal-related legal actions continued in the United States until August 2016.[130]

Ethical Analysis

The discussion of this case will focus on an important engineering-related phenomenon: the transfer of technology from a more developed country to a less developed one. The ethical issues in this case stem from characteristics of the process by which the technology for producing MIC was transferred from UCC in the United States and established in UCIL's Bhopal plant in India. Depending on the relative adequacy of that process, the transfer process foreseeably could—and in the event did—have an enormous negative effect on the health and safety of UCIL's MIC workers and people living near the plant and in Bhopal.

At the outset of Chapter 4, a distinction was made between the foreground "precipitating cause" of a phenomenon of interest and various

---

[126]See, for example, Varma and Varma (2005), p. 37, and Edwards (2002).

[127]The Government of India provided one estimate. In October 2006, it filed an affidavit in the Indian Supreme Court stating that 558,125 injuries had been established as due to the release of MIC gas at the Bhopal plant. See http://storage.dow.com.edgesuite.net/dow .com/Bhopal/Aff%2026Oct06%20of%20UOI%20in%20SC%20in%20CA3187%20n %203188.pdf.

[128]Hazarika (1989).

[129]Polgreen and Kumar (2010).

[130]See "Bhopal Litigation in the U.S.," http://www.bhopal.com/Bhopal-Litigation-in-the-US.

background "contributory causal factors" that, although they did not themselves trigger the phenomenon, helped set the stage for another factor to precipitate its occurrence.

UCC has always contended that the Bhopal disaster was caused by single factor: the action of an alleged saboteur who was a disgruntled UCC employee. If true, this deed was the precipitating causal factor of the incident. UCC has also claimed that such an action is the only conceivable explanation for how water got into MIC liquid holding tank E610.

> Shortly after the gas release, Union Carbide launched an aggressive effort to identify the cause. . . . Investigations suggest that only an employee with the appropriate skills and knowledge of the site could have tampered with the tank. An independent investigation by the engineering consulting firm Arthur D. Little, Inc. determined that the water could only have been introduced into the tank deliberately, since process safety systems—in place and operational—would have prevented water from entering the tank by accident.[131]

In light of the distinction between foreground precipitating and background stage-setting causal factors, the concept of "cause" UC employed in this passage is arguably extremely reductive; it limits itself to an alleged precipitating causal factor.

Another candidate for the precipitating cause of the incident is implicit in an MIC worker's claim that during the evening of the disaster he was told by a "novice" MIC supervisor,[132] in his position for only two months, to wash out with water a pipe that led from "a device that filtered crude methyl isocyanate before it went into the storage tanks."[133] In addition, the worker claimed that inside that pipe was a valve that had not been closed.[134]

It will probably never be known with certainty whether the precipitating cause of the MIC emission was the action of a saboteur, the above-mentioned washing out of the pipe, or some other single foreground factor. However, a number of background stage-setting factors had been at work in the MIC plant long before the night of the disaster. These factors enabled, facilitated, were conducive to, and/or helped induce the occurrence of the toxic emission. They arguably set the stage for a possible

---

[131] http://www.bhopal.com/Cause-of-Bhopal-Tragedy.
[132] Diamond (1985a). This individual's immediately prior position was allegedly at a UCI battery factory. See also Diamond (1985b).
[133] Diamond (1985b).
[134] Ibid.

saboteur or operational mistake of some sort to trigger the deadly emission. At least eight such table-setting factors merit scrutiny.

1. *Plant location.* UCC chose Bhopal for its pesticide factory for several reasons. There were rail connections to the north, south, east, and west of India from Bhopal; new employment opportunities were desperately needed in the poor, underindustrialized state of Madhya Pradesh; the Indian government was willing to rent land on favorable terms to companies that agreed to build there; and there were ample supplies of water from Bhopal's lakes for use in the plant's chemical processes.[135] However, the factory location turned out to be a noteworthy contributory causal factor in the MIC disaster for two quite different reasons.

Chemical engineer Kamal Pareek started working at UCIL's Bhopal plant in 1971, served as senior project engineer during the creation of the MIC plant in the late 1970s, and left the company in December 1983.[136] Speaking about the MIC plant shortly after the disaster, Pareek said:

> When we set up this plant, we used workers just out of the agricultural age. . . . What we needed were people who had grown up with technology, not workers who have had technology imposed on them by economic circumstances.[137]

In effect, he acknowledged that workers raised in an agricultural society had been hired to work in an industrial-society factory. Consequently, there was a mismatch between the entrenched behavioral and mental habits of many of the new workers and the mental and behavioral requirements for operating such a factory safely. This is the first reason why the location of the Bhopal plant was a contributory causal factor in the Bhopal disaster.[138]

Pareek's remarks raise a question: when did he recognize the disconnect between the agricultural cultural background of many of the workers being hired to work in UCIL's MIC plant in Bhopal and the kinds of mental and behavioral orientations (such as executing tasks strictly in accordance with specified safety procedures) that are essential for safe operation of an MIC plant? Was it while he was senior project manager for

[135] "Pesticide Plant Started as a Showpiece but Ran into Troubles," *New York Times*, February 3, 1985.

[136] Diamond (1985a) and "U.S. Company Said to Have Control in Bhopal," *New York Times*, January 28, 1985.

[137] Diamond (1985c).

[138] Ibid. Pareek also pointed to the culture of broader Indian society as a contributory causal factor: "We Indians start very well, but we are not meticulous enough to follow up adequately."

the construction of the Bhopal MIC plant or only in retrospect, after he left the company? If the former, did he go to UCIL top management and let its members know unambiguously that hiring such workers was creating a risky situation vis-à-vis the safe operation of the plant? In light of FERE2 and the nature of the manufacturing process being carried out in the plant, Pareek arguably had a derivative ethical responsibility to do so.

To his great credit, Pareek apparently went to UCIL management to warn them while he was with the company. In a 2004 interview he stated:

> I told them that the way things are going it wouldn't really surprise me if we have a huge tragedy on our hands. I did not use the word "tragedy," I said "disaster." . . . They wished me good luck and threw me out.[139]

If that quote accurately represents what he did, engineer Pareek arguably fulfilled his FERE2-based ethical responsibility to try to prevent an unreasonable risk of harm from continuing to loom over the plant.

The second reason why locating the pesticide factory in Bhopal proved to be a contributory causal factor to the disaster involves ineffective zoning. As noted, originally UCIL did not manufacture Sevin at the Bhopal plant. It imported "pure Sevin" and "diluted, packaged, and shipped" it for sale.[140] Then, in line with an Indian government drive to achieve agricultural self-sufficiency and stem currency outflows by making the components of pesticides locally, in 1972, at the behest of UCC, UCIL proposed to "manufacture and store MIC at the plant,"[141] just as it had recently done with alpha-naphthol.

In August 1975, public authorities issued a Bhopal development plan that required that "obnoxious industries," "including plants that manufactured pesticides and insecticides, be relocated to an industrial zone 15 miles away" from the city of Bhopal. The belief was that in that zone, the prevailing winds would "carry obnoxious gases away from the city area."[142] However, while the commissioner and director of town and country planning "ordered the Carbide plant to be relocated in 1975,"[143] that directive was never implemented. The factory stayed where it was. Shantytowns and middle-income housing grew up adjacent to it, and the population of Bhopal increased rapidly.[144] Thus, when, in 1979, the MIC

---

[139] Lewis (2004), 03:38–55.
[140] "Pesticide Plant Started as a Showpiece," *New York Times*, February 3, 1985.
[141] Reinhold (1985).
[142] Ibid.
[143] Ibid.
[144] Ibid.

production plant was finally built, a toxic-substance-producing facility was situated adjacent to a large and rapidly growing group of people living just outside and nearby the factory. The risk to public safety that this facility posed should have triggered a serious effort to mitigate it, but apparently that did not occur. That so many people were living so close to a plant making toxic materials strongly suggests that Madhya Pradesh lacked the ability to establish and maintain a suitable buffer zone between where the toxic material MIC was being produced and where increasing numbers of people were living. This lack of buffer zone was a background contributory causal factor to the disaster.

2. *Worker training.* Many workers in the Bhopal MIC plant that came online in 1980 appear to have had insufficient training and experience to be able to work safely and effectively with that extremely dangerous chemical. Many knew only their own microtasks and nothing of how those tasks fit into the larger production system. As one of the MIC workers said about the pipe he claimed the novice supervisor asked him to wash out with water, "I didn't check to see if that one was leaking. It was not my job."[145] He continued:

> I was trained for one particular area and one particular job. I don't know about other jobs. During training they just said, "These are the valves you are supposed to turn, this is the system in which you work, here are the instruments and what they indicate. That's it."[146]

"Previously, operators said, they were trained to handle all five systems involved in the manufacture and storage of methyl isocyanate. But at the time of the accident, they said, only a few of about 20 operators at Bhopal knew the whole methyl isocyanate plant."[147] Yet the Bhopal operating manual states, "To operate a plant one should have an adequate knowledge of the process and the necessary skill to carry out the different operations under any circumstances."[148]

Along with the narrow training provided, the duration of training decreased. "In Bhopal, operators hired for the new plant in 1979 received three weeks of training. By 1984, that training had practically disappeared."[149]

---

[145] Diamond (1985b).
[146] Diamond (1985a).
[147] Ibid.
[148] Ibid.
[149] Willey (2014). On the decreasing duration of training given to new operators, see Chouhan (2005), p. 207, fig. 6.

An inspection team from UCC's technical center in South Charleston, West Virginia, visited the Bhopal plant in 1982.[150] Its report stated: "There was also concern that sufficient 'what if' training was not conducted to provide personnel with a basic understanding of the reasoning behind procedures, as opposed to rote memorization of steps."[151] Concurrently, according to workers, employment qualifications were being reduced at the Bhopal plant. MIC operator jobs that once required college science degrees were being filled with high school graduates.[152] M. K. Jain, a high school graduate and an operator on duty the night of the disaster, claimed to have had three months of instrument training and two weeks of theoretical work. This left him, he said, unable to understand large parts of the plant.[153] Moreover, with declining pesticide sales and factory losses starting in 1982, employees said that the Bhopal plant had "lost much of its highly trained staff, its morale and its attention to the details that insure safe operation."[154] Chemical engineer Pareek summed up the consequences of the inadequate worker training:

> The whole industrial culture of Union Carbide at Bhopal went down the drain . . . The plant didn't seem to have a future and a lot of skilled people became depressed and left as a result.[155]

It may be that the drive to cut company operating losses had led to diminished commitment to keep worker training levels where they had been and to willingness to hire less qualified workers who lacked the background needed to understand the workings of the plant as a whole. Whatever the reason, it is remarkable that, according to Pareek, the workers who worked at the MIC facility did not even realize how lethal MIC was. "No one at this plant thought MIC could kill more than one or two people."[156] Many workers thought that MIC's chief effect was to irritate skin and eyes. In short, the training of operators at the MIC plant seems to have been inadequate and getting worse. Any engineer at UCIL or UCC who believed that the worker-training level at the Bhopal MIC plant was inadequate had a specific FERE2-based derivative ethical responsibility to make that known to top management as soon as she

[150] Stix (1989), p. 50.
[151] Ibid.
[152] See Chouhan (2005) p. 207, fig. 5.
[153] Diamond (1985a).
[154] Ibid.
[155] Ibid.
[156] "Most at Plant Thought Poison Was Chiefly Skin-Eye Irritant," *New York Times*, January 30, 1985.

or he recognized it, to try to prevent an unreasonable risk of harm from increasing and going unaddressed.

3. *Compensatory technology*. Given the modest and apparently declining level of training of workers at the MIC plant, advanced technological systems could have been used to compensate. The UCC MIC plant in the U.S. state of West Virginia had a computer-and-sensor-based system that monitored the production process and quickly alerted staff to leaks. No such system existed at UC's Bhopal factory.[157] According to the UCC technical manual, "Although the tear gas effects of the [MIC] vapor are extremely unpleasant, this property cannot be used as a means to alert personnel." Nevertheless, as allegedly happened the night of the accident, workers' eyes detected leaks in the MIC production process. In the words of MIC worker Suman Dey, "We were human leak detectors."[158]

UCIL engineers likely knew—or arguably should have known—about the primitive way leaks were being detected. Did any who saw what was going on make any effort to warn top management at the Bhopal MIC plant that it was relying on a method of leak detection prohibited in the UCC technical manual? Did any engineers at Bhopal strongly urge UCIL top management to request that UCC provide a computer-based system for the Bhopal plant that would monitor nascent toxic leaks? Any engineers that took one or more such initiatives were arguably fulfilling FERE2-based derivative ethical responsibilities to try to prevent unreasonable risks of harm.

4. *Maintenance*. Maintenance was problematic at the UCIL Bhopal plant during the period leading up to the incident. First, preventive maintenance was apparently not a high priority there. Not only had the number of MIC operators per shift been cut from 12 to 6 in 1983, likely meaning further delays in completing maintenance tasks,[159] but, allegedly, the number of "maintenance supervisors" had also fallen significantly between 1979–80 and November, 1984.[160] Engineer Pareek claimed that "[the plant] cannot be run safely with six people."[161] These cuts might have stemmed from top managers' concern with cutting costs to reduce losses from operating the Bhopal plant.

However, there was another, more subtle side of the maintenance matter. As Indian chemical engineer Kiran Rana put it,

[157]Diamond (1985a).
[158]Diamond (1985b).
[159]Diamond (1985a).
[160]Chouhan (2005), p. 207, fig. 5.
[161]Ibid.

The idea of spending money now to save money later is a concept completely alien in what is basically a subsistence economy.[162]

In other words, *the idea of preventive maintenance* had not yet put down deep roots in Indian society or at the MIC plant. This situation was conducive to allowing MIC plant safety systems to go unrepaired for long periods of time.

Another factor antithetical to good maintenance was that instruments were often not trusted because they so often gave inaccurate readings. Consequently, opting not to carry out safety-preserving interventions that would have been prompted by anomalous but credible instrument readings became accepted practice at the plant rather than violations of plant cultural norms. Under FERE2 and FERE3, any engineer who observed declining preventive maintenance at the MIC plant, and/or instrument readings being distrusted and dismissed, arguably had a derivative ethical responsibility to warn Bhopal top management of the risk. Moreover, if doing that failed, such an engineer arguably had a derivative ethical responsibility to alert and inform Indian public safety officials about the additional risk that endemic distrust of instrument readings was creating.

The MIC plant had three main safety systems: a water spray (to contain a chemical leak), a gas scrubber (to neutralize toxic release material), and a flare tower (to burn off escaping gases). However, the MIC disaster occurred in the midst of a maintenance backlog. The gas scrubber had been shut down for maintenance 12 days just before the disaster, even though the "plant procedures specify that it be 'continuously operated' until the plant is 'free of toxic chemicals.'"[163] According to workers, the flare tower had also been out of operation for six days at the time of the deadly emission. "It was awaiting the replacement of a four-foot pipe section, a job they estimated would take four hours."[164] Other safety-related systems had been turned off, for example, the refrigeration unit or chiller, which, had it been active, would have greatly slowed the reaction of MIC with water.[165] In short, maintenance at the Bhopal MIC plant at the time of the disaster seems to have been in disarray.

If that was the case, it is difficult to believe that engineers working at the Bhopal MIC plant were unaware that key safety systems were defective or inoperable. Perhaps some engineers who worked there were like the

[162]Diamond (1985c).
[163]Diamond (1985a).
[164]Ibid.
[165]Ibid.

MIC operators who had narrow notions of their jobs and did not perceive sounding the alarm about inoperable safety systems as something they had a derivative ethical responsibility to do. Or perhaps some knew they *should* have warned about the inoperable safety systems but chose to adopt a passive posture and kept silent. Such behavior by any engineers at the Bhopal MIC plant would have been at odds with FERE2. Whether any engineers at UCC's Bhopal plant explicitly warned plant management about serious problems with the plant's safety systems but were rebuffed, as Pareek claimed he was, is unclear. If any did so and were rebuffed, did they have an FERE2-based ethical responsibility to take their concerns public?[166]

5. *Public education.* Although with the benefit of hindsight, the author's view is that UCIL should have made a much stronger proactive and concerted effort to educate the public about the hazardous chemical being produced at the MIC plant and about what to do in case of an emergency. Much of the public did not even know that one could avoid or mitigate the effects of MIC gas simply by putting a wet towel over one's face.[167] Moreover, there was no distinctive warning horn that would be sounded to alert the public *only* in an actual emergency.[168] "Several Bhopal residents said that many of the people living near the plant thought [the plant] produced 'Kheti Ki Dawal,' a Hindi phrase meaning 'medicine for the crops.'"[169] Even some doctors who helped care for victims of the emission were allegedly unaware of the toxicity of the escaped gas.[170]

6. *Regulation.* The Madhya Pradesh labor department had an inadequate number of inspectors to check out factories in the state regarding worker safety. Each inspector was responsible for more than 150 factories and had a quota of 400 inspections per year.[171] Moreover, the inspectors had primitive measuring instruments and were expected to travel to the factories by public transportation. Deficiencies found in the factories allegedly triggered only minimal fines, and effecting remedies of unsafe conditions took a long time. More fundamentally, the regulatory rules pertaining to industrial and environmental safeguards that were applied to UCIL's Bhopal plant and other "obnoxious industries"

[166] For detailed discussion of the conditions under which an engineer has a presumptive ethical responsibility to go public with her or his safety concerns, see Case 11.

[167] Reinhold (1985).

[168] Diamond (1985a) and "Slum Dwellers Unaware of Danger," *New York Times*, January 31, 1985.

[169] "Most at Plant Thought Poison Was Chiefly Skin-Eye Irritant," *New York Times*, January 30, 1985.

[170] Reinhold (1985).

[171] Ibid.

were outdated. In the wake of the disaster, Arjun Singh, Chief Minister of Madhya Pradesh state, made an astute comment, albeit at a point when it was too late to be acted upon: "Most of these rules were framed a long way back. They certainly need updating in view of new processes."[172]

7. *Operator manuals.* The native language of most of the workers at the MIC plant was Hindi, but the operating manual used at Bhopal was in English. Five Indian engineers had adapted it from one written in English for UC's West Virginia plant.[173] However, one wonders whether the adapted version, including detailed technical warnings, rules, and safety procedures, was comprehensible to the increasingly modestly educated MIC operators who were supposed to absorb and follow it, especially as there was considerable operator turnover after the MIC production facility was launched. The ability to speak and understand basic conversational English by no means guarantees that individuals who can do so also comprehend the prescriptions and proscriptions of a dense English-language technical manual.

8. *MIC Storage.* UCC decided that at the new MIC facility it would be "more efficient" for MIC to be stored in large tanks instead of in drums, "so that a delay in the production of methyl isocyanate would not disrupt production of the pesticides of which it is a component."[174] "Many plants store methyl isocyanate in 52-gallon drums, which are considered safer than large tanks because the quantity in each storage vessel is smaller."[175] In fact, MIC had been stored in such drums at Bhopal when it was imported from the United States. "Tank storage began in 1980, when Bhopal started producing its own methyl isocyanate" in the recently completed MIC facility.[176] In a legal deposition after the disaster, engineer Edward A. Munoz, former managing director of UCIL, claimed that in the mid-1970s, he had recommended that

> for safety and cost reasons, MIC be stored in small containers, but that he was overruled by Union Carbide's design engineering group in Charleston, W. Va. Union Carbide officials have denied this version of events, adding that Munoz was fired from his job because of differences with management over business strategy.[177]

---

[172] Ibid. It is worth noting that it can take some time for a building code to be updated to take into account structural innovations.

[173] Diamond (1985a).

[174] Ibid.

[175] Ibid.

[176] Ibid.

[177] Stix (1989), p. 48. See also Tyabji (2012), p. 49, n. 14: "When the plant was first designed, Edward A. Munoz, the managing director of UCIL, took the position that large-

It is curious that the MIC produced at Bhopal was stored in tanks, since the UC technical manual stated that storage of MIC is riskier, and contamination of it more likely, in large tanks than in small containers. Because of this, the manual stated that refrigeration of the MIC was necessary for bulk storage.[178] Yet, according to the Indian government's chief scientist, on the day of the accident, the refrigeration unit was not connected, "because managers had concluded," allegedly "after discussions with American headquarters, that the device was not necessary."[179] In my view, any UCC or UCIL engineer or engineer-manager who participated in the process that culminated in UCC's decision to have the MIC produced at Bhopal stored in large (15,000 gallon) tanks rather than in small drums or containers, or who was involved in the alleged UCC decision to disconnect the MIC refrigeration unit, had a derivative ethical responsibility, under FERE1, to oppose both decisions. Whether any did is unclear.

Much of the foregoing can be summarized by saying that neither the culture of the UCIL MIC plant nor that of broader Indian society in which it was embedded was conducive to safe operation of the Bhopal facility. In fact, the opposite was arguably the case; that is, the cultures of the Bhopal MIC plant and of the surrounding Indian society as a whole were conducive to operating the plant in ways that posed high degrees of risk to workers and nearby residents. Let us examine the key of idea of "culture" in greater detail.

* * *

Engineering professionals involved in transferring technology from a more developed country (MDC) to a less developed country (LDC) must grasp and internalize the concept of *culture* as many social anthropologists conceive of it. That is to say, the culture of a social unit refers to *the total, inherited, typically slowly changing way of life of the members of that social unit*. Put differently, the culture of a social unit refers to its interactive complex of characteristic background societal elements of four types: *mental, social, material,* and *personality/behavioral*. The cultural system of a social unit is embedded in a particular environmental context

---

volume storage of MIC was contrary to both safety and economic considerations. In a sworn affidavit to the Judicial Panel on Multidistrict Litigation considering the Bhopal case, Munoz said that he had recommended, on behalf of UCIL, that the preliminary design of the Bhopal MIC facility be altered to involve only token storage in small individual containers, instead of large bulk storage tanks. However, UCIL was overruled by the parent corporation, which insisted on a design similar to UCC's Institute, West Virginia, plant."

[178] Diamond (1985a).
[179] Ibid.

and forms an interactive system that is transmitted, sometimes in modified form, to newcomer members of that social unit.

Just as it is not fruitful to view a basketball team as a *thing*, but rather as a *system* of interactive components, it is helpful to view the culture of a social unit as a "cultural system" comprising four interacting subsystems (Figure 4). The **ideational subsystem** (I) of society's cultural system consists of the social unit's characteristic ideational constructs (ideas, values, core beliefs, ideals, worldviews, etc.). Its **social subsystem** (S) consists of the social unit's characteristic societal forms (social institutions, social structure, social status system, social groups, societal traditions, social roles, etc.). Its **material subsystem** (M) consists of the social unit's characteristic technological artifacts and characteristic ways of making them. Its **personality/behavior subsystem** (P/B) consists of the social unit's characteristic personality traits and related behavior patterns. The word *characteristic* is critical here. For something to be part of a social unit's culture, it must be sufficiently widespread in that unit to qualify as characteristic of it. No isolated, unique, one-off elements can be part of a social unit's culture.

Most of the eight stage-setting factors in the Bhopal case, discussed earlier, relate to one or more of these subsystems of the MIC plant's or the broader society's cultural system. Thus, the lack of an *idea of preventive maintenance* pertains to the Indian ideational subsystem (of both the plant and Indian society); the lack of effective *social institutions of zoning and industrial regulation* pertains to Indian society's social subsystem; the lack of an *advanced computer system* to monitor incipient problems in the MIC production process pertains to the plant's material subsystem; and the lack of *safety-related and what-if training* of both workers and the public pertains to the personality/behavior subsystem of the plant and wider Indian society.

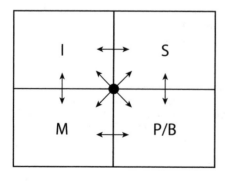

Fig. 4. The cultural system of a social unit.

More generally, one thing that seems to have gone wrong in the Bhopal incident was that an advanced technological production system—a product of work done in UCC's engineering design division, a social unit with its own distinctive cultural system and that was embedded in the culture of American society of the day—was introduced into two quite different and overlapping cultural systems: those of India and of the UCIL MIC plant in Bhopal. Such an introduction sounds harmless enough. However, it should never be assumed that the safety-preserving elements that were present, functioning, and reasonably taken for granted in the MDC cultural system where the design was created will also be present and function smoothly in the plant of the recipient LDC where the designed system will be used to manufacture product.

Unfortunately, this appears to be what happened in the Bhopal case: an *assumed match and compatibility* of the two cultural systems as regards elements related to the safe operation of the transferred MIC plant technology apparently blocked recognition by managers and engineers of a *profound actual mismatch and incompatibility* between the safety-related elements of the respective cultural subsystems of donor-designer and recipient-operator. The Bhopal disaster shows that doing so was devastating to LDC health and safety.

This brings us to a key ethical idea of the Bhopal case: the idea of an **ethically responsible process of international technology transfer** (ER-PITT). How can a process of transferring a technological product or system from one country to another for that process qualify as an ethically responsible one? Among other things, the process must be such that the engineers and managers involved fulfill the following FERE-based derivative ethical responsibilities:

(i) to ensure that all conditions for the safe and effective operation of the transferred technology are satisfied before the production process is allowed to commence;

(ii) to try to temporarily suspend the production process if any of the conditions for the safe and effective operation of the production cease to be satisfied; and

(iii) to try to prevent resumption of the production process until all unmet conditions for its safe and effective operation of the process are satisfied.

It is essential that the engineers and managers involved recognize that *the conditions referred to in these three ethical responsibilities are not all technical*; some are *cultural* in nature. For example, in the Bhopal

MIC case the operational existence of key ideas and values (preventive maintenance, holding worker and citizen safety paramount), the existence and effective operation of key social institutions and groups (such as zoning, industrial regulation, and a well-qualified, well-trained workforce), and the existence of key behavioral patterns (emergency preparedness, worst-case-scenario drills, and habitual rigorous adherence to specified operational procedures) are all important nontechnical conditions for the safe and effective operation of the MIC production process. *To the extent that a particular process of international technology transfer neglects these nontechnical elements, it falls short of qualifying as an ERPITT.* Any engineer-manager or engineer who pushes forward (or acquiesces in others pushing forward) a process of MDC-LDC technology transfer when any of those conditions is not satisfied is arguably negligent.

Conceptually speaking, conceiving "technology transfer" as "moving hardware from A to B" is a deep and dangerous misunderstanding of what is involved. The fact is that *the transfer of technology from an MDC to an LDC is a complex process of cultural transfer.* That is, a technological system developed in a MDC that is to be transferred to a LDC typically *embodies and presupposes* certain values, bodies of knowledge (implicit as well as explicit), organizational resources, ways of making things, personality traits, and behavior patterns that reflect the cultural system of the MDC in which it was designed and produced. These cultural elements must exist, be readily replicable, or have functional equivalents in the recipient LDC culture if safe and effective operation of the transferred technological system is to be assured. If that is not the case, then plant safety and effectiveness are apt to be jeopardized.

The critical mistake of cultural mismatch or incompatibility in processes of international technology transfer *is* avoidable. For example, consider the transfer of the technology for a rural water-supply system from Britain to Senegal that occurred in the 1980s.[180] The self-conscious goal in that case on the part of the MDC—in this case, a particular MDC engineering company in the UK—was "robust" international technology transfer, not just the transfer of hardware. The British company's team realized its goal of effecting "the maximum transfer of knowledge" by deliberately inducing active local participation in the project at all levels: central government, local authority, and village. The purpose of doing so was

[180] Wenn and Horsfield (1990).

so that the existing Senegalese organizations should not only gain the ability to operate and maintain the facilities provided through the project, but be able to construct new facilities to serve other populations in the future as aid funds allowed.[181]

One way the British firm tried to ensure an ERPITT was to form a multidisciplinary team consisting of engineers, hydrogeologists, an economist, and, revealingly, *a social anthropologist*. The latter was responsible for investigating existing rural Senegalese cultural infrastructure, practices, and preferences so that they could be taken into account in designing an effective and responsible transfer process.

> The most obvious technology to be transferred was that involved in the water-engineering works themselves, i.e., basic design and construction knowledge. Less obvious but equally important was the transfer of skills associated with other aspects of project life, such as management and accounting, language ability, and the use of computers.[182]

Ronald Dore made a related point about the breadth of the knowledge that must be transferred along with the hardware:

> If I.G. Farben build[s] an ethylene glycol plant in Indonesia which uses the latest chemical technology and is wholly owned by Indonesian capitalists, that is only the beginning of technology transfer. It is only the beginning as long as there are German technicians there supervising the running of the plant, as long as crucial spare parts can only be imported from Germany, and as long as German experts have to be called in when plant refurbishing and maintenance are required. Only when there are Indonesians who understand enough of the workings to deal with all production and maintenance difficulties; only when some of them have the underlying theoretical knowledge used in the plant's processes, and when that knowledge is incorporated into university engineering courses so that it can be locally reproduced— only, in short, when the knowledge which went into the making and the maintaining of the plant is transferred fully from German heads to Indonesian heads—can one say that technology transfer has taken place.[183]

Dore's implicit point about ensuring the transfer of the *full range* of knowledge about the making and maintaining of the plant is valid and

[181] Ibid., p. 148.
[182] Ibid.
[183] Dore (1989), p. 1673.

important. However, a complementary point needs to be added, not about "technology transfer" per se, but about an ethically responsible process of international technology transfer. An ERPITT requires that another body of knowledge be obtained and acted upon: namely, *knowledge of all conditions for the safe and effective use/operation of the transferred technology*. That body of knowledge encompasses not only knowledge related to Dore's focus on plant construction and maintenance but also knowledge of the prevailing culture at the plant and of the prevailing culture of the local society, systems that might contain elements that if overlooked or tolerated, could impede or endanger the plant's safe and effective operation.

The vital importance in processes of technology transfer of paying close attention to the culture of the local recipient society cannot be overemphasized. If the well-intentioned organizations that attempted to introduce biogas technology installations into 20 Papua New Guinea villages in the 1970s had recognized that

> most Papua New Guineans have very strong superstitions concerning the use by others of their excrement in any form, and that this cultural taboo mitigates against its collection for use in biogas production,

then they would have realized that this cultural mismatch precluded the effective operation of the new energy system.[184]

Another example in which the profound mistake of viewing technology transfer as "moving hardware" was avoided involves a U.S. engineer whom, for reasons of confidentiality, I shall call "Tom Turner."[185] Turner designed an athletic footwear manufacturing process to be implemented in an Asian country. In stark contrast to the Bhopal example, his technical competence was exercised with discerning "cultural systems vision." He recognized that the workers who would be making the athletic shoes on the assembly line would not be experienced in working with chemicals of the sort routinely used for such processes in MDCs like the United States. Therefore, in his process design he specified that the substances used in the manufacturing process be ones that would not cause harm to the workers *even if* they were not fastidious in handling them. He chose thus, rather than selecting chemicals that *unless* they were handled fastidiously *would* pose a risk to the health or safety of the workers who

---

[184]Newcombe (1981), p. 501.
[185]Turner related what he had done on this project during a visit to my engineering ethics seminar in the mid-1990s.

used them in risky fashion shaped by the local LDC culture. Turner thus drew upon cultural knowledge about the recipient society in making sure that the establishment of the athletic footwear factory was an example of an ERPITT. In this case, rather than trying to change the local culture to make it compatible with the cultural requirements embedded in his process design, he adapted the process design to make a salient difference in the recipient culture—low worker knowledge about and familiarity with handling toxic chemicals—not harmful to worker safety.

While the two cases differ in some noteworthy respects, both involved international technology transfer from MDCs to LDCs, and both involved potential risks to local people. In the athletic footwear plant case, the engineer went to great lengths (e.g., by considering the nature, training, and local practices of the workers who would be carrying out his process) to ensure that the manufacturing process introduced into the LDC would be such that the conditions (including the cultural ones) for the safe and effective operation of the process were satisfied. The Union Carbide managers and engineers do not appear to have done so. If they actually *did* do so, their list of conditions for the safe and effective operation of the Bhopal MIC plant was probably too narrow. While they may have considered technical requirements for the safe and effective operation of the plant, their vision probably did not extend far enough. It is highly unlikely that it included consideration of whether the cultural requisites for the safe and effective operation of the plant in India were satisfied, including ones related to worker mentalities and training, social institutions and groups, behavioral patterns, and common practices, at both the plant and local cultural levels.

Did the UCC and UCIL managers (whether engineers or nonengineers by training) and engineers substantively involved in the transfer of MIC technology to India fulfill their FERE-related ethical responsibilities? If the facts are as related earlier, it seems likely that some did, while others did not. But, in fairness, other parties arguably deserve slices of the blame pie for the cognitive myopia that underlay Union Carbide management's incomplete checklist of safety-related conditions. Consider, for example, MDC and LDC schools of business and engineering, and the professions of business and engineering.

It is virtually certain that few if any managers or engineers involved in this devastating incident were ever taught to conceive of technology transfer as a complex cultural process, or to conceive of and treat cultures as *complex systems* with interacting *ideational, social institutional, material cultural,* and *personality/behavioral* subsystems that evolve and are embedded in natural environments.[186] Hence, it is not surprising that even if the engineering professionals involved were well intentioned, technically competent, and gave serious consideration to the safe operation of the plant in techno-economic terms, none appears to have perceived the safety situation *through a cultural systems lens*. Hence, it is unlikely they ever inquired seriously into what would be required for the transfer of MIC technology to be an ERPITT, by pondering the full set of cultural conditions for the safe/low-risk operation of the complicated new production system introduced into that part of India. While many engineering-centric firms have rapidly become globalized in the scope of their practice, their leaders, including engineers, sometimes fail to be sensitive to cross-national differences in plant/corporate cultures and broader societal cultures and to how those differences, if uncompensated for, can compromise safety in MDC-LDC technology transfers.[187]

Lessons

When dealing with the transfer of a potent, potentially risky technology from a MDC to a LDC, the would-be ethically responsible engineer must ensure that all conditions, techno-economic *and* cultural, for the safe and effective operation of the technology in the recipient country are satisfied. If he or she fails to do so, the engineer is akin to an experienced parent who gives an uneducated adolescent a sophisticated, powerful weapon without ensuring that the recipient will receive adequate training, not only about how to use the weapon but also about the conditions under which it is safe and unsafe to use it.

A MDC technology donor's **cultural resources for safety** may well differ greatly from those of the LDC technology recipient. Assuming that technology transfer means just moving hardware, and that the mental, social, material, and personality/behavioral resources for safety that exist in the MDC culture also exist or will automatically emerge in the LDC culture, can create a major risk of significant harm in the LDC, as happened in the Bhopal case. Under FERE1, engineers involved with the process of international technology transfer have a weighty derivative ethical responsibility not to make such risk-of-harm-engendering assumptions. Under FERE2, they have an ethical responsibility to try to prevent existing

---

[186] On this important notion, see Kaplan and Manners (1972), and McGinn (1990), ch. 4.

[187] When dividing the blame pie, significant slices should also arguably be reserved for the national government and relevant state government of India, for shortcomings in awarding building permits, enforcing development plans, and regulating risky industries.

culturally induced safety risks from going unidentified and unaddressed. Even if the unreasonable risk of harm cannot be prevented, involved engineers have the FERE3-grounded derivative ethical responsibility to attempt to alert and inform those at risk of being harmed by the engineers' work that they are vulnerable.

The human toll of the 1984 MIC emission at the UCC/UCIL Bhopal MIC plant was enormous. Unfortunately, the author sees little evidence that the education of U.S. engineering students over the last three decades has changed such that recent graduates are now significantly better prepared to participate, as engineer-managers or practicing engineers, in MDC-to-LDC technology transfers of risky technologies in a more comprehensively responsible way than were most of their counterparts who were involved with the Bhopal plant in the run-up to the disaster.

Discussion Questions

1. When technology is being transferred from a more developed country to a less developed one, what must be the case for that process to be an ethically responsible one?
2. Are schools of business and engineering partly to blame for technology transfer processes, such as in the Union Carbide case, that go awry and kill and injure people? Why or why not?
3. What is meant by the "culture" or "cultural system" of a society?
4. How can the culture of a country and the culture of a manufacturing facility in that country that is the recipient of a risky technology transferred from a MDC affect whether that technology transfer process is an ethically responsible one?
5. Is there anything problematic about citing an alleged saboteur as the sole cause of the Bhopal pesticide plant disaster? Explain.
6. What, in your view, is the most interesting or intriguing nonobvious "contributory causal factor" that played a role in the occurrence of the Bhopal plant disaster? Explain your choice.

CASE 8: THE SPACE SHUTTLE *CHALLENGER*

Background

On January 28, 1986, at 11:38 a.m. EST, the U.S. space shuttle *Challenger* took off from Kennedy Space Center in Florida on Mission 51L.

Fig. 5. Crew members of space shuttle *Challenger*, Mission STS-51-L.

Seventy-three seconds after liftoff, a fire that broke out on the right solid rocket booster led to an explosion of the adjacent external fuel tank. The shuttle system disintegrated, and the orbiter plummeted into the Atlantic Ocean. All seven crewmembers perished (see Figure 5).

The immediate cause of the disaster was the failure of a seal at the joint between two segments of the right solid rocket booster. That seal failed because the two O-rings that were supposed to prevent hot gas from escaping were not seated in the groove between the booster segments. Unusually cold weather had made the O-rings insufficiently flexible to seat properly and seal the joint. Hot gas escaped through the compromised seal and was ignited, leading to an explosion of the external fuel tank and the disintegration of the shuttle system.

Ethical Analysis

The focus in this case discussion will be on three ethics-related issues: whether and under what conditions engineers should resist the politicization and economization of engineering decision-making; how engineers should respond when their efforts to prevent harm from resulting from a project they've been involved with are rebuffed by management; and whether and under what conditions the normalization of engineering risk is ethically problematic.

1. *The politicization and economization of engineering decision-making.*[188] The *Challenger* case provides several examples of decisions about critical technical matters that appear to have been made on the basis of considerations having little to do with technical factors and much to do with political and/or economic factors.

Some historical background is relevant here. By inspecting recovered boosters from earlier shuttle flights, Morton Thiokol Inc. (MTI)[189] engineers who specialized in solid rocket boosters (SRBs) had discovered evidence that hot gas had partially burned through the O-rings meant to seal adjacent rocket booster segments. Moreover, the extent of the O-ring erosion was growing, to the point that it had recently been discovered in the *secondary* O-rings.

After four postponements, *Challenger*'s launch was set for January 28, 1986. The preceding afternoon, five MTI engineers responsible for the SRBs concluded that the company should recommend against launching the next morning. Besides being concerned about the partial destruction of the O-rings, the MTI engineers were worried that the unusually cold weather forecast for the next morning would make the O-rings so stiff that they would not sit properly in the grooves and seal the rocket-booster joints. The resulting escape of hot gases would pose a risk of explosion.

The evening before launch, a teleconference took place between NASA managers, a number of MTI engineers, and four MTI senior managers, all engineers by training. One of the latter, Robert Lund, vice president of engineering, conveyed MTI's initial recommendation to NASA. Specifically, following their specialist engineers' position, MTI recommended to NASA officials that *Challenger* not be launched below 53°F—the lowest temperature at which the shuttle had previously been safely launched and recovered. This limit would have precluded launching early next morning, when the temperature was expected (and proved) to be much colder.

[188]The decisions in question may be made by politicians or nontechnical managers and acquiesced in and/or implemented by engineers. Such decisions can also be initiated by engineers in response to political and/or economic pressures.

[189]Starting in 1981, Morton Thiokol Inc. provided the solid-fuel rocket boosters NASA used for space shuttle launches.

NASA officials, some of whom were engineers, were extremely irritated by, even angry about, this recommendation. George Hardy, deputy director of science and engineering at NASA's Marshall Space Flight Center, exclaimed, "I am appalled. I am appalled by your recommendation."[190] Lawrence Mulloy, solid rocket booster program manager at Marshall, responded, "The eve of a launch is a hell of a time to be inventing new criteria." He went on to ask, sarcastically, "My God, Thiokol, when do you want me to launch, next April?"[191] In response, MTI senior management requested an offline caucus, ostensibly to reevaluate the data it had. At the meeting, which lasted about half an hour, engineer Jerald E. Mason, vice president of MTI's Wasatch Operations, told engineer Robert Lund, "Take off your engineering hat and put on your management hat."[192] That statement merits scrutiny, for it seemed to urge Lund (and two other MTI vice presidents) to not prioritize safety concerns based on technical considerations but rather, economic considerations, in the form of continued profit for MTI.

Up to that time, MTI had been the sole source of rocket boosters for NASA. Making and selling them to NASA was a major source of MTI's income and profit. MTI top management was concerned that if it did not reverse its initial recommendation against launching *Challenger*, it might alienate NASA further and run the risk of losing its status as the sole provider of the rocket boosters to NASA. To the extent that concern over that possible loss was *a* motive or *the* motive that drove MTI's engineer-managers to reverse the initial recommendation, that would be a classic (and tragic) example of the economization of technical judgment. In the words of the presidential commission:

> The Commission concluded that the Thiokol Management reversed its position and recommended the launch of 51-L, at the urging of Marshall and contrary to the views of its engineers in order to accommodate a major customer.[193]

After the *Challenger* disaster, former U.S. astronaut Col. Frank Borman, commander of *Apollo 8* and later CEO of Eastern Airlines, offered the following view:

---

[190] Berkes (2012).
[191] Bell and Esch (1987), p. 37.
[192] Ibid., p. 38.
[193] Rogers (1986), vol. 1, ch. 5, p. 104.

When NASA came out with the request for a proposal from airlines to run the shuttle like it was a [Boeing] 727, I called them and I told them they were crazy. The shuttle is an experimental vehicle and will remain an experimental, highly sophisticated vehicle on the edge of technology.[194]

The decision to request proposals from commercial airlines to operate the shuttle routinely, "like it was a 727," was arguably not based primarily on technical factors. It appears to the author to have been driven primarily by the economic priority of quickly making the shuttle operation financially self-sustaining. To that end, NASA attempted to attract more paying shuttle customers. This led, in the early 1980s, to a decision to increase the size of shuttle crews from two to between four and seven, including mission and payload specialists, to meet customer needs. To the extent that the orbiter ejection seats installed for the shuttle's test flights were removed because of the need for the new larger crews, the decision to remove them would arguably qualify as a technical decision driven more by economic considerations than by belief that the shuttle was now safe enough to make ejection seats unnecessary.[195]

As regards the politicization of technical judgment, some have suggested that the pressure NASA put on MTI to recommend launching on the morning of January 28 reflected pressure to launch that had been put on NASA by the Reagan administration White House. There was clearly strong interest in the executive branch in getting the delayed *Challenger* launched, allegedly so that the president could speak in real time with Christa McAuliffe, "the first teacher in space," from the rostrum of the House of Representatives during his 1986 State of the Union address.[196] However, it could also be that the White House did *not* deliberately exert such pressure but that NASA officials *perceived* pressure to launch as soon as possible, especially in light of the four previous postponements.[197] If and to the extent that NASA put any pressure on MTI to authorize launch because of actual or perceived political pressure from the White House, that would arguably be an example of the politicization of technical decision-making.

---

[194] Bell and Esch (1987), p. 45.

[195] Ibid.

[196] Ibid., p. 48. NASA had proposed remarks to the White House about the *Challenger* flight for Reagan's 1986 State of the Union address. See Hunt (1986).

[197] Ibid., p. 37. On whether the Reagan White House pressured NASA to launch *Challenger* on January 28, 1986, see Cook (2006) and Boyd (1986).

The key ethics-related point here is that an engineering judgment made primarily on the basis of political and/or economic considerations is apt to subject people affected by it to greater risk of harm than they would have been exposed to if the judgment had been made strictly on the basis of technical considerations. Under FERE1, an engineer has an ethical responsibility not to cause harm or create an unreasonable risk of harm. For an engineer to either politicize or economize a technical judgment, or to acquiesce in the politicization or economization of such a judgment and act accordingly, may well effectively put people affected by that judgment at a significantly higher risk of harm than need be. That arguably qualifies as creating an unreasonable risk of harm. When people's lives are at stake, it is difficult to imagine circumstances under which an engineer would *not* have a derivative ethical responsibility to oppose or rebut an engineering judgment made primarily or exclusively on political or economic grounds.

2. Responses of engineers to the rebuffing of their efforts to prevent work-related harm. The individual widely credited with having tried the hardest to prevent the disaster that resulted from launching *Challenger* "outside the database," that is, under conditions significantly different from those encountered previously, was Roger Boisjoly, an MTI mechanical engineer and aerodynamicist.[198]

Boisjoly was an expert on the rocket booster joint O-rings. Not only did he make the case, during the offline caucus of MTI senior managers and engineers, against MTI's recommending to NASA that it launch the next morning, but he had also taken other preventive measures long before the eve of the fateful launch. In July 1985, he had written a memo to Robert Lund, MTI's vice president of engineering, about "the seriousness of the current O-ring erosion problem in the SRM joints from an engineering standpoint."[199] Boisjoly was admirably explicit about the possible consequences of erosion in both the primary and secondary O-rings:

> If the same scenario should occur in a field joint (and it could), then it is a jump ball as to the success or failure of the joint because the secondary O-ring cannot respond to the clevis opening rate and may not be capable of pressurization. The result would be a catastrophe of the highest order—loss

---

[198] "I fought like hell to stop that launch." Roger Boisjoly, quoted in Berkes (2012). Two other MTI engineers also fought to prevent the *Challenger* from being launched were Arnold Thompson and Allan McDonald. See Rogers (1986), ch. 5, "The Contributing Cause of the Accident," http://history.nasa.gov/rogersrep/v1ch5.htm.

[199] Boisjoly (1987), p. 4.

of human life. . . . It is my honest and very real fear that if we do not take immediate action to dedicate a team to solve this problem, with the field joint having the number one priority, then we stand in jeopardy of losing a flight along with all the launch pad facilities.[200]

Boisjoly's candid memo failed to persuade MTI management to stop the launches of the space shuttle until the joints were fixed, which is what he believed should have happened. But through his various actions before the launch, Boisjoly arguably fulfilled various derivative ethical responsibilities under FERE2.

Besides laying out upstream his best estimate of the possible grim consequences of not solving the O-ring erosion and seal-preservation problem, Boisjoly also took steps to fulfill his ethical responsibilities *after* the accident took place. In testimony before the presidential commission investigating the accident, Boisjoly refused to follow the instructions given to him by MTI attorneys "to answer all questions with only 'yes' and 'no' and not to volunteer anything freely."[201] Later, at a closed meeting with two commission members, in response to a question from one of them, Boisjoly gave him

a packet of memos and activity reports . . . to clarify the true circumstances leading to the disaster. . . . I thought it was unconscionable that MTI and NASA Management wouldn't tell the whole truth so that the program could go forward with proper corrective measures.[202]

This post facto deed was also in line with FERE2, although in this instance it was aimed at preventing a *repetition* of harm or unreasonable risk of harm in the future.

Because of his actions, Boisjoly became persona non grata at MTI and was ostracized by some of his colleagues. He eventually resigned from the company. In 1988, Roger Boisjoly received the Award for Scientific Freedom and Responsibility from the American Association for the Advancement of Science.[203] The award citation read as follows:

For his exemplary and repeated efforts to fulfill his professional responsibilities as an engineer by alerting others to life-threatening design problems

---

[200] Ibid.

[201] Ibid., p. 9.

[202] Ibid.

[203] To the best of my knowledge, Boisjoly never received a comparable major award from any professional engineering society.

of the *Challenger* space shuttle and for steadfastly recommending against the tragic launch of January 1986.[204]

While designating Boisjoly "a hero for arguing against the launch, and then publicly calling attention to what happened," and stressing that he "deserve[s] credit for his effort to stop the launch, and later for revealing what happened," Stephen Unger suggests that in one respect Boisjoly may not have done enough:

> [I]t should be recognized that he (and his colleagues) did not violate the chain of command when their soundly based objections to the launch were overridden by management. They remained silent, rather than going public, or trying to warn the astronauts.[205]

Did Boisjoly have an ethical responsibility to go beyond what he did during the telecaucus and to "go public" and/or "try to warn the astronauts"?

Regarding the "go public" option, given NASA's determination to launch, MTI's decision to accommodate NASA, and the fact that the launch (11:38 a.m. EST) took place only about 12 hours after the end of the teleconference (11:15 p.m.), it is unclear whether contacting a newspaper, television reporter, or public official immediately after the teleconference and warning that *Challenger* was at serious risk of an accident if launched the next morning would have stopped the launch.

More fundamentally, *prior to the accident*, Boisjoly, like many or most engineers, did not believe it was proper to violate the chain of command and go public. As he told the presidential commission:

> I must emphasize, I had my say, and I never [would] take [away] any management right to take the input of an engineer and then make a decision based upon that input, and I truly believe that. . . . [S]o there was no point in me doing anything any further than I had already attempted to do.[206]

Boisjoly believed that once he had warned MTI management of his concerns about risk, as he had done both far upstream and the evening before launch, he had fulfilled his ethical responsibilities. It was only *after the accident* that he changed his view about not violating the chain

[204]http://archives.aaas.org/people.php?p_id=331.
[205]Stephen Unger, email to the author, September 22, 2014.
[206]Kerzner (2013), p. 479.

of command.[207] It may be that even if there was only a small chance that contacting the press would have made a difference, Boisjoly had an FERE2-based derivative ethical responsibility to go public and *try* to get the launch postponed.[208] However, it is doubtful he believed that he had an ethical responsibility to do so at that point. In warning management as he did, Boisjoly felt he had done all he had an ethical responsibility to do.

What about the option of warning the astronauts? Whether or not the *Challenger* crew members were actually reachable by phone after the teleconference, FERE3 suggests that Boisjoly had a derivative ethical responsibility to do what he could to *try to alert and inform* the astronauts and the teacher that launch the next morning would carry a greater risk of harm than usual. However, it is doubtful Boisjoly recognized FERE3 as an ethical responsibility that is incumbent upon engineers. Hence, he did not recognize that he had the aforementioned derivative ethical responsibility in that situation.

It is highly unlikely that the day before or the morning of launch NASA managers, some of whom were engineers, informed the astronauts and the teacher about (a) the increasing O-ring seal erosion and blowby, (b) the unprecedented weather conditions under which they were to be launched, and (c) the fact that MTI line engineers had serious questions about the safety of launching under those conditions. Hence, the seven *Challenger* crewmembers were probably deprived of a chance to give their informed consent to being launched under those circumstances. That all seven passengers had probably signed a generic risk-disclosure consent form weeks or months earlier does not imply that their boarding of *Challenger* and willingness to be launched under the unusual circumstances that prevailed in the run-up to launch on the morning of January 28 was done with their informed and voluntary consent.

The importance of acting in accordance with FERE3 needs greater emphasis in engineering education.

3. *Normalization of risk.* One of the more remarkable things about the *Challenger* disaster is the fact that it was *known* that there was erosion in the O-rings since early 1984, almost two years before flight 51L.

---

[207] As noted earlier, after being told by MTI lawyers not to "volunteer anything freely," he gave the "packet of memos and activity reports" to a commission member because he thought that was the only way to ensure that "proper corrective measures" would be taken to prevent the repetition of harm.

[208] For more detailed discussion of the conditions under which an engineer has a presumptive derivative ethical responsibility to "go public" or "publicly blow the whistle," see Chapter 4, Case 11.

But that year "Thiokol ran a computer analysis based on empirical data from subscale tests that indicated that O-rings would still seal even if eroded by as much as 0.09 inch—nearly one-third of their diameter."[209] This led Thiokol to conclude that "[O-ring erosion] is not a constraint to future launches."[210] Engineer Lawrence Mulloy, manager of NASA's Marshall Space Flight Center's solid-fuel rocket booster project, then introduced "the concept that a certain amount of erosion was actually 'acceptable,' along with the notion that the O-rings carried a margin of safety."[211] Even though the frequency and depth of the O-ring erosion increased in 1984 and 1985, Mulloy used the expression "allowable erosion," and 11 months before the *Challenger* accident he and Thiokol personnel "spoke of the observed blow-by of O-rings in two field joints . . . as an 'acceptable risk.'"[212] In April 1985, the worst seal failure to date occurred, with significant erosion of the primary O-ring *and* the first recorded erosion of the secondary O-ring.[213] However, in July 1985, "Mulloy presented earlier nozzle-joint erosion problems as 'closed.'"[214] In December 1985, he held that the "SRM joint O-ring performance [was] within [our] experience base."[215] He later stated, "since the risk of O-ring erosion was accepted and indeed expected, it was no longer considered an anomaly to be resolved before the next flight."[216] Finally, five days before *Challenger*'s fatal voyage on January 28, 1986, "the entire problem of erosion to both nozzle joints and the field joints had been officially closed out."[217]

In other words, the risk that erosion of the O-rings posed, one that had been growing over time, was no longer viewed as concerning. Instead, it was *normalized*, that is, regarded as expected, usual, and acceptable, and hence deemed no cause for concern, since all the shuttle flights made

[209]Bell and Esch (1987), p. 42. It is worth adding that reliance on data from subscale model tests can be problematic unless all similitude criteria are satisfied. See "Similitude (model)," http://en.wikipedia.org/wiki/Similitude_(model).

[210]Ibid.

[211]Ibid.

[212]Ibid.

Boisjoly wrote that a quality assurance manager had once advised him that "when faced with a tough question of whether a product was acceptable," he should ask himself, "'Would you allow your wife or children to use this product without any reservations?' If I could not answer that question with an unqualified 'Yes,' he said, I should not sign off on the product for others to use. That is what ethical analysis of acceptable risk should be." Boisjoly (1987), p. 11.

[213]Bell and Esch (1987), p. 43.

[214]Ibid.

[215]Ibid.

[216]Ibid.

[217]Ibid.

so far, which had generated erosion of the O-rings, had been completed successfully. Looking back on the accident, Mulloy offered the following FERE2-related advice: "Be careful in rationalizing the acceptance of anomalies you didn't expect."[218] If unexpected anomalies are rationalized, the risks they pose are likely to be viewed as acceptably small. Depending on how meticulous the process is by which a certain risk comes to be regarded as normal and acceptable in a particular episode, designating a risk of harm as normal and acceptable can have the same effect as opting not to try to eliminate or mitigate it, something that, if the risk is substantial enough, is arguably incompatible with FERE2.

Mulloy made another interesting observation after the fatal accident. Looking back on the changing perception and official assessment of the risk posed by the observed erosion, he stated, "In retrospect, that is where we took a step too far in accepting the risks posed by the O-ring erosion."[219] To characterize normalizing and accepting the risk posed by the O-ring erosion as "taking a step too far" is arguably an understatement. Whatever the nature of the actual process through which the *Challenger*'s O-ring erosion risk came to be deemed as normal and acceptable, as a general proposition I propose the following: normalization of the risk associated with an endeavor in which human lives are at stake is a strong candidate for being regarded as a negligent process to the extent that it is not searching, probing, systematic, and meticulous, for normalization of a risk effectively leaves it intact, sanctions not doing anything to eliminate or mitigate it, and thereby sustains it. Normalization of a risk is an act of commission that sanctions an act of omission: namely, not doing anything to address that normalized risk of harm. For that reason, the normalization of a risk can be at odds with FERE1 and FERE2.

The pressure to increase the number of shuttle flights may have subtly pressed some of those in decision-making positions to regard the anomalous as normal and acceptable, and hence as not being an unreasonable risk. One would have thought that the increasing depth of erosion in the primary O-rings *and* the first appearance of erosion in the secondary O-rings would have triggered alarms and impelled those in charge to take serious precautionary steps. That arguably did not happen, although engineer Boisjoly tried to bring that about. Once an engineering risk has been judged and designated as "acceptable," it can be difficult to revoke that judgment even when new data call it into question. This is especially

---

[218] Ibid., p. 50.
[219] Ibid., p. 43.

true if making the judgment that the risk is acceptable facilitates getting on with an important organizational objective.

In an organization with a dominant "gung-ho culture," when a large-scale engineering project has built up a certain momentum, the culture is apt to exert pressure on the people involved to keep the project on track. This can induce technical personnel who do not want to be perceived and treated as pessimists or as "sticks in the spokes of the wheel of progress" to alter their perceptions and conceptions of what is anomalous and unacceptable and what is normal and acceptable. I suspect that some engineers working on the *Challenger* launch, perhaps under pressure from managers, prematurely stopped treating the shuttle like the novel, experimental technological system it still was, and started treating it like an established, stable, and well-understood system that it was not.

Like the Citicorp Center, *Challenger* should not have been treated in the same way that familiar, well-understood, stable technological artifacts properly could. Instead, in my view, the engineers involved, both at MTI and NASA, had ethical responsibilities to diligently check out the anomalies and novel design aspects and possible connections between them, and to recheck those aspects to make sure they were reliable and stable. Any *assumptions* that *Challenger* was reliable and stable, and any declaration that its recurrent risks were acceptable, would arguably have been at odds with FERE1.

In late 2012 and early 2013, the then-new Boeing 787 "Dreamliner" had problems involving its innovative lithium-ion batteries, leading to malfunctioning electrical systems. But in this case, unlike with *Challenger*, the entire fleet of in-service 787 aircraft was grounded (by the FAA), until a solution for the battery fire problem was devised (even if the cause of the original fires was still not completely understood when the planes were allowed to fly again). Unlike NASA, civilian airline companies were not permitted to operate their planes while a solution was being sought. Only when a solution was found and tested were the planes allowed back into operation. NASA appears to have done otherwise: the shuttle continued to operate in the face of knowledge of O-ring erosion but without any urgency on the part of MTI or NASA senior managers, some of whom were engineers, to come up with a genuine solution to the problem, such as a new joint design.

When there is anything innovative that involves radical engineering design, the engineers involved have a derivative ethical responsibility to be sure they do not treat the innovative artifact in a conventional or normal way. In the case of *Challenger*, as regards the anomalous erosion data, a novel experimental aircraft was treated in a way that would be appropriate only for a conventional aircraft that was well understood and whose critical components were known to be reliable based on a substantial body of empirical evidence.

Lessons

What are the most important ethics-related lessons to be learned from the *Challenger* case? First, engineers should be wary of accepting employment in organizations that have a gung-ho culture, that is, a culture in which strong emphasis is put on moving ahead with a project regardless of expressions of concern by engineer-employees that doing so carries a serious risk of harm. In gung-ho cultures, precautionary action may be viewed as something that would retard the organization's work. A major problem of an organization with such a culture is that its engineers and other workers may be deterred from speaking up and urging, at some level in the organizational hierarchy, a slowdown in the firm's work on the project because of concern over risk of harm. It is often difficult enough for the engineer to fulfill all her or his ethical responsibilities under normal circumstances; it is even harder to do so in an organization with a gung-ho culture. It is interesting that in the *Challenger* case, top executives of NASA claimed they never heard anything about the concerns of MTI's engineers that it would be dangerous to launch *Challenger* under the conditions expected to exist on the morning of January 28, 1986. A gung-ho culture sometimes induces employees at lower organizational levels to filter out engineers' disturbing concerns in their communications with upper management.

Second, engineering students and professional engineers should beware of any engineering firm that is willing to politicize or economize (or acquiesce in the politicization or economization of) technical decision-making. Doing so is apt to engender unreasonable risks to life and property that are inimical to FERE1.

Third, engineers should be suspicious of impatient efforts to "normalize risk," for succumbing to the temptation to do so, or going along with a decision to do so, may keep an engineering organization from seriously addressing problems that pose significant risks to human safely and threaten major financial harm. Normalizing risk can effectively sustain it over time and deter an organization from seriously attempting to diminish or eliminate it. Consequently, depending on how meticulously it is done, normalizing risk can be incompatible with FERE1 and FERE2 and may induce complacency that is unlikely to encourage parties to ensure that FERE3 is fulfilled in the context in question.

Discussion Questions

1. After the caucus between MTI senior management, MTI engineers who were specialists on the O-rings, and NASA managers the evening before launch ended as it did, did Boisjoly have an ethical responsibility to go public with his concerns in an attempt to prevent the launch? Why or why not?
2. Did Boisjoly have an ethical responsibility to try to warn the astronauts about the conditions under which they would be launched and about the concerns that he and other engineers had about the safety of launching under conditions expected for the next morning? Why or why not?
3. Based on what you read in the case discussion, was it ethically responsible of Lawrence Mulloy to designate the risk of an explosion from the escape of hot gas through adjacent rocket booster joints as an "acceptable risk"? Why or why not?

CASE 9: A COMPOSITE-MATERIAL BICYCLE PROJECT

Background

In the 1980s, interest grew in the possibility of developing strong, lightweight, corrosion-resistant bicycle frames made of new, exotic composite materials.[220] In response, in 1989, Velo, a leading U.S. bicycle manufacturer, contracted for consulting services with an independent engineer, Steve Smith.[221] Smith claimed he had a workable method of designing and manufacturing composite-material bicycle frames.

---

[220]The following case description is based on an unpublished 1995 account that engineer "Brown" wrote about his real-life experience. Unless otherwise noted, all quotations in the case description are from Brown's original account. Brown was a guest in my ethical issues in engineering seminar in 1995 and again in 1997. Remarks quoted from tape recordings of those class sessions are referenced as "S95" and "S97" ("S" stands for "seminar"). For further discussion of this case, see McGinn (1997). When I sent Brown a draft version of the essay just cited, he responded with three pages of comments. Quotations from that document are referenced as "C97" ("C" stands for "comments").

[221]Individual and company names have been changed to preserve confidentiality.

When Smith could not convince Velo that his approach was workable, he engaged another engineer, Bill Brown (the head of his own very small company), as a secondary engineering consultant. Smith enlisted Brown to "validate his methodology and technical approach," that is, to confirm to Velo the viability of Smith's plan for designing and manufacturing composite-material bicycle frames. To make his case to Velo, Smith set up a meeting including himself, Brown, and Velo's technical staff, whose members "knew little of composite structures" other than that, at the behest of Velo's marketing division, "they wanted one in their product line." During Smith's presentation at the meeting, it became "quite obvious" to Brown that Smith's approach "created far more new problems than it attempted to solve." At one point in the meeting, Smith asked Brown to "validate or endorse his structural concept," one which, to Brown, "appeared workable by surface cosmetic and general conceptual standards, but did not have the underlying reinforcement fibers oriented to react to any of the critical load conditions this bicycle frame would see in use." In response, Brown stated that he "apparently did not have the same insight and awareness of the problem as [Smith] and that [Smith's] understanding of the dynamics and intricacies of his concept surpassed his own." Brown "offered to refrain from commenting" until he "understood the process as well as its author."

Later in the meeting, Jim Jones, Velo's director of R&D, asked Brown "point blank" whether he thought Smith's approach was workable. Brown responded that his relationship to Velo was only through Smith, and that Brown's company "would work with [Smith] to address issues which were important to the client's [Velo's] requirements."

The limitations of Smith's design and manufacturing concept became clearer to Velo over time. Eventually a Velo representative approached Brown without Smith's knowledge and confided that it wished to "begin a direct relationship" with Brown's company and "bypass" Smith altogether. Brown responded that he would "continue to provide our services to [Smith,] as we had agreed, so long as [Smith's] relationship existed with [Velo]; and that if ever that relationship should terminate equitably and [Smith] be compensated for bringing our company to the problem, we would then[,] with [Smith's] knowledge and consent[,] negotiate a direct contract with [Velo] for the services sought."

Eventually, Velo terminated its relationship with Smith. When Smith told Brown that he was now free to begin a direct relationship with Velo, Brown negotiated a new agreement with the bike company. The result for Brown's firm was "a $720,000 drafted contract that included royalties on each composite bicycle sold. . . . This was a very good deal for us, and as it turned out for the client as well."

After developing "a workable manufacturing plan for composite bicycles," but prior to deciding on a composite frame design, Brown felt it important to see the strength and stiffness of Velo's existing 4.5 pound aluminum-alloy production frame could be duplicated at a lighter weight. He asked Jones if he would agree to "some finite-element computer runs aimed at optimizing the structure of the existing metal frame." Jones replied, "in no uncertain terms, 'Do not do that!'" Since, however, Brown believed that doing so was "a vitally important step," he decided to optimize the structure at his own company's expense. He determined that "with no loss of stiffness or strength, the weight of the existing aluminum frame could be reduced from 4.5 to 3.1 pounds," bringing it within 3 ounces of the weight of the composite frame we had designed."

With more than $300,000 of the contract's $720,000 unspent, Brown contacted Charles Clark, president and CEO of Velo, and informed him that "the cost to reduce the weight of the existing aluminum frame to 3.1 pounds [would] be about $6.35 per bicycle with no additional investment in facilities or personnel." In contrast, deciding to manufacture a composite-material bicycle would mean "building a new facility, hiring and training a second production staff at an initial cost of $2.6 million[,] and a unit cost of $97.00." Clark decided that "his marketing staff's insistence on having a carbon-fiber bicycle frame had more to do with incorporating the 'buzz-word technology du jour' than [with] relying on fundamental applied engineering to achieve the desired weight reduction deemed necessary to maintain market share." He therefore asked Brown to complete any task in progress and close the composite-bicycle effort. The improved aluminum-frame bicycle was a successful part of Velo's product line for several years in the early 1990s.

Clark's decision to close the composite-bicycle project saved Velo "a great deal of money." It also "meant the loss of the remaining $300,000 on the contract" and that Brown's company "was going [to] have a lot of unbillable time for a few months." To Brown, this case demonstrated that while "ethics have a price, . . . integrity has a reward," for, it turned out, the referral business he got as a result of his work on the composite-bicycle project "returned the lost revenue several times over."

Ethical Issues

I shall explore ethical issues raised at three points in this episode: (1) Brown's response to Smith at the meeting with Velo, when Smith asked Brown to endorse his composite-frame manufacturing plan;

(2) Brown's action when told by Jones, Velo's vice president of R&D, not to do the optimization studies Brown thought it advisable to do; and (3) what Brown did when he reached his conclusions about the carbon-fiber bike-frame production option.

Analysis

Episode 1: Smith Asks Brown to Endorse His Plan

When Smith asked Brown, at the meeting with Velo's technical staff, to "validate his [Smith's] methodology and technical approach," Brown faced a "multi-faceted . . . ethical quandary." Since he had been introduced to Velo as a consultant of Smith's, Brown felt "ethically bound not to undermine [Smith's] business with this client"; however, since he also felt "bound to contribute both experience and technical expertise with integrity to [his] company and its reputation," Brown felt he "could not endorse the technically flawed program which was the subject of the meeting" without harming his own company and its employees. As noted, Brown claimed that he "apparently did not have the same insight and awareness of the problem" as Smith and that "[Smith's] understanding of the dynamics and intricacies of his concept surpassed [Brown's] own." Brown "offered to refrain from commenting on Smith's approach "until [he] understood the process as well as its author."

Brown's conduct here was mostly laudable. However, two aspects of his behavior seem ethically problematic. First, as the preceding quotation indicates, in responding to Smith's endorsement request at the meeting, Brown dissembled. One reason he did so was that he adhered to a "simple business ethic"; namely, he believed himself "ethically bound" not to do anything that would undermine Smith's reputation with *his* client, Velo. Brown seems to have regarded that ethical responsibility as *absolutely* rather than prima facie binding. He realized that endorsing Smith's approach "could cost [Velo] many hundreds of thousands of dollars before they understood its weaknesses." That Brown nevertheless feigned ignorance and was evasive with Velo suggests he felt his responsibility not to undermine his client [Smith] was categorical.

However, Brown did *not* have an ethical responsibility to refrain from critical comments on Smith's method and approach. The NSPE *Code of Ethics for Engineers* states that engineers "shall not attempt to injure, maliciously or falsely, directly or indirectly, the professional reputation, prospects, practice, or employment of other engineers, nor untruthfully criticize other engineers' work."[222] But it does *not* hold that engineers are obliged to refrain from criticizing their professional colleagues. It holds only that such criticism of the work of fellow engineers must be truthful, nonmalicious, and not intended to injure. The "ethic" embodied in the NSPE code appears, in this instance at least, to be more nuanced and conditional than Brown's seemingly categorical "simple business ethic."[223]

If Velo had not eventually come to recognize the flaws in Smith's plan and had proceeded on the basis of his proposal, the company might not only have wasted much money but have reached a point at which it was on the verge of manufacturing structurally flawed bicycles. Whether Brown's adherence to his "simple business ethic" would have impelled him to continue to dissemble or be evasive if Velo was poised to make a decision that could have resulted in a product that risked its users' physical safety is unknowable. Fortunately, Brown's initial lack of candor was sufficiently upstream in the product development process that it did not create an unreasonable risk of (unjustifiable) harm to members of the public.[224] However, his choice to remain publicly noncommittal, even though Velo stood to lose a lot of money if it implemented ideas of Smith that Brown knew to be structurally flawed, is troubling from the point of view of respecting Velo's financial interests. It is hard to see how Brown's evasiveness can be squared with his FERE4-based ethical responsibility to do his best to serve the legitimate interests of his indirect client, here, Velo.

This first episode had a second ethically questionable aspect. By his own admission, during Smith's presentation at the meeting with Velo it "became obvious" to Brown that Smith's "proposed solution created far more new problems than it attempted to solve." Nevertheless, Brown's decision to be a consultant for Smith and apparent willingness to attend the meeting with Velo in the role of endorser of Smith's method and approach *before checking on their validity and viability* invites criticism. Brown seems to have been ethically irresponsible for not having investigated the workability of Smith's plan *before* agreeing to put himself in a

---

[222]National Society of Professional Engineers (2007), sec. III.8.

[223]When Brown visited my seminar in 1997, I asked him about his seemingly categorical position. At first, he indicated a strong preference for simply withdrawing from any project he deemed seriously and irredeemably flawed. But when pressed, he reluctantly acknowledged that there could be circumstances in which he would feel ethically compelled to "sabotage the relationship" with the client who brought him into a relationship with a company, e.g., if the client was trying to "foist something on the marketplace that we knew was bad" (S97), e.g., unsafe.

[224]The latter, of course, is something that engineers have an FERE1-based ethical responsibility not to do.

situation in which he would be subject to strong pressure to dissemble or prevaricate (because of his "simple business ethic"). He knew or should have known that he would be expected to support his client's proposal at the meeting and, given his "simple business ethic," should have realized that, independent of the proposal's merit, he would be extremely reluctant if not unwilling to say anything that would cast doubt upon his client's plan and thereby on his and his firm's reputation. In short, Brown seems to have negligently put himself in a position in which he was foreseeably confronted by a classical conflict of interest: respecting the legitimate financial interest of his indirect client, Velo, and the safety interest of the bicycle-riding public background client could easily come into conflict with protecting the professional reputational interest of his direct client, Smith.

It turns out, however, that this account of events in and around this first episode, one based on Brown's initial written case study and what he said in my ethical issues in engineering seminar in 1995, is too black and white. It does not do justice to the complexities and dynamics of the sociotechnical engineering situation Brown actually faced. Although consistent with Brown's original document, the foregoing account and analysis rest upon several important assumptions. For example, they assume that Brown was unaware of drawbacks of Smith's concept prior to the meeting with Velo and that Brown knew or should have known he would be expected to endorse or validate Smith's plan at the meeting. However, both assumptions proved to be incorrect. Let us briefly consider each in turn.

First, when he came to my seminar in 1997 to discuss this case, Brown explained he had had several meetings with Smith about the latter's proposed manufacturing approach *before* the meeting with Velo's staff, at which he had "expressed his concern that for fundamental reasons his solution did not appear to be workable."[225] Indeed, Brown stated that prior to the meeting with Velo, he had already presented Smith "with several optional approaches to [solving] Velo's problem which did address the unworkable aspects of his initial concepts."[226] What Brown realized at the meeting with Velo was not what he already knew—that Smith's approach was structurally flawed—but that Smith's concept "created far more new problems than it attempted to solve." *But since Brown had conveyed his concerns to Smith before the meeting with Velo, Brown's*

[225] S97.
[226] Ibid.

professional ethical responsibility not to undermine his client's reputa-tion and credibility with critical comments on Smith's plan was signifi-cantly diluted if not completely dissolved by Smith's public request for validation from the very person who had previously told him in private that his proposed solution idea was structurally flawed.

Second, far from understanding that he was attending the meet-ing with Velo to endorse or validate Smith's ideas, Brown was in fact "surprised"[227] that Smith called upon him during that meeting to validate what Brown regarded as "a very creative, clever, but highly unworkable solution."[228] Given the early stage of development of Smith's plan, Brown regarded it as distinctly "premature" for endorsement or adoption[229] and claimed he was surprised by Smith's request to validate it. He saw his role entering the meeting as that of a consulting engineer with expertise in composite-material structures who had been engaged by a client and in-vited by him to participate in a three-way meeting, the purpose of which, he assumed, was to jointly "explore problems" with Smith's embryonic ideas. That he was taken aback by Smith's endorsement request is clear from Brown's statement:

> What I don't know to this day is whether [Smith] felt that he was under pres-sure to perform on this contract and needed to produce a positive response in the manufacturer as to his performance. Those are things I don't know, but they could easily have been motivational factors that would have influ-enced why he put us [Brown and his company] on the spot.[230]

This less tidy situation calls for reassessment of the initial ethical judg-ments about Brown's conduct.

Was Brown's initial evasiveness ethically irresponsible? While he did dissemble and was evasive, Brown "respected, and had confidence in, Velo's ability to discover the potentials and pitfalls of Smith's concept."[231] He felt it would "serve no purpose to deny Smith an opportunity to pres-ent and defend his recommendations to Velo," especially since Brown's "assessment of the situation and personal dynamics [at the meeting] was that Velo's funding commitments would only be toward engineering [and] technology validation at this time, and that production funding

[227] Ibid.
[228] Ibid.
[229] Ibid.
[230] Ibid.
[231] C97.

would be based on demonstrated merit."[232] Whether this more complex state of affairs absolves Brown of being ethically irresponsible for feigning ignorance depends on *whether his confidence was well grounded that Velo would in fact discover by itself the pitfalls of Smith's concept and do so in a timely manner.*

If Brown had good reason to think Velo would discover these pitfalls on its own before losing significant money (by adopting Smith's plan), much less producing unsafe bicycles for the market, then his evasive behavior would seem to be (barely) ethically justifiable in spite of FERE1. But to the extent Brown lacked good reason for thinking Velo would discover the flaws by itself before spending much money on Smith's unworkable ideas or producing structurally flawed bicycles, then his lack of candor would be in tension with FERE1 and FERE4 (vis-à-vis Velo). As argued previously, Smith's request for Brown's endorsement of his approach at the meeting with Velo's officials released Brown from any FERE4-derived ethical responsibility he had to Smith to avoid comments critical of Smith's plan that might damage his professional reputation.

But what of Brown's "apparent willingness to attend the meeting with Velo in the de facto role of endorser of Smith's method and approach *before checking on their validity and viability*"? It appears that, to the contrary, Brown clearly recognized and explicitly told Smith *prior* to their meeting with Velo of serious flaws in his method and approach and did *not* accept before the meeting the role of endorser or rubber-stamp validator of Smith's supposedly flawed solution idea. If true, this revised idea of Brown's state of knowledge and role understanding absolves him of any charge that he negligently violated FERE1 by being willing to attend the Velo meeting as endorser without first checking out the validity of Smith's ideas. As Brown put it, his notion of the purpose of the meeting was that it was to be devoted to jointly "exploring problems" with Smith's ideas, not to "foisting [flawed] solutions" on a skeptical manufacturer by allowing his expertise in composite structures to be exploited by his client.[233]

EPISODE 2: TO OPTIMIZE OR NOT TO OPTIMIZE

When Brown entered into a contract with Velo, the "original problem statement" was "to design a composite frame that was equal in strength

---

[232] Ibid.
[233] S97.

and equal in stiffness to the existing [tubular metallic] frame."[234](S95) After developing a working manufacturing plan for composite bicycles, Brown made an important decision. Because the composite analysis was so much more complex and therefore more costly, Brown suggested that his firm should "make certain that the [aluminum-alloy] frame [Velo] had given us as a baseline was in fact as good as it could be before we departed [from it] and tried to make that same frame out of some . . . advanced material."[235](S95) But, to his surprise, when he asked Jones, the head of Velo R&D, for permission to optimize the existing frame, Jones told him, unambiguously, not to do so, something Brown felt was "a very shocking thing for him to say until I [learned] subsequently that [Jones] was its designer"[236] and that his "design ego" was the main factor driving Jones's denial of Brown's request.[237]

Brown could easily have taken that no as a definitive answer and acted accordingly. However, he "decided to go off the customer's book and if we found something that was of note on our own invested nickel, then we would bring it to the customer's attention."[238] In fact, he found that the design of the existing tubular metallic frame was "nowhere near where it should have been to be the appropriate baseline to solve the [composite] problem."[239] The reason why Brown undertook to optimize the frame on his own time and resources was that he regarded optimization of the existing frame as "an essential, sound engineering practice."[240] As he put it:

> [I]f you're going to design an experiment and the objective is to make something the best you can, and you have a baseline to start with as your point of reference, your control if you will, and you find that there's something wrong with that control that's going to throw off the whole experimental base that you're trying to accomplish, . . . so we wanted to make certain that the control was in fact as refined as it could possibly be so that all of the rest of the work that we had to accomplish would be based on the best answer [for] their existing technology.[241]

[234] S95.
[235] S95.
[236] Jones did a potential disservice to his employer by trying to prevent the use of certain methodologies that might have improved (and, in fact, did lead to improvements in) a product he had designed. This posture is at odds with FERE4, engineers' ethical responsibility to do the best they can to serve the legitimate business interests of their employer or client.
[237] S97.
[238] S95.
[239] Ibid.
[240] Ibid.
[241] Ibid.

Given the terms of his contract, optimization was not legally binding on Brown. Indeed, Jones told him in no uncertain terms not to do it. Nevertheless, in my view, in this case optimization was more than ethically permissible; it was something that Brown had a derivative ethical responsibility to do. His ethical responsibility to optimize the alloy frame did not stem from the fact that optimization is an element of good engineering practice as such. After all, not everything that is "good engineering practice" is always morally obligatory for engineers, such as benchmarking, careful record keeping, and fastidious literature searches. Rather, it stemmed from two other considerations.

First, engineers have a fundamental ethical responsibility to act so as to serve the legitimate interests of their clients or employers as well as they can (FERE4). For the design engineer, this general obligation ramifies into a more specific ethical responsibility to design the best product or process that meets the given specifications under the specified constraints.[242] This, Brown reasonably believed, required that the existing frame be optimized. To proceed without optimization would be akin to a surgeon's performing an important, innovative operation on a patient after having acquiesced in the plea by a member of the patient's family, unbeknown to the patient, not to carry out an exploratory diagnostic procedure that could significantly enhance the prospects for a successful surgical outcome. The specific ethical responsibility to optimize the existing frame derives, at two removes, from FERE4, the engineer's general ethical responsibility to serve the legitimate interests of the employer or client to the best of his or her ability.

Second, this derivative ethical responsibility was strengthened by the fact that Brown's task called for him to go beyond what Walter Vincenti, extending Thomas Kuhn and Edward Constant, called "normal [engineering] design" and to enter the realm of "radical [engineering] design."[243] One enters this realm when the configuration, the operational principle, or, as in this case, the structural material of an item of technology constitutes a marked departure from the reigning, well-understood normal design paradigm. The responsibility to assure that the baseline aluminum-alloy frame was optimal in design stems from the fact that "composites are unique" and "tricky" (S95) and sometimes behave in ways that are not yet adequately understood. Moreover, composites,

[242] This obligation holds as long as the constraints specified are not such that observing them jeopardizes the safety of the ensuing product or process.
[243] Vincenti (1992), pp. 19–21.

although very strong and very stiff, are "very brittle and don't give much external evidence of internal failure."[244] Therefore, optimizing the structure of the old frame in terms of strength and stiffness would provide a baseline reference for evaluating the behavior of the composite frame under critical-load conditions.

Without this baseline, it would have been impossible for Brown to know whether he was designing the best product for his client and hence serving the client's interests to the best of his ability. Without optimization, the new composite frame might have been lighter than the existing frame but, unbeknown to the designer, not appreciably lighter (and perhaps less reliable) than the optimized, more cost-effective tubular metal frame. In carrying out the "essential, sound engineering practice" of optimization, even at his own expense, Brown exhibited exemplary moral character and ethically responsible engineering judgment.

EPISODE 3: OPTION DISCLOSURE AND PROBLEM REDEFINITION

As a result of his optimization study, Brown determined that for an incremental manufacturing cost of $6.35 per bicycle, and with a change in only a couple of materials, the existing metal-alloy frame could be reduced in weight by a third, bringing it within 3 ounces of the target weight of the projected composite-material bicycle. Moreover, this could be done

> without adding about $2 million in capital equipment, without adding a total secondary staff of people who [would have] had to learn new skills . . . not totally available in the . . . marketplace, apply these new skills to the new equipment, go through the learning curve headaches, and come up with an optimally designed and manufactured corporate product.[245]

Given these findings, Brown felt a responsibility to disclose this new option to Clark, Velo's CEO. Why? As he put it:

> [I]t is my job to provide [the people employing us] with full and complete information so that the conditions under which they employ us are honest and viable. . . . [B]y withholding information that's essential to their decision-making process, I'm not being honest and complete in the service I provide. . . . It's simply our job to present alternatives that are either known in advance or that come up in the course of our investigation.[246]

[244] S97.
[245] S95.
[246] S95.

Brown's remarks here conflate two considerations: (1) the consulting engineer's moral character and (2) the employer's best interests, here in making sound business decisions. The ethical responsibility to disclose the alternative production option stems not just or primarily from the fact that honesty is a moral virtue but from the fact that option disclosure was necessary for Clark to make a sound decision by being informed, rather than making him vulnerable by keeping him uninformed.[247] For Velo to make an informed production decision about whether to proceed with development of a composite-material bicycle, the company had to have knowledge of the range of options available to it, including the improved noncomposite option. Consulting engineers have an ethical responsibility to serve their client's or employer's legitimate interests to the best of their ability (FERE4), and making important business decisions on an informed basis is clearly a legitimate client or employer interest. Since Brown had acquired information bearing substantially on his client's ability to make such decisions, he had FERE4-based derivative ethical responsibilities to optimize the existing aluminum-alloy frame *and* to disclose to Clark the manufacturing option for the resultant, significantly lighter aluminum frame.

But for Brown, the responsibility to work to the best of his ability to serve the legitimate interests of the client went a step beyond *optimization* and *option disclosure*. It encompassed a third derivative ethical responsibility: *problem redefinition*. Instead of just making the client a carbon-fiber bicycle, Brown helped Velo CEO Clark see that the company's goal was "[to make the] best possible product, given all possible perspectives of the problem."[248] (S95) This in turn involved helping the client see that "the real problem" was "to build a lightweight bicycle frame that was equivalent in strength to their existing frame,"[249] whether made out of a fiber-reinforced composite material or out of a less exotic metallic material.

This is a robust and ethically admirable notion of what it means for an engineer to serve the legitimate interests of her or his client to the best of her or his ability. It enabled the client to choose between a new, expensive-to-manufacture composite-material bicycle of uncertain structural reliability and a redesigned, much-cheaper-to-produce reliable metal-tube bicycle as strong and almost as light as the composite. Had Brown's idea of serving the legitimate interests of a client to the best of his ability taken the

---

[247]Ibid. Brown believed that manufacturing 300 to 350 bicycles a day out of fiber-reinforced composite materials could "expose the company . . . to significant product liability claims."

[248]S95.

[249]Ibid.

form of uncritically following the client's marching orders, the sole light-weight bicycle option available to Velo would have been an expensive-to-manufacture and decidedly suboptimal carbon-fiber bicycle. Paradoxically, FERE4, the general ethical responsibility of engineers to serve the legitimate interests of the client as well as they can, here imposed upon engineer Brown a derivative ethical responsibility to challenge and attempt to revise the problem formulation originally given him by his client.

As in the case of William LeMessurier and the Citicorp Building, the composite-material bicycle case suggests that fulfilling FERE4 is not always simply a matter of deploying state-of-the-art technical competence to achieve what is requested or demanded of the engineer. Sometimes, fulfilling that responsibility requires having the courage of one's convictions—in this instance of undertaking to do exactly what one was told *not* to do—and of deftly exercising various diplomatic, organizational, and communications skills. In taking the path of optimization, option disclosure, and problem redefinition, engineer Brown again exhibited exemplary ethical conduct. It would have been far easier for him to have kept knowledge of the new, lighter metal-alloy option to himself or to have conveyed it only to Jones, with whom, for obvious reasons, it would probably have remained a closely held secret.

Lessons

Two important lessons can be drawn from this case. First, a secondary engineering consultant, A, should be extremely careful about the conditions under which she or he agrees to serve as a consultant to a primary engineering consultant B in a meeting arranged by B with its client company C, whose business B is trying to secure. If A is not careful about those conditions, then she or he could find herself or himself in a complex conflict-of-interest situation, one in which she or he has potentially conflicting ethical responsibilities: to B, to company C, and/or to the public, the ultimate end users of C's products. This could lead the engineer to act in a way incompatible with FERE1.

Second, it is important to realize that there are several quite different models of the ethically responsible engineering professional in the world of engineering practice. One is someone who is diligent and competent in carrying out the instructions/marching orders/tasks she or he has been given by the employer or client. The other is someone who, besides being competent, keeps the FEREs clearly in mind, treats society as the omnipresent background client whose interests take precedence over any conflicting private interests, and shapes his or her practice to be consistent with those basic ideas as they apply to the situation at hand, even if it means opposing or going against specific instructions from the employer or client. In this case, Brown admirably exemplified the second, robust ideal of the ethically responsible engineering professional.

Discussion Questions

1. Was Brown's conduct at the meeting with Smith and Velo's technical staff ethically responsible? Why or why not?
2. When Jones told Brown not to perform the optimization studies on the existing aluminum-alloy bike frame, and Brown decided to do them "on his own time and nickel," was he a consultant irresponsibly going against the wishes of a client or a consultant doing something he had an ethical responsibility to do? Justify your position.
3. Did Brown have an ethical responsibility to inform Velo's CEO that he had the option of building a new, lighter-weight aluminum-alloy bicycle frame that was cheaper and almost as light as the projected composite-material frame, or should Brown have simply proceeded to develop and implement his proposed plan for the desired composite-material frame? Explain.
4. If Brown had simply carried out his contract and proceeded to develop the desired composite-material frame, would his actions have been in accord with FERE1 and FERE4? Why or why not?
5. Can an engineering consultant engaged by a client have an ethical responsibility to go *against* the instructions given her or him by the client? Justify your answer.

CASE 10: NANOTECHNOLOGY R&D

Background

Most of the cases considered thus far have revolved around specific existing engineering products or systems or those under development. This case is different; it is about an emerging, flourishing *area* of technological activity, namely, nanotechnology (NT). Because NT is still in a relatively early phase of development, the nonobvious effects on society of its novel end products are largely matters of speculation. Given this situation, what kinds of ethical issues can be identified as ones likely to arise in this emerging area of technology?

To begin with, "nanotechnology" is a misleading expression. While it seems to refer to a particular technology, the term is actually linguistic shorthand for "scientific and engineering work carried out at the nanoscale."[250] Nanoscale scientific and engineering research is currently burgeoning, and some projected applications of the knowledge being generated and the nanomaterials and nanodevices under development are highly controversial. As increasing numbers of engineers get involved in nanotech R&D, they would do well to be aware of the ethical issues related to that work, including but not limited to those that will arise downstream when its fruits are applied in society. It would be even more desirable if they gain that awareness *before* those applications are diffused into society at large.

Why is NT a controversial area of research in the eyes of the public, at least the small part of the public who have heard of and know a bit about it? Many perceive NT as controversial for three reasons: its novel properties, product size, and disaster scenarios.

1. *Novel properties.* The properties of materials at the nanoscale often differ dramatically from their properties at the macro- and microscales, including optical, chemical, electromagnetic, and mechanical properties. This is so for two reasons. First, a relatively high percentage of a nanomaterial's constituent atoms are at its surface rather than within its volume, compared with the relatively smaller percentage of atoms at the surface of a bulk material. This greater ratio of surface area to volume of a nanomaterial compared with a bulk material engenders properties and phenomena called *surface effects*, such as increased strength and increased chemical reactivity. Second, intra- and interatomic relationships at the nanoscale particle size can also significantly affect a material's properties and sometimes give rise to novel electrical, optical, magnetic, and chemical characteristics and phenomena, often dubbed *quantum effects*.

That, for these two reasons, materials at the nanometric scale can exhibit quite different and unexpected properties from those of the same materials at larger scales—for example, the micrometer, millimeter,

---

[250] *Nanoscale* normally refers to structures with a length scale of 1–100 nanometers, where a nanometer is one-billionth of a meter (10^{-9} meter). To get an idea of how extraordinarily small a nanometer is, consider the following facts: "A sheet of paper is about 100,000 nanometers thick," and "There are 25,400,000 nanometers in 1 inch." National Nanotechnology Initiative (2016).

or bulk scales—bestows on nanomaterials an aura of the bizarre and unpredictable.

2. *Product size.* Many of the end products of nanotech R&D will share the nanoscale's property of being extraordinarily small. Thinkers have imagined diverse applications of nanotech research that exploit that extraordinary smallness, such as nanomedications and nanodevices. Although such products could bestow significant benefits on those able to access them, other nanotech applications could pose enormous risks or inflict major harms if used in certain ways. To offer a few examples of both kinds, one can imagine and anticipate effective, targetable anticancer nanomedications; nanomaterials that find their way into the natural environment, to problematic effect; affordable nanofilters able to efficiently clean up environmental pollution in less developed countries; and nanodevices that when secretly deployed by those that control them pose major risks to individual privacy and anonymity. It is likely that some future nanoproducts will be both impressively efficient and incompatible with one or more moral rights of individual humans. This prospect is disquieting to contemplate.

3. *Disaster scenarios.* A third factor that has contributed to the view that NT is a controversial area of research is that imaginative and well-informed authors have created narratives of nanodevices that self-replicate and learn. In some such visions, those who created such devices to serve their purposes lose control over them, sometimes to great human and/or environmental harm.[251]

Ethical Issues

What ethical issues can confront researchers in an emerging area like nanotechnology? The following remarks fall into three parts, corresponding to three domains or societal levels: the *microsocial*, the *mesosocial*, and the *macrosocial*. The "microsocial" domain or level is the realm of R&D laboratories and production facilities. The "mesosocial" domain is the intermediary realm of interactions between laboratory researchers and production facility engineers, and individuals from institutions that represent or affect society at large. The "macrosocial" domain refers to the domain of society as a whole, at the national or global level.

---

[251] See, e.g., Joy (2000) and Crichton (2002).

Analysis

1. *Microsocial.* An earlier case discussed in this book that dealt with engineering research—Case 3: the Schön affair—focused on research misconduct and several ethically questionable practices linked to collaborative research. Here I shall examine ethical responsibilities related to several different ethically questionable research practices, the first of which is specific to nanotechnology.

The cardinal fact that nanotech researchers must always keep clearly in mind is that the properties a material exhibits at the nanoscale are often quite different from those it exhibits at larger scales. Consequently, although a material is known to be safe for humans to work with in the lab at the metric and millimetric scales, that fact *by no means implies or guarantees* it is safe for them to work with the material in the lab at the nanoscale.

Surprisingly, in a survey of Stanford nanotech researchers, carried out between late 2010 and mid-2012, about a fifth (19.4%) of the 215 respondents thought it would *not* be irresponsible for a nanotech researcher who knew it was safe to work with a particular material in the lab at the metric and millimetric scales to *assume* the same was the case at the nanoscale.[252]

To the contrary: since one cannot validly infer the behavior of a material at the nanoscale—a nanomaterial—from its behavior at the metric or millimetric scales, FERE1, engineers' fundamental ethical responsibility not to cause harm or create an unreasonable risk of harm through their work (or work about which they are technically knowledgeable) implies a derivative ethical responsibility to abstain from any mode of thought or practice that would cause such harm or create such an unreasonable risk of harm. In particular, engineers have a derivative ethical responsibility to avoid making and/or operating on the *assumption* that a material known to be safe at the scales of meters and millimeters is therefore also safe at the nanoscale. That 19.4% of the researcher respondents indicated it would be responsible to make that assumption shows that rather than *assuming* graduate students and postdoctoral scholars understand that cardinal fact about nanotechnology, researchers and teachers need to take proactive steps to underscore the *irresponsibility* of making that assumption. Nanotech researchers must universally and explicitly recognize they have an ethical responsibility to *not* assume the safety of a nanomaterial

[252]McGinn (2013), pp. 98–99.

based on its observed safety at larger scales. Nor should they base their research practices on such an invalid assumption.

Given the pressures under which many nanotech researchers work, it is not surprising that they occasionally resort to taking shortcuts in doing their work. Sometimes these shortcuts violate published laboratory safety rules. How would other nanotech researchers react if they observed a researcher in their lab taking a prohibited shortcut?

A 2005–2006 survey of more than a thousand researchers from 13 U.S. NT research labs included the following scenario:

> For several weeks, a nanotech lab researcher has been taking a relatively safe, timesaving shortcut in doing her/his work. This shortcut clearly violates published laboratory procedures. So far, no unfortunate results have occurred because of this behavior. Other lab users know that s/he is taking the shortcut.[253]

The respondents were then asked what reaction they thought would be the most common response of users in their respective nanotech labs who observed the individual taking prohibited shortcuts. Respondents were given six options:

(a) users would report the individual to lab management;
(b) users would cease having professional contact with the individual;
(c) users would try to persuade the individual to stop shortcutting;
(d) users would start taking rule-violating shortcuts of their own;
(e) users would take no action and the situation would continue unchanged;
(f) users would bring up the prohibited shortcutting for debate at a lab meeting.[254]

Encouragingly, 43.8% of the respondents indicated they thought option (c)—trying to persuade the researcher to stop taking prohibited shortcuts—would be the most common response in their labs. However, surprisingly, just short of a quarter (24%) of the respondents indicated they thought option (e)—no action would be taken by the user observing the short-cutter; situation unchanged—would be the most common response in their labs. Interestingly, the 13 laboratories differed widely in the percentage of their respondents who chose the "no action taken, situation unchanged" option. The percentage of respondents in a single NT

[253] McGinn (2008), p. 108.
[254] Ibid.

lab who chose option (e) ranged from 12% at the low end to about 40% on the high end. This finding suggests that NT research laboratories differ significantly in the strength of their respective laboratory safety cultures.

Given the FEREs, the properties of nanomaterials, the pressures under which many nanotech researchers work, and the importance of a research lab having a strong safety culture to deter prohibited shortcutting, several derivative ethical responsibilities follow:

- nanotech researchers have an ethical responsibility to not take prohibited shortcuts in their laboratory work;
- if less coercive measures do not succeed in stopping prohibited shortcutting, nanotech researchers have an ethical responsibility to report such behavior to laboratory managers;
- top managers of a nanotech lab have an ethical responsibility to actively promote a strong safety culture in their facilities;
- researchers in nanotech labs have an ethical responsibility to help socialize new researchers to do things in ways consistent with maintaining a strong safety culture in their labs; and
- nanotech researchers have an ethical responsibility to combat in their labs any manifestations of a laissez-faire culture in the actions and practices of their fellow researchers, whether newcomers or veterans.[255]

2. *Mesosocial*. The mesosocial domain is the realm of interactions between research (and development) engineers and representatives of pertinent societal institutions, such as government, law, regulation, and communication. Without entering into the fine details of the arguments, nanotech researchers active in this domain also have several derivative ethical responsibilities: (i) to avoid engaging in hype, for example, by exaggerating the benefits of research projects for which they are seeking funding, and (ii) to avoid helping legitimize distorted mass media coverage of their work or achievements by participating in or endorsing such coverage when they have reason to suspect it will be significantly flawed.

Nanotech researchers' ethical responsibilities to avoid engaging in hype and to avoid cooperating with distorted mass media coverage of nanotech achievements stem from the fact that failure to do so can help cause long-term harms to the research enterprise. Such harms could result from the failure of publicly funded research to measure up to hyped accounts of its projected benefits. This would risk diluting public willingness to initiate or continue supporting such research projects in the

---

[255] McGinn (2010), pp. 1–12.

future. Contributing to the public's acquiring a distorted understanding of the nature, limitations, and constraints of nanotech research endeavors could also harm the NT research enterprise, for example, by eroding the motivation of young researchers considering entering the field.[256]

3. *Macrosocial.* As regards the macrosocial realm, several matters give rise to ethical responsibilities incumbent on NT researchers.

The aforementioned 2005–2006 survey yielded two surprising findings about ethics and nanotechnology research at the macrosocial level.

Many researchers have long believed their only research-related ethical responsibilities have to do with three general topics: lab safety, data integrity (e.g., avoiding fabricating and falsifying data), and publication proprieties (e.g., not appropriating others' ideas or words without credit or plagiarizing their work). Many researchers have long denied having any ethical responsibilities related to the downstream applications and uses of their work in society at large. As noted, physicist Leon Lederman used a seductive metaphor to refute the notion that researchers can have ethical responsibilities to society: "We [scientists] give you powerful engines[;] you [society] steer the ship."[257] One implication of that statement is that, for Lederman, only those who choose, shape, implement, and regulate the eventual applications of nanotech research have ethical responsibilities to society, not those who, through research, create the malleable knowledge base or experimental devices usable for many applications.

In the same 2005–2006 survey, NT researchers were invited to indicate their levels of agreement/disagreement with the following claim:

Nanotechnology researchers should always strive to anticipate ethical issues that may arise out of future applications of their work.

About two-thirds (66.5%) strongly or somewhat agreed that NT researchers have a responsibility to try to anticipate ethical issues that may arise from future applications of their work. In contrast, only 13.6% strongly or somewhat disagreed with that notion.[258]

The same survey questionnaire also contained this item:

If a nanotech researcher has reason to believe that her/his work will be applied in society so as to pose a risk of significant harm to human beings, s/he has an ethical responsibility to alert appropriate parties to the potential dangers.

---

[256] For more detail on these two ethical responsibilities, see McGinn (2010), pp. 7–8.
[257] Lederman (1999), p. A15.
[258] McGinn (2008), p. 114.

Slightly more than three-quarters (76.3%) *strongly agreed* with that proposition and another 13.4% *somewhat agreed,* while only 4% strongly disagreed, and 0.8% somewhat disagreed. Summing these pairs of figures, nearly 9 of every 10 respondents (89.7%) *strongly or somewhat agreed* that under the stated condition researchers have an ethical responsibility to alert appropriate parties to the potential dangers, while only about 1 in 20 (4.8%) somewhat or strongly disagreed with that claim.[259] That is a striking and encouraging finding.

What might account for this apparent sea change, whereby the great majority of *researchers* agree that they can have derivative ethical responsibilities to act preventively (FERE2) and to alert the public about risks of downstream harm (FERE3) based on grounded concerns about projected end uses of their findings in society? One possibility is that in contemporary societies, researchers can no longer reasonably claim to be unaware of the risks associated with the powerful "engines" they make available to society. This is especially understandable when dominant social groups with recognized track records pilot the "ships" that researchers' "engines" will power, and when regulators and policy makers, influenced by parties standing to gain economically by proceeding with dispatch, readily give the newly equipped ships green lights. In the contemporary era, engineering and scientific researchers diffuse the fruits of their labor in societies whose characteristics are often well known to them. Consequently, while it is *not always* foreseeable that particular research outcomes will be turned into ethically problematic applications, *sometimes* it will be. This is apt to happen when there are substantial military or economic advantages to be gained by proceeding, even if ethically problematic issues or effects are foreseeable as by-products.

In such cases, the researcher cannot avoid an ethical responsibility simply by washing his or her hands à la Pontius Pilate, or by citing the alleged neutrality and merely enabling character of research knowledge. What seems clear is that most of the respondents in the 2005–2006 survey included NT researchers among the groups comprising NT development engineers, manufacturers, regulators, and end users, whose members, they believe, have *ethical responsibilities toward society.* In short, according to most surveyed respondents, *responsibility to anticipate* and *conditional responsibility to alert* are two ethical burdens incumbent upon NT researchers.

[259] Ibid.

A third source of ethical responsibilities for nanotech researchers involving the macrosocial realm has to do with an important projected application area of nanotech research: human-implantable nanodevices (HINDs).

NBIC convergence refers to the projected confluence of streams of research in four areas: nanotechnology, biotechnology, information science, and cognitive science. One expected result of this convergence is a new type of HIND: one able to gather and transmit information about human brain–related conditions and processes. HINDs may also eventually offer the ability to monitor and wirelessly communicate the conditions or states of internal body organs or systems, such as body temperature, pulse, blood sugar, blood glucose, and heart activity. The medical benefits of such innovations could prove enormous. For example, HINDs are envisioned that could alert to imminent epileptic episodes and help restore vision and correct hearing dysfunctions.

While the potential benefits of biomedical nanodevices (and nanomedications) are impressive, they are also likely to raise significant ethical issues. These will invite ongoing ethical reflection by, among others, *the practitioners of nanoscale engineering and science involved in achieving them.* At least five facets of biomedical HIND research will raise ethical responsibility issues for engineers.

1. New nanodevices will eventually be tested experimentally on humans. The substantial, possibly enormous financial interests at stake in bringing such products to market quickly could easily be in tension with the ethical responsibility of involved technical professionals to ensure that humans give their informed voluntary consent to being test subjects. In light of FERE1 and FERE3, as such testing draws near, involved (and noninvolved but knowledgeable) NT researchers will have an ethical responsibility to oppose any plans to implant experimental HINDs in patients without their being candidly and comprehensively informed beforehand about the full spectrum of risks involved. This is true even if the test subjects are among the intended beneficiaries of such devices.

2. Suppose that in the next quarter century, medical doctors, engineers, and scientists are able to devise and ethically test a HIND that can transmit wirelessly, to doctors or technicians monitoring the patient, information indicating whether any internal physical condition is changing in ways that suggest the emergence or imminent occurrence of a significant health problem. Such a device might enable certain risky medical

problems to be nipped in the bud, with preventive rather than curative care, hence at much less cost than would be involved had the medical condition reached a more advanced and risky stage before being detected and treated.

Under such circumstances, it would probably not be long before some individual or organization had the idea that such a device (or range of such devices) should be routinely implanted in all newborns to spare them the serious medical problems that can result when nascent abnormal somatic conditions go undetected and evolve into more advanced disease stages. If it were to turn out that upstream preventive care enabled by information transmitted by HINDs typically costs much less than downstream curative care, it is likely, at least in a country like the United States, that medical insurance for young children who have had the devices implanted in them would cost much less than if they have not. One can further imagine that parents of modest economic means who wished to avoid paying the much higher medical insurance costs for children without such devices might feel pressured to have their children get early-warning HINDs.

Indeed, such devices (and the prevailing U.S. insurance market) could alter the very notion of responsible parenthood. Any parents who elect *not* to have these devices implanted in their young child might be regarded and treated as *irresponsible* parents. They would be regarded much as are parents who elect not to have their young children vaccinated against diphtheria, pertussis, and tetanus. One can hope that at least some engineers involved in the R&D work on such HINDs will do more than simply strive to meet the design specifications stipulated by the doctors with whom they are collaborating.

3. Nanoengineers will be well positioned to know whether devices with certain feature sets or specifications might compromise device safety or durability. Engineers working on HINDs will also have a derivative ethical responsibility thoughtfully to address design questions having to do with issues such as the ability of unauthorized parties to gain access to information transmitted by the devices, and the built-in security of the devices against malevolent attempts to hack and disable them. Under FERE1 and FERE2, it would arguably be ethically irresponsible of engineers working on biomedical HINDs to remain oblivious to, indifferent to, or silent about such matters.

4. Engineers designing or testing such devices might be able to make an informed case that in addition to a presumably costly HIND able to transmit information about the greatest number of key inner states or

conditions, a more affordable model could and should also be developed, one that would provide early information about a "core set" of such variables. In addition, a would-be ethically responsible nanoengineer who tries to be sensitive to the context in which her or his devices will be used could make a strong case for designing one or more "orphan devices," reminiscent of "orphan drugs." Such HINDs, although not pertinent to the health needs of all or most young children, would be extremely pertinent to the special health needs of smaller groups of at-risk would-be users, for example, children of certain races, ethnicities, and genetic endowments.

5. It is highly likely that early NBIC nanodevices will be therapeutic in intent, for example, aimed at addressing problems in brain functioning. However, they could eventually be developed for purposes of cognitive enhancement. Since the human mind and human autonomy are extremely sensitive ethical concerns to most people, it is not surprising that the U.S. public *and* nanotechnology researchers are deeply divided about the ethical acceptability of human cognitive enhancement. In the aforementioned 2005–2006 survey, nanotech researchers responded to the following item: "How morally acceptable is the following nanotechnology project goal to you: to increase human mental abilities." The responses diverged enormously:

- 18.4% replied "very" morally acceptable;
- 16.4% "quite" morally acceptable;
- 19.5% "somewhat" morally acceptable;
- 20.1% "slightly" morally acceptable;
- 18% "not at all" morally acceptable, and
- 7.4% "do not know."[260]

To note that there is no consensus on the ethical acceptability of this nanotech-related goal is an understatement. In fact, there is major dissensus.

Like researchers working with human embryonic stem cells, researchers working on NBIC HINDs for cognitive enhancement would do well to become sensitive to and learn to discuss thoughtfully the issue of the ethical acceptability of their work. Not only is doing so in the researchers' own self-interest, they may have an FERE1-derived ethical responsibility to do so. For if they do not, the fate of such work, and hence of the

[260] McGinn, 2005–06 NNIN Ethics Survey Data File (unpublished): frequency distribution for nanotechnology researcher responses to question A13.2. This item asked how morally acceptable the respondent would find the future nanotechnology R&D project goal of enhancing human mental abilities.

beneficial innovations it might yield, could well depend by default on the positions of politicians, corporate representatives, and clerics. Such individuals often have potent political-economic interests and/or ideological commitments that tend to take precedence over available empirical evidence and incline those under their sway to adopt rigid, uncritical, and parochial points of view. This could block or delay the delivery of major benefits from such work.

In short, engineers working on biomedical nanodevices, perhaps in collaboration with doctors, are likely to have a number of macrolevel ethical responsibilities, ones connected with the design, clinical testing, and eventual uses of various models of the devices. Educating themselves to participate thoughtfully in discussions about such innovations could itself become an FERE1-based derivative ethical responsibility of NT researchers. Remaining ignorant or casually dismissive of ethical issues and responsibilities in their work could make the persistence of eliminable harms more likely.

Lessons

An emerging area of technology is apt to raise a range of ethical issues, not just ones spawned by controversy over its projected downstream social consequences. Working in an emerging technology area, even in the research phase, can impose ethical responsibilities on engineers involved in it.

Researchers in an emerging area of technology have clear ethical responsibilities related to the creation and maintenance of a robust safety culture in the lab or manufacturing facility. Researchers involved in competitive fund raising and in promoting accurate public understanding of engineering and science through mass media also have ethical responsibilities, ones often overlooked. Happily, researchers in some emerging fields of technology, such as nanotechnology, are beginning to recognize that engineers working downstream on applications of knowledge and their regulation are not the only ones with responsibilities to society as a whole. Biomedical nanodevice R&D is likely to prove a burgeoning, fruitful, and ethically charged area of engineering in the next half century. Engineers working in this area will be empowered, more widely respected, and more influential if they are sensitive to, literate about, and committed to fulfilling the ethical responsibilities likely to pervade all phases of such work.

Discussion Questions

1. Can it be ethically acceptable or responsible for an experienced researcher to take prohibited shortcuts in a nanotechnology research laboratory when under time pressure? If not, why not? If it can be ethically acceptable, give an example of conditions under which it would be.

2. Is it ethically irresponsible for a nanotechnology researcher to hype her or his work in the mass media? Why or why not?

3. An engineer testifying as an expert witness in a trial hears the judge make a statement about nanotechnology that he or she knows is wrong. The engineer believes that, left uncorrected, the statement will probably change the outcome of the trial. Does he or she have an ethical responsibility to inform the judge about the incorrect statement? Why or why not?

4. Does a nanotechnology researcher who has good reason to believe that a government regulatory agency's decision about a nanomaterial was made on political grounds have an ethical responsibility to bring that situation to the attention of the public? Why or why not?

CASE 11: THE FORD PINTO

Background

In the 1960s, Ford Motor Company was concerned about the influx of small, cheap Japanese and German cars, especially those made by Toyota, Datsun (now Nissan), and Volkswagen. Ford feared it could suffer a significant loss of share in the U.S. subcompact market. Hoping to avoid such an outcome, Lee Iacocca, named Ford's executive vice president in 1967, pushed the development of a new model to compete with the Japanese and German imports. Iacocca stipulated that the Ford Pinto should cost at most $2,000, weigh at most 2,000 pounds, and be ready for sale in September 1970, the start of the 1971 model year.

After its debut, the Pinto was involved in a number of accidents. In some of them, when the car was hit from behind, the fuel tank was punctured or ruptured, or the filler neck became detached from the fuel tank. If that happened and a spark occurred, leaking fuel could catch fire, risking passenger injury or death from burning.[261] In 1977, *Mother Jones* magazine published a controversial article[262] contending that Ford had brought the Pinto to market knowing that its fuel tank was liable to rupture and its filler neck liable to become detached from the fuel tank when the car was hit from behind at as little as 20 miles per hour. The article also claimed that Ford knew how to make the fuel tank safer but decided not to do so. In June 1978, in response to a National Highway Traffic Safety Administration study of the Pinto, and the agency's ensuing threat to impose a mandatory Pinto recall, Ford agreed to recall 1.5 million 1971–76 Pintos and 30,000 1975 and 1976 Mercury Bobcats with similar fuel-tank systems. Ford installed polyethylene shields in hundreds of thousands of recalled Pintos to protect the fuel tank against rupture, lengthened the filler pipe to make it less likely to disconnect from the fuel tank when the car was hit from behind, and strengthened the car's weak bumper.[263] Production of the Pinto ended with the 1980 model year.

Ethical Analysis

The Pinto case raises several noteworthy engineering-related ethical issues. Three will be explored here. The first involves a key feature of the Pinto's design, the second pertains to its product-development period, and the third to public whistleblowing.

   1. *Design.* A critical element of the Pinto's design was the nature and placement of its fuel tank. Like other American automakers in that time period, Ford chose to place the Pinto's fuel tank *behind* the car's rear axle and *below* the floor pan of the trunk.[264] The rear-facing side of the fuel tank was "six inches in front of the rear bumper,"[265] while the front-facing side of the tank was "3 inches from the [car's rear] axle housing"[266] and close to a shock absorber bracket. Thus positioned, the fuel tank was vulnerable to rupture in a rear-end collision.

   For several reasons, it is initially puzzling that Ford chose to put the subcompact Pinto's fuel tank in that location. First, Ford was aware that

[261]The number of burn deaths from car fires precipitated by collisions that damaged Pinto fuel tanks has long been contested. Estimates range from 27 to 500. See Schwartz (1991), p. 1030, and Dowie (1977), p. 1.

[262]Dowie (1977).

[263]Strobel (1980), p. 89.

[264]Graham (1991), p. 129.

[265]Viscusi (1991), p. 111.

[266]Strobel (1980), p. 82, and Baura (2006), p. 41.

fuel tanks could be designed for location in other positions. In fact, in 1961, Ford had secured a U.S. patent on a "saddle-type" tank designed to sit *over* the rear axle.[267]

Second, in 1967, Ford had purchased an English-built Rover subcompact, whose fuel tank was *over* the rear axle. Ford crash-tested it with a moving barrier at 30 m.p.h. and found that "there was no deformation, puncture, or leakage of the fuel tank."[268]

Third, in February 1968, UCLA engineers had published a summary of the findings of several experiments they had conducted involving automobile rear-end collisions, including one involving collision-induced passenger-car fires.[269] The research, funded in part by Ford, which also donated some vehicles for crash-testing[270], concluded, among other things, that "[f]uel tanks should not be located directly adjacent to the rear bumper . . . as this location exposes them to rupture at very low speeds of impact."[271] The article stated that "[c]ollision studies to date tend to support relocation of fuel tanks to the [over-the-axle] area, but further research is needed before this location can be recommended."[272] The authors observed that a fuel tank placed "above the rear axle and below the rear window" is "least often compromised from collisions of all types."[273]

Fourth, in early 1969, more than a year before the 1971 Pinto went into production, Ford introduced the Mercury Capri in Europe. Unlike in the Pinto, the Capri's fuel tank was located *above* the rear axle. In the advertising campaign for the new Capri, the location of its fuel tank was touted as a safety feature. Its gas tank was said to be "safely cradled between the rear wheels and protected on all sides."[274]

Why, then, in spite of knowing "before it began producing its controversial Pinto model that the car's gas tank was more susceptible to fire and explosion in low-speed crashes than other available designs,"[275] did Ford put the Pinto fuel tank behind the car's rear axle and under the floor pan of the trunk? Ford appears to have done so primarily for marketing reasons.

[267]Strobel (1994), p. 43.
[268]Ibid.
[269]Severy, Brink, and Baird (1968).
[270]Ibid., p. 3106.
[271]Ibid.
[272]Ibid., p. 3101.
[273]Ibid., p. 3106.
[274]Strobel (1980), p. 81.
[275]Strobel (1979), p. A1.

Placing the fuel tank behind the rear axle and under the trunk floor panel had the advantage of "keeping the tank outside of the passenger compartment (thereby preventing gasoline vapors from entering it)." Moreover, "[p]lacement behind the rear axle was necessary if Ford was to add a hatchback and station wagon to the Pinto line."[276] However, an internal Ford document of January 31, 1969, indicates that Ford engineers in the Product Development Group had evaluated the over-the-rear-axle design. Their conclusion: "Due to the undesirable luggage space attained with these proposals [to place the tank over the rear axle], it was decided to continue with the [behind-the-rear-axle] strap-on tank arrangement as base program."[277] Ford's upper-level managers, not the Pinto engineers, may have made the initial decision to put the Pinto fuel tank behind the rear axle and close to the car's weak, "essentially ornamental"[278] rear bumper. However, as the "undesirable luggage space" quote shows, Ford engineers working on the Pinto, knowing that substantial luggage space was an important priority for the new car, decided against putting the fuel tank over the axle, as at least one Ford engineer had proposed,[279] in order not to undermine that priority. To ascertain how their conduct relates to the FEREs, let us explore what Ford engineers knew firsthand about the relative vulnerability of the Pinto's fuel tank in low-speed collisions.

In mid-1969, while the Pinto was under development, Ford engineers modified three Mercury Capris to make their rear ends similar to that of the Pinto as then designed.[280] They tested one Capri by backing it into a wall at 17.8 mph, while the other two were rear-ended by moving barriers at 21 mph. In all three cases, gas leaked, either because the filler pipe came out of the fuel tank, the tank's welds split open, the tank fell out of the car, or sharp objects punctured the tank. Let us call this Test 1. The results of Test 1 should have strongly suggested to Pinto development engineers who knew its results that putting the car's fuel tank behind the rear axle would make it vulnerable to fuel spillage and possible fire when hit from behind at 18–21 mph, especially given the weakness of the Pinto bumper.

The Pinto was introduced on September 11, 1970. About a month later, on October 9, 1970, Ford engineers crash-tested an *unmodified* Capri and found that, in contrast with the three modified Capris, it survived a

[276] Huber and Litan (1991), pp. 129–130.
[277] Strobel (1980), pp. 81 and 279.
[278] Huber and Litan (1991), p. 130.
[279] Ibid., p. 129.
[280] Ibid., p. 81.

31 mph rear crash into a wall *without* leaking any fuel. This should have driven home to the Ford engineers involved with this test, and those who learned of its results, that the Pinto as it was currently being sold was more vulnerable to fuel spillage and possible explosion at low speeds than was the Capri. Call this Test 2.

It appears that Ford engineers did not crash-test actual production-model Pintos until October 10, 1970, a month after the car's debut. Over the next 15 months, Ford engineers carried out a number of crash tests of Pintos at various speeds. Some involved running a Pinto backward into a fixed wall, others a moving wall hitting a Pinto from behind. Most of the tests resulted in fuel leakage, either from a ruptured or punctured fuel tank or from filler-pipe problems, such as its disconnecting from the fuel tank. However, in some tests, the addition of some kind of protective device to the car preserved the integrity of the fuel tank.

One such device was a nylon-reinforced rubber bladder made by Goodyear Tire and Rubber Company. Put inside the metal fuel tank, the bladder kept fuel from spilling out of the tank when Pintos moving at 20 and 26 mph were crashed into a cement wall and their tanks were punctured.[281] The bladder's estimated unit cost was $6.00.[282] Another change that was tested was the installation of "two longitudinal side rails," called "hat sections," at the rear of the car, aimed at strengthening it. One Ford engineer estimated that "the design cost of adding rails to both sides of the Pinto would have been $2.40 per car."[283] Engineers also considered adding a "flak suit," akin to "a rubber tire that fit around the [fuel] tank to cushion it from impact and prevent puncture."[284] The estimated design cost of the flak suit was $8 per car. Call this set of tests and design alternatives Test 3.

The modest costs of the tested and considered changes notwithstanding, Ford apparently believed that shutting down production and retooling to implement changes would be too expensive. Shortly after the Pinto went on sale in September 1970, the U.S. Department of Transportation had proposed that, as of 1972, cars be required to withstand a 20 mph fixed-barrier rear crash without gasoline leakage, and, by 1973, a 30 mph fixed-barrier rear crash (equivalent to a 44 or 45 mph car-to-car crash). Nevertheless, Ford executives decided in late 1971 that, to save money, no additional "fuel system integrity" changes would be incorporated into

[281] Strobel (1980), p. 85.
[282] Ibid., p. 83.
[283] Ibid., p. 87.
[284] Ibid., p. 88.

the 1973 and later Pinto models until "required by law."[285] A confidential Ford memo in late April 1971 stated that "the design cost of the flak suit or the bladder would be $8 per car." The memo went on to state that based on this design change cost, executives recommended that Ford "defer adoption of the flak suit or bladder on all affected cars until 1976 to realize a design cost savings of $20.9 million compared with incorporation in 1974."[286]

At this remove, it is difficult to know which Ford engineers were familiar with the results of which crashworthiness tests. However, any Ford engineer who was familiar with the results of Test 1, knew about the fuel-tank integrity tests of the Rover and the Capri, and yet endorsed, signed off on, or acquiesced in the design and placement of the more risky Pinto fuel tank and fuel pipe "packages" arguably did not act in accord with FERE1 and probably also FERE3, for such actions arguably helped engender an unreasonable, easily avoidable, undisclosed risk of harm to Pinto drivers and passengers. It is equally hard to see how any Ford engineer who knew of the results of Test 3 and yet made no attempt to persuade management to adopt one of the inexpensive tested or considered protective devices can be said to have acted in accord with FERE2.

One Pinto engineer *did* make such an effort. She or he wrote a memorandum to management informing it that if the fuel tank were positioned behind the rear axle, spending $6.65 per car—perhaps for a nylon-reinforced rubber fuel-tank bladder-liner—would "increase the fire safety level of the 1973 Pinto from a 20-mile-per-hour moving-barrier standard to 30 miles per hour."[287] Laudably, and in accordance with FERE2, that anonymous engineer tried to prevent an incremental risk of harm to drivers and passengers by giving management an option that, for little additional cost, would allegedly have made the fuel tank safe in rear car-to-car impacts up to 45 mph. However, according to former Ford engineering executive Harley Copp, Ford chose not install the $6.65 part. According to Copp, Ford did so because not installing the safety-enhancing part "improved the profitability of the vehicle."[288]

[285] Ibid., pp. 88–89.
[286] Ibid., p. 88.
[287] Strobel (1980), p. 179.
[288] Ibid. Ford's decision to not install any of the tested or considered protective devices in the 1973 and later models of the Pinto until "required by law" held until 1977, when the federal government adopted its first "safety requirement designed to protect against gas leakage and fire risk in rear-end crashes." Ibid., p. 89.

Other Ford engineers had a different view of the Pinto's safety. In early 1980, Ford Motor Company was unsuccessfully prosecuted for reckless homicide. The case, *State of Indiana v. Ford Motor Company*, arose from the deaths in August 1978 of three young women in a 1973 Pinto that burst into flames when it was hit from behind by a van on an Indiana highway. Several "Ford engineers testified at the trial that they had used Pintos for their own families."[289] For example, Francis G. Olson, a Ford design engineer, testified that he considered the Pinto "so safe that he bought one for his 18-year-old daughter."[290] The implication was that these engineers regarded the Pinto as safe. It would be interesting to know whether any of Ford's engineer-witnesses knew, when they testified, of the results of Tests 1, 2, and/or 3, of the Ford engineer's memorandum regarding the $6.65 part, and/or of Ford's decision to not make further "fuel system integrity" changes in the 1973 and later Pinto models until "required by law."

In short, when a decision about where to locate a key safety-related part of a car is influenced primarily by marketing/"styling" considerations— and decisions whether to mitigate risk by adding inexpensive protective devices are influenced primarily by considerations of unreduced profit— rather than by technical engineering considerations, an engineer working on the car may be faced with a difficult ethical issue: how should she or he act in response to such a management decision? The answer to that question depends on her or his role in the product-development process, and on what she or he knew or should have known about the relative safety of alternative part locations at various points in time. Depending on an involved engineer's state of knowledge about the safety risks involved—and/or on what knowledge she or he *should* have had about those risks—she or he might have an FERE1-based ethical responsibility to not sign off on a part location that she or he believes is significantly riskier than one or more known viable alternatives that are not significantly more expensive. One reason that responsibility is important is that an engineer signing off on such a location may enable management to justify its economically driven choice by pointing to the engineer's presumably safety-driven sign-off approval. The engineer might also have an FERE2-based ethical responsibility to try to persuade management to change the location or to modify the part so that it poses a smaller risk. Finally, the engineer might have an FERE3-based ethical responsibility to

[289]Hoffman (1984), p. 418.
[290]Stuart (1980b), p. A12.

try to alert and inform people who she or he knows would be unknowingly put at incremental risk of incurring significant harm by putting the safety-related part in question in the location chosen by management for marketing or cost reasons.

2. *Production planning period.* A second important ethical issue raised by the Pinto case has to do with the vehicle's "production planning period," that is, the time engineers are given for going from the birth of a new conceptual design to production. To ensure that the Pinto made it to market by the time Iacocca had stipulated, namely, the start of the 1971 model year, the new-car production planning period was slashed from the usual 43 months to 25 months.[291] This meant that tooling assembly had to be started before the Pinto development process had been completed. While some Ford engineers may have protested this major compression of the product development process, there is no evidence of which I am aware that engineers involved did so. If they did not, they arguably should have.

The rushed product development process and the need to start tooling before detailed design and testing were completed *may* also have contributed to the decision to stick with the riskier fuel-tank location, because any changes in the fuel-tank location made late in the product development process not only might have been costly but might have ensured failure to meet Iacocca's tight deadline. In the words of Richard DeGeorge, "The engineers were squeezed from the start. Perhaps this is why they did not test the gas tank for rear-end collision impact until the car was produced."[292]

If Pinto safety engineers were unable to seriously test the car's fuel-tank system for rear-end collision impact integrity until the car was already being produced, as seems to have been the case, then the development engineers' upstream willingness to go along with the prescribed, substantially compressed Pinto development cycle arguably contributed to putting into the marketplace a car that was inadequately tested for safety and known to be riskier than it needed to be. Unless the development engineers had good reason to think that going along with the compressed production-planning period would not compromise or preclude adequate safety testing before production, that upstream acquiescence is difficult to square with FERE1.

3. *Public whistleblowing.* While writing the abovementioned memo to management about the $6.65 part was in line with FERE2, it does not

---

[291] Strobel (1979), sec. 2, p. 12, and DeGeorge (1981), p. 9.
[292] DeGeorge, ibid.

appear that any then currently employed Ford engineer familiar with the Pinto's development decided to blow the whistle publicly on the incremental risk the tank's location created.[293] Did any Ford Pinto engineer with such knowledge have an ethical responsibility to do so?

In an influential article[294], Richard DeGeorge argued that if the following three conditions are satisfied, then it is ethically *permissible* for an engineer to publicly blow the whistle on a product that he or she believes is unsafe:

C1. "if the harm that will be done by the product to the public is serious and considerable";

C2. "if [the engineers] make their concerns known to their superiors; and

C3. "if, getting no satisfaction from their immediate superiors, [the engineers] exhaust the channels available within the corporation, including going to the board of directors."[295]

I have concerns about two of these conditions. First, I do not believe that C1's scope should be limited to cases where the serious and considerable harm in question would be visited on *the public*. In the *Challenger* case, the engineers involved arguably had an ethical responsibility to blow the whistle after the fact when the most important harm in question was done to the six astronauts and the first teacher in space, not to the public at large. If NASA had been a private company, I fail to see that the engineers involved would not have had an ethical responsibility to publicly blow the whistle just because the public was not a directly affected party. Second, also involving C1, I contend that there can be an ethical responsibility to blow the whistle to prevent an *unreasonable risk of harm*, not just to prevent harm per se. Third, the requirement, stated in C3, that the engineer go up the corporate chain of authority *as far as the firm's board*

---

[293] The Ford engineer who came closest to doing so, although ex post facto, was Harley Copp. Copp was "a . . . retired Ford executive who was one of the company's highest-ranking engineers. Ford dismissed Mr. Copp, a critic of the Pinto, in 1976, citing poor performance." (Stuart [1980a]). After Ford forced him to retire, Copp became an independent engineering consultant. (Strobel [1980], p. 35). In that role, he testified for plaintiffs about Ford's approach to the Pinto's safety in relation to cost in the reckless homicide lawsuit *State of Indiana v. Ford Motor Company*, as well as in an important product liability case, *Grimshaw v. Ford Motor Company* (1981), brought against Ford by a severely burned survivor of another Pinto rear-end collision. See 174 Cal. Rptr. 348,119 Cal. App. 3d 757 at 361–362, https://h2o.law.harvard.edu/cases/50.

[294] DeGeorge (1981). Put differently, DeGeorge held that the three conditions are jointly sufficient for an engineer's act of public whistleblowing to be ethically permissible.

[295] Ibid., p. 6.

of directors seems too onerous. I would set the bar lower and require only that the engineer go to her or his boss and to her or his boss's boss.

DeGeorge next argued that for an engineer to be ethically *obligated*, or, in other words, for her or him to have an ethical responsibility or duty to "bring his case for safety to the public, . . . two other conditions have to be fulfilled, in addition to the three mentioned above."[296]

> C4. "He must have documented evidence that would convince a rea-sonable, impartial observer that his view of the situation is correct and the company policy wrong."
> C5. "There must be strong evidence that making the information pub-lic will in fact prevent the threatened serious harm."[297]

I also have concerns about C4 and C5. Regarding C4, requiring that the engineer have "documented evidence" is, in my view, too strong a requirement, especially since engineers in corporations are sometimes instructed not to put things down on paper precisely to avoid having relevant documents captured by plaintiffs in the discovery phase of trials of product-liability lawsuits. Moreover, even if the engineer in question does have documented evidence, it might be so technical that it would not "convince a reasonable, impartial observer" that the engineer's view of the situation is correct and the company's policy is wrong, simply because it is incomprehensible. That fact in itself should not release an engineer from an ethical responsibility to try to prevent the projected harm or unreasonable risk of harm. I would replace C4 with a revised condition:

> C4*: The engineer must possess either (a) sufficient evidence re-garding the harm in question or (b) a combination of technical expertise, authority, and insider knowledge that would persuade impartial observers literate in the field in question that the engi-neer's view of the situation should be taken seriously.

Regarding C5, it is not clear that an engineer could ever *know* in ad-vance that the evidence he or she has is so strong that making it public "will in fact prevent the threatened serious harm." The engineer could have strong technical evidence of imminent or eventual harm but, for one or another reason, be quite unsure whether, given current political-economic

---

[296] In other words, DeGeorge viewed C1–C5 as individually necessary conditions for an engineer to have an ethical responsibility to blow the whistle publicly when safety is jeopardized. It is unclear whether he regarded conditions C1–C5 as jointly sufficient for an engineer to have an ethical responsibility to blow the whistle publicly.

[297] DeGeorge (1981), p. 6.

reality, releasing it would actually have the desired effect of preventing harm. Requiring that the engineer have strong evidence that making her or his information public "will in fact prevent the threatened harm" from occurring seems to me to set the bar too high for blowing the whistle to become an engineer's ethical responsibility. Maintaining that requirement would mean that the circumstances in which an engineer had an ethical responsibility to publicly blow the whistle would be extremely rare.

I propose that condition C5 be revised as follows:

> C5*: The engineer must have reason to believe that making public the information she or he has, taken together with her or his credibility, authority, and insider knowledge, will create at least a *decent chance* that the threatened serious harm will be prevented, significantly lessened, or stopped, and/or that its recurrence will be made less likely.

The greater the magnitude of the pending harm, the lower should be the level of certainty that engineers should be required to have (about their chances of being able to prevent the harm by publicly blowing the whistle) before doing so becomes their FERE2-based derivative ethical responsibility.

Given the strength of conditions C4 and C5, it is not surprising that DeGeorge concluded that Ford engineers did *not* have an ethical responsibility to engage in public whistleblowing in the Pinto case. To the extent that the reader finds merit in conditions C4* and C5*, she or he is invited to reflect and draw her or his own conclusion about whether any Ford engineers involved with the Pinto's development might have had an ethical responsibility to blow the whistle publicly on Ford management for making certain decisions about the Pinto's fuel-tank system design, and if so, at what juncture(s).[298]

Lessons

The Pinto situation illustrates an important recurrent problematic pattern in engineering practice. I refer to this as *the upstream acquiescence, downstream misfortune* pattern of practice. In this case, the apparent

---

[298]Noteworthy decision junctures arguably include when the Pinto's development period was set at 25 months, when its fuel-tank-system design and location were finalized and production tooling was set, and when Ford management decided in late 1971 not to add further protective "fuel-system integrity" devices to 1973 and later Pinto models until "required by law," regardless of their cost or efficacy.

willingness of Pinto engineers to go along upstream with a substantially shortened new-product-development cycle arguably helped make it more likely that downstream the safety engineers would be unable to do a thorough job testing the safety of the car (including the fuel-tank system) before the design was fixed and tooling was set. Testing would be compromised by being done less rigorously, less comprehensively, less extensively, and/or more rapidly than usual before production had begun, or by being performed only after production had begun.

Given the limited evidence available about what specific Ford engineers who worked on the Pinto actually knew, or should have known, about fuel-tank safety in the Pinto and Capri (and at what points), and about the specific management-imposed constraints under which Ford engineers worked on the Pinto, it is impossible to make categorical ethical judgments about the conduct of engineers involved in the early stages of the car's development. Instead, to provoke discussion of whether and how the FEREs applied to engineers involved in this and similar cases, I shall limit myself to offering three provisional claims.

1. If and to the extent that Pinto development engineers who acquiesced upstream in Ford's substantially compressed development period for the car realized, suspected, *or should have realized or suspected,* that having the car ready by September 1970 would make it unlikely the safety engineers would have enough time to do a thorough job testing the safety of the car before tools had to be set, then that acquiescence would be incompatible with FERE1.

2. If and to the extent that any Pinto safety engineers *knew or should have known* that the preproduction time available for testing the vehicle's fuel-tank system—to ensure that it was acceptably safe under various crash scenarios—was insufficient, then, under FERE1, they had a derivative ethical responsibility not to attest that the Pinto fuel-tank and fuel-pipe systems were production-ready.

3. If and to the extent that any Pinto safety engineers knew, in the first few years the Pinto was for sale, inexpensive fuel-tank protective devices were available that would significantly enhance its safety level, they had an FERE2-based derivative ethical responsibility to try to persuade management to order that one be adopted. Further, if any Pinto safety engineers who were knowledgeable about this situation were unsuccessful in persuading management to do so, they arguably had an FERE3-based derivative ethical responsibility to try to alert the public about the safety risk of the vehicles and inform it about how little it would cost to significantly enhance the safety of the vehicle.

Discussion Questions

1. Was it ethically responsible for Ford engineers involved in working on the Pinto's development to go along upstream with the shortening of the vehicle's product development cycle from 43 to 25 months? Why or why not?

2. Could it be ethically responsible for a Ford engineer working on the Pinto to recommend, make, or approve a decision about where to locate the car's fuel-tank system based primarily on stylistic or marketing rather than safety considerations? Why or why not?

3. After the Pinto debuted in September 1970, Ford engineers identified a range of inexpensive protective devices able to improve the integrity of the vehicle's fuel-tank system. When Ford management took the position that no additional "fuel system integrity" changes would be incorporated into the 1973 and later Pinto models until "required by law," did the engineers who knew of the range of such devices have an ethical responsibility to try to persuade management to order that one or more of them be incorporated into the car? Why or why not?

4. If Ford engineers familiar with such protective devices failed to persuade management to order that one or more be incorporated, and if the existence of such devices remained unknown to the public, did such engineers have an ethical responsibility to try alert and inform prospective Pinto customers of such options? Why or why not?

5. To what extent do you agree or disagree with the author's revision of DeGeorge's five criteria for public whistleblowing to be an engineer's ethical responsibility? Why or why not?

CASE 12: TOPF & SONS: CREMATORIUM OVENS FOR THE NAZI SS

Background

In World War II, the Nazis murdered millions of Jews, as well as many thousands of Sinti and Roma, Soviet prisoners of war, non-Jewish Poles, homosexuals, communists, socialists, trade unionists, mentally and physically disabled people, Jehovah's Witnesses, and other groups.[299] Much of

---

[299] http://www.ushmm.org/wlc/en/article.php?ModuleId=10005143 and http://www.topfundsoehne.de/download/medienmappe_ef_en.pdf, p. 3.

Fig. 6. Topf and Sons cremation ovens, Nazi concentration camp, Weimar (Buchenwald), Germany, April 14, 1945.

the murder was carried out between 1942 and 1945 in six concentration camps in German-occupied Poland, in which "Krema" (crematoria) had been built. These structures contained rooms in which victims were poisoned to death,[300] and furnaces in which their corpses were reduced to vapor, ashes, and bone fragments.[301] A German engineering firm, Topf & Söhne, produced 66 cremation ovens (Figure 6) for the Nazi SS for use at concentration and extermination camps.[302]

[300] While carbon monoxide was used initially, cyanide-based Zyklon B later became the primary poison utilized. See http://www.pbs.org/auschwitz/40–45/killing/.

[301] http://www.topfundsoehne.de/download/medienmappe_ef_en.pdf, p. 5.

[302] "Topf and Sons," http://en.wikipedia.org/wiki/Topf_and_Sons. The first crematorium ovens that Topf & Söhne made for the SS, in 1939, were mobile and not linked with gas chambers. The use of stationary ovens to incinerate the bodies of those murdered in gas chambers began in 1942–43. In 1939, Topf ovens were used to incinerate corpses of those who had died in a typhus outbreak at Buchenwald. See ibid.

Master brewer Johannes Andreas Topf founded the company in Erfurt in 1878 to manufacture "combustion equipment and systems."[303] By 1889, the firm specialized in "heating systems, brewery and malting facilities."[304] Two decades later, Topf was "one of the world's major producers of malting equipment for breweries. The product line grew to encompass boilers, chimneys and silos, then ventilation and exhaust systems and, beginning in 1914, incineration ovens for municipal crematoria."[305] Topf was the market leader in the latter field in the 1920s.[306] During WWII, the firm's co-owners and co-managing directors were the founder's grandsons, brothers Ludwig Topf Jr. and Ernst-Wolfgang Topf.[307]

Of the 66 cremation furnaces Topf & Sons built for the Nazi SS, 46 were located at the Auschwitz-Birkenau extermination camp.[308] It has been estimated that at least 1.1 million people were "gassed to death at Auschwitz, 90 percent of them Jews."[309]

On May 31, 1945, fearing arrest by the U.S. Army, Ludwig Topf Jr. committed suicide.[310] In March 1946, the Soviet Army arrested four key Topf engineers[311] involved with the firm's work for the Nazi SS on crematorium ovens and ventilation systems. Soviet authorities interrogated three of them the same month, and one of those three again in March 1948.[312]

Before proceeding, a comment is in order. I included this case in this work despite the fact that it involves a situation of the sort that contemporary engineers are highly unlikely to encounter in their careers. I chose to include it for several reasons. First, as will become clear, some justifications that Topf engineers gave for cooperating with the Nazi SS cited reasons that apply to normal contexts in which engineers face ethical challenges. Learning that these specious justifications were given to

---

[303] "Topf & Söhne: Die Ofenbauer von Auschwitz," http://www.topfundsoehne.de/cms -www/index.php?id=94&l=1.

[304] Ibid.

[305] Knigge (2005), p. 4, http://www.topfundsoehne.de/download/medienmappe_ef_en .pdf, p. 4.

[306] http://www.topfundsoehne.de/cms-www/index.php?id=94&l=1.

[307] Ibid.

[308] "Topf and Sons," http://en.wikipedia.org/wiki/Topf_and_Sons.

[309] Volkhard Knigge gives a figure of 1,086,000, including "more than 960,000 Jews." See Knigge (2005), p. 3.

[310] http://www.topfundsoehne.de/cms-www/index.php?id=295&l=1.

[311] Kurt Prüfer, Fritz Sander, Karl Schultze, and Gustav Braun. See Fleming (1993) and http://www.topfundsoehne.de/cms-www/index.php?id=294&l=1.

[312] Ibid. A few weeks after his arrest, Fritz Sander died from heart failure. See also Fleming (1930).

excuse engineering misconduct in such an extreme case may disincline engineers from using them in more normal situations.

Second, another valuable aspect of this case is that transcripts exist of Soviet Army interrogations of key Topf engineers. They enable one to see how the engineers attempted to justify their cooperation with the Nazi SS in their own words.

Third, this case vividly illustrates how important it is for engineers to pay careful attention to the environments in which their creations are commissioned and will be used.

Ethical Analysis

Starting in 1939, the Nazi SS commissioned many works from Topf & Sons. Topf's contracts with the SS involved them in designing, producing, installing, and maintaining incineration ovens, initially mobile one-muffle models, later stationary multimuffle models, as well as ventilation systems for gas chambers and incineration ovens. The key question to be explored here is whether in carrying out their wartime work for their employer, any Topf & Sons engineers violated any ethical responsibilities of engineers.

To answer this question, let us examine the activities of four key Topf personnel: the firm's nonengineer co-owner and co-managing director, Ernst-Wolfgang Topf, and three key Topf engineers: Kurt Prüfer, Fritz Sander, and Kurt Schultze.

1. *Ernst-Wolfgang Topf* (EWT). On June 21, 1945, about two weeks before the Soviet occupation of Erfurt began, EWT left Erfurt and fled to what became West Germany.[313] He was not permitted by the Soviet authorities to return to what soon became East Germany. EWT was put on trial before an American denazification tribunal but released for lack of evidence.[314] He blamed his brother and Kurt Prüfer for the firm's work for the Nazi SS and claimed he did not know what the Nazis were doing with the ovens.[315] For the rest of his life he insisted that furnaces of Topf & Sons had been "misused (*missbraucht*) in an unforeseeable manner" and "clung to his concept of 'innocent ovens.'" (*unschuldigen Öfen*).[316]

---

[313] Wahl (2008).

[314] "Topf and Sons," http://en.wikipedia.org/wiki/Topf_and_Sons.

[315] Ibid.

[316] "Ernst-Wolfgang Topf in the West," http://www.topfundsoehne.de/media_en/en_ct _spurensichern_imwesten.php?subnav=imwesten.

EWT maintained that during the Hitler Reich, "in our company, no-one was guilty, morally or objectively," and held that the furnaces were not wrong or evil in themselves.[317]

Citing the "misuse of a product in an unforeseeable manner" is an argument sometimes used to try to deny that technical practitioners are ethically accountable for what is done by others with the fruits of their work. As seen in Case 10 (nanotechnology R&D), Leon Lederman made the same argument about the misuse by others of potent knowledge developed by scientists. Lederman denied that scientists who generate potent scientific knowledge have any ethical responsibility for devastating applications of it by others. Similarly, EWT denied that he (and, presumably, the firm's engineers) bore any responsibility for the supposedly unanticipatable and horrible use by the Nazis of the "neutral" crematorium ovens his firm designed and produced. If valid, this argument would mean that no Topf employee bore any ethical responsibility for contributing to the Nazi killing program.

But this argument fails to take into account that sometimes the generators of critical scientific knowledge, and the designers and producers of potent items of technology, are either familiar with or could easily have made themselves familiar with the characteristics of the social contexts and the agendas of key actors in the contexts into which the knowledge or items of technology they generated would be introduced and used. Thus, not surprisingly, in arguing that he was not ethically responsible, EWT claimed he was completely unaware of the uses the Nazis made of his firm's ovens. Probably because the Americans accepted this line of defense, they released him.

However, as will shortly become clear, there is reason to suspect that EWT knew what the Nazi SS was doing with his firm's ovens starting in 1942. A statement he made in denying accountability suggests he might have known what the Nazi SS was doing with the ovens his firm produced for them:

> What burned in those ovens was already dead. . . . You can't hold the builders of the ovens responsible for the deaths of the people who were burned in them.[318]

As noted at the beginning of Chapter 4, a background enabling or facilitating factor can contribute to bringing about an outcome as much

[317]Ibid.
[318]Walsh (1994).

as a foreground precipitating action. As will be argued shortly, this fact undermines EWT's claim of nonresponsibility as expressed in this quote.

Consider the fact that Zyklon B, carbon monoxide, and the gas chambers enabled the killing of massive numbers of innocents without their having to be murdered directly by Nazi soldiers.[319] The use of poison gas eliminated a roadblock in the form of serious psychological problems that had plagued those who had been killing innocents directly, by shooting and/or burying them.[320] *Topf & Sons' furnaces removed another kind of roadblock or bottleneck to more efficient killing/murder*: the accumulation of poisoned corpses. It enabled the Nazi SS to make "disappear" the corpses of murdered innocents ever more rapidly and in ever greater numbers. That increased disposal ability stimulated the Nazi SS to murder more and more innocents by poison gas, for instead of quickly accumulating and causing a bottleneck, the resultant corpses could be rapidly destroyed in the furnaces, eliminating the need for mass burials and pyres in huge earthen pits.[321]

Thus, I submit that EWT's claim that "[y]ou can't hold the builders of the ovens responsible for the deaths of the people who burned in them" is wrong. The "builders of the ovens" *can* be held *partly* responsible for the deaths of at least some of those whose corpses burned in them, because without the furnaces' large and increasing cremation capacity, it is unlikely that anywhere near as many innocents would have been gassed to death. To the extent that he knew or should have known what was being done with his "innocent ovens," EWT was complicit in the murder carried out with the aid of his firm's cremation furnaces and ventilation units, both commissioned by the Nazi SS.[322] I shall return to EWT's state of knowledge about this matter below.

---

[319] See "Transition to gassing," in "Einsatzgruppen," http://en.wikipedia.org/wiki /Einsatzgruppen.

[320] "In 1941, SS General Eric von dem Bach-Zelewski told his superior Heinrich Himmler that the Nazis had been murdering Jews, including women and children, at close range and in cold blood all summer. Bach-Zelewski was worried about this method's traumatizing effects on his men. . . . Himmler realized he had to find new methods that would spare his troops the psychological strain of killing human beings at close range." http://www.pbs.org /auschwitz/40–45/killing/.

[321] "The issue of how many people could be gassed is somewhat misleading, . . . as the bottleneck to the gassing operations at Birkenau was the incineration of the victims." Van Alstine (1997).

[322] It should be noted that the Topf brothers and Kurt Prüfer joined the Nazi party at the end of April 1933. See http://www.topfundsoehne.de/cms-www/index.php?id=102&l= 1 and http://www.topfundsoehne.de/cms-www/index.php?id=99&l=1.

2. *Kurt Prüfer*. Prüfer had studied structural engineering for six semesters in Erfurt[323] and was a senior engineer in the Topf firm. He specialized in the construction of cremation furnaces.[324] Prüfer played a leading role in designing and building the incineration ovens at Auschwitz-Birkenau from 1940 to1944 and admitted that "[f]rom 1940 to 1944, 20 crematoriums for concentration camps were built under my direction—for Buchenwald, Dachau, Mauthausen, Auschwitz, and Gross-Rosen."[325] When Ludwig Topf Jr. told him in 1940 that the SS Command urgently wanted more powerful incineration ovens, Prüfer responded by inventing a "triple-muffle oven."[326]

Prüfer was arrested and interrogated by officers of the American Third Army on May 30, 1945, the day before Ludwig Topf Jr. committed suicide.[327] Prüfer was released shortly thereafter, apparently because he persuaded the Americans that he believed the crematoria he had designed and built "were in [the Nazi] concentration camps for health reasons only."[328] But, in 1946, when interrogated by the Soviet Army, which had captured the files of the Auschwitz Central Building Administration of the Waffen SS and Police, Prüfer told a different story. He acknowledged that he had visited Auschwitz five times in 1943.[329] When asked whether he knew what was happening with the ovens he had built he was quoted thus:

> I have known since spring 1943 that innocent human beings were being liquidated in Auschwitz gas chambers and that these corpses were subsequently incinerated in the crematoriums.[330]

Asked what his motivations were for continuing to build the cremation ovens, Prüfer (allegedly) answered,

> I had my contract with the Topf firm and I was aware of the fact that my work was of great importance for the national socialist state. I knew

---

[323] Ibid.

[324] Ibid.

[325] Fleming (1993). "Altogether, Topf & Sons delivered at least twenty-five stationary ovens with seventy-seven incineration chambers to the Buchenwald, Dachau, Mauthausen, Gusen, Auschwitz, Gross-Rosen and Mogilev Concentration Camps. The firm also repeatedly supplied the SS with mobile cremation ovens upon request." http://www.topfundsoehne.de/cms-www/index.php?id=123&l=1.

[326] Fleming (1993).

[327] http://www.topfundsoehne.de/cms-www/index.php?id=295&l=1.

[328] Fleming (1993).

[329] Ibid.

[330] Ibid.

that if I refused to continue with this work, I would be liquidated by the Gestapo.[331]

Thus, Topf senior engineer Kurt Prüfer admitted knowing what was being done with the ovens by early 1943.[332] This admission by itself renders co-owner and codirector Ernst-Wolfgang Topf's claim that he had no idea what the Nazis were doing with Topf & Sons' cremation ovens highly suspect.

3. *Fritz Sander.* Prüfer reported to the chief engineer of Topf & Sons: Fritz Sander. Sander's most notorious "achievement" was probably the invention, in 1942, of a more efficient "crematorium for mass incineration."[333] He did so in response to learning, probably from a September 1942 in-house memorandum from Prüfer, who was in contact with the SS, that the latter regarded the current incineration rate at Auschwitz-Birkenau of 2,650 corpses per day, or 80,000 per month, insufficient for their purposes.[334] Sander seems to have viewed this situation as posing a technical challenge that needed to be met. He chose to use his engineering skills to devise a "continuous-operation corpse incineration oven for mass use" that used "the conveyer belt principle."[335] In such a cremation unit, for which Topf & Sons submitted a patent application in 1952, corpses were "brought to the incineration furnaces without interruption"[336] and flames from previously burning corpses ignited other corpses above them on grates.

> Fritz Sander designed his four-story oven in the manner of an enormous refuse incineration plant. The corpses would be introduced sideways on a kind of conveyor belt. . . . They would then slide down along the zig-zag array of slanted grates, catching fire from the burning corpses already below them. After a preheating period of two days, the oven would thus be capable of remaining in operation "continually," without the introduction of further combustion material. The burning corpses alone would keep the fire going.[337]

[331] Ibid.
[332] As the following remarks about Fritz Sander suggest, Prüfer probably knew by fall 1942.
[333] Ibid.
[334] Kurt Prüfer (1942), http://www.holocaust-history.org/auschwitz/topf.
[335] http://www.topfundsoehne.de/cms-www/index.php?id=94&l=1 and Fleming (1993).
[336] Fleming (1993).
[337] http://www.topfundsoehne.de/cms-www/index.php?id=120&l=1.

Sander's design remained a state secret until after the war and was not carried out. His double-edged ability to focus on the technical details of his engineering work without considering its gruesome human implications is apparent from a memorandum he wrote to the Topf & Sons "Geschäftsleitung," that is, the Topf brothers, on September 14, 1942:

> The high demand for incineration ovens for concentration camps—which has recently become particularly apparent in the case of Auschwitz . . . has prompted me to examine the issue of whether the present oven system with muffles is suitable for locations such as the abovementioned. . . . In my opinion, with regard to the design of an incineration oven for the purposes of a concentration camp, the ideal solution would be an oven that could be continuously loaded and likewise operated . . . , i.e., the corpses for burning would be loaded on at respective intervals—without interrupting the incineration process—and on their way through the oven would catch fire, burn, burn out and reduce to ash, and then land in the ash-chamber beneath the grate in the form of burnt-out ash. *Here I am quite aware that such an oven must be regarded purely as a facility for extermination, so that concepts of reverence, the separation of ashes and emotions of any kind must be dispensed with entirely.* All of this is probably the case already, however, with the operation of numerous muffle furnaces. Special war-related circumstances prevail in the concentration camps, making such methods indispensable. . . . Bearing in mind the remarks I made above, it must be assumed that the authorities in question are also approaching other oven construction companies with regard to the supply of efficient, fast-functioning cremation ovens. In these companies, the question of the cheapest method of constructing such ovens for the abovementioned purposes will be examined as well. . . . For this reason I consider it urgently necessary to have my suggestion patented, so that we can secure priority.[338] (emphasis added.)

On January 3, 1953, "J. A. Topf & Söhne" was awarded German patent #861731 for a "Process and Apparatus for Incineration of Carcasses, Cadavers, and Parts Thereof."[339]

Sander was interrogated by the Soviets on March 7, 1946.[340] When it came to justifying his Topf engineering work on projects initiated by

[338] Ibid.
[339] Rosenbaum (1993).
[340] Fleming (1993).

the SS, he did not deny that he knew what was being done with the crematoriums.

> I was a German engineer and key member of the Topf works and I saw it as my duty to apply my specialized knowledge in this way in order to help Germany win the war, just as an aircraft construction engineer builds airplanes in wartime which are also connected with the destruction of human beings.[341]

Thus, Sander invoked a would-be exculpatory analogy, claiming that what he had done was similar to what is done by "aircraft construction engineers" who build items of technology that often kill innocents in wartime but whose designers are not thereby deemed ethically irresponsible.

However, Sander's analogy is flawed. From 1942 on, Topf ovens were knowingly designed, built, and optimized for use directly on the corpses of murdered innocent civilians, whereas military planes were not. The latter typically kill innocent civilians indirectly, as by-products of efforts to destroy targets with military value. Only if aircraft construction engineers built airplanes in wartime specifically to kill innocent civilians more efficiently might there be a valid analogy between them and the Topf crematorium engineers. Hence, Sander cannot plausibly rebut the claim he acted incompatibly with FERE1 by analogizing his work to that of aircraft construction engineers.

4. *Karl Schultze.* The focus of senior engineer Karl Schultze's work was on designing and building ventilation systems for the gas chambers and the ovens. "I was responsible for the ventilation systems and for its air injection into the muffles."[342] Schultze maintained that he was *unaware* of what was being done to innocents in the crematoriums.

> It did not become known to me what was the purpose of installing these ventilation systems, nor did it become known to me that these morgues were to be used as gas chambers. . . . I did not know that in the crematoriums in Auschwitz-Birkenau innocent human beings were being liquidated. I thought criminals were being killed there who had partly been sentenced to death because of the crimes they had committed against the German army in Poland and other occupied territories. I am a German and supported and am supporting the Government in Germany and the laws of our Government. Whoever opposes our laws is an enemy of the state, because our laws

[341] Ibid.
[342] Ibid.

established him as such. I did not act on personal initiative but as directed by Ludwig Topf. I was afraid of losing my position and of possible arrest.[343]

Thus, Schultze cites several would-be exculpatory factors: ignorance and false belief, loyalty to the state, loyalty to his employer, and fear of arrest. However, he did admit to being in Auschwitz three times in 1943, and acknowledged that he "personally led the installation work in Auschwitz crematoriums and gas chambers." Also, in four other death camps, he "supervised the gas injection into the Krema furnaces for the purpose of increasing the incineration capacity."[344] It is, however, difficult to believe that Schultze and Prüfer were both senior engineers at Topf & Sons, that both worked at Auschwitz after 1942, and that while Prüfer knew innocent people were being murdered at Auschwitz-Birkenau, Schultze was unaware of the liquidation of innocents being carried out there with the help of the systems he installed.[345]

Schultze's attempt to justify his activities by stating that he was an employee of Topf and that he was ordered to build the ventilation systems is not compelling. As we have seen, FERE4 imposes on employees an ethical responsibility to work to the best of their ability to serve their employer or client only when doing so would be serving the *legitimate* interests of that party. Topf & Sons had no legitimate interest in designing, optimizing, making, and installing systems that at least some, and probably all, of the firm's senior decision-makers and engineers knew were being used by the SS to liquidate innocents. So, in spite of being an employee of Topf & Sons, Schultze had no ethical responsibility under FERE4 to design and install those ventilation systems to the best of his ability. On the contrary, under FERE1 he had an ethical responsibility to *not* do so. His work in ventilating the gas chamber rooms actually facilitated the gassing to death of more innocents. That it might well have been risky for Schultz to have opted out of such work does not negate the fact that he had an ethical responsibility to not carry on with his work, assuming that, contrary to what he claimed, he realized (or should have realized) what his work products were actually being used to do.

* * *

---

[343] Fleming (1993).
[344] Ibid.
[345] Schultz claimed he believed the Krema were being used to kill criminals who had been convicted of crimes against the German occupying forces. See Fleming (1933).

Ernst-Wolfgang Topf was fortunate he was not permitted to return to the part of Germany, including Erfurt, then controlled by the Red Army.[346] He went on to start a firm in West Germany that designed and constructed refuse incineration ovens. His enterprise went bankrupt in 1963, and he died in Brilon, West Germany, in 1979.[347] Prüfer and Schultze, along with Topf & Sons' production manager, engineer Gustav Braun,[348] were sentenced by the Soviets to "25 years deprivation of liberty and sent to a Special Penal Camp of the Ministry of the Interior."[349] As noted, a few weeks after he was arrested, Fritz Sander died in Red Army custody from a heart condition.[350] Prüfer died "in a prison camp hospital of a brain hemorrhage on Oct. 24, 1952."[351] Schultze and Braun were released from prison on October 5, 1955, after spending about 7 years in a gulag.[352]

Lessons

Regarding the strategies they adopted to avoid being judged ethically irresponsible for the work they did for the Nazi SS, the Topf engineers discussed can be divided into two categories. However implausibly, Karl Schultze (and Ernst-Wolfgang Topf) claimed ignorance of the fact that the Topf furnaces were being used to kill innocents. In contrast, Kurt Prüfer and Fritz Sander acknowledged they knew what the SS was doing with the Topf ovens but denied any ethical responsibility for what happened, citing various would-be exculpatory factors: fear of being killed by the SS, loyalty to their employer, loyalty to the state in time of war, and the fact they were only doing what many other engineers do during wartime: designing technologies of human destruction.

It is hard to imagine a case that could underscore more strongly than this one does the cardinal point that engineers must always be attentive

---

[346] EWT was also fortunate that he was not arrested and put on trial by the British. Consider the case of Dr. Bruno Tesch, the German chemist and industrialist who was the head and owner of the company, Tesch und Stabenow, that made Zyklon B gas and sold it to the Nazis. He was put on trial by the British in March 1946, charged with "supply[ing] poison gas used for the extermination of allied nationals interned in concentration camps well knowing that the said gas was to be so used," found guilty, and sentenced to death by hanging. See British Military Court (1946).

[347] http://www.topfundsoehne.de/cms-www/index.php?id=296&l=1.

[348] Fleming (1993).

[349] Ibid.

[350] http://www.topfundsoehne.de/cms-www/index.php?id=294&l=1.

[351] Fleming (1993).

[352] Ibid.

to not just the design features, performance, and likely profitability of the products and systems they design/produce/install/maintain but also to the political and cultural *contexts* within which those products will be deployed and used. Attending to that broader context may give engineers a fairly good idea whether their engineering product or system is likely to be used for ethically justifiable purposes or to cause unjustifiable harm to those targeted by its controllers. Decontextualizing one's engineering work, whether in design or production, can be extremely dangerous and ethically irresponsible, through negligence, and hence inimical to FERE1.

Regarding the important factor of context, this case recalls the Cadillac engine-control-chip case. Both involved apparent contextual myopia. In the Topf case, however, a much broader context should have been taken into consideration by the engineers involved. The Topf engineers should have paid careful attention not just to the local sociotechnical-environmental system context, as should have happened in the Cadillac case, but to *the macro political-economic-cultural-environmental context of the engineering work being carried out by Topf & Sons personnel.*

As happened in this case, a point can be reached when it no longer rings true for an engineer to claim ignorance of the fact that her or his work product was being used to cause massive unjustifiable harm to innocents. To the extent that an engineer realizes or should have realized the magnitude of the harm her or his work product was being used to inflict, to continue with that engineering activity, not to mention using one's engineering skills to *augment* the human destructive power of one's products, is profoundly antithetical to FERE1. The evidence suggests that the three Topf engineers discussed failed to fulfill their ethical responsibilities under FERE1 and FERE2. Moreover, they could not validly point to FERE4 to justify their work on contracts with the Nazi SS.

To conclude our discussion of this case, Volkhard Knigge's thoughtful ethical assessment of Topf & Sons' collaboration with the Nazi SS is worth pondering:

Topf & Sons' unquestioning cooperation with the SS is especially disturbing. For neither the company owners nor the participating employees correspond to the image of the fanatical National Socialist or the radical anti-Semite. They were neither mere "cogs in the wheel" nor "desk murderers," nor did they act under force or on command. They knew the precise purpose of the technology they were developing and they could have discontinued their business relations with the SS without suffering serious consequences. Yet their willingness to cooperate was apparently motivated merely by the fact that extermination and mass murder were the will of the state and supposedly served the interests of Germany—and by the related technical challenges, which spurred the ambitions of the engineers. The absence of a sense of humanity towards the "natural enemies" of the "people's community" sufficed for complicity in mass murder.[353]

Discussion Questions

1. Senior engineer Karl Schultze argued that because he (i) did not know what his ventilation systems were being used for, (ii) did what he did because he was directed to do so by company director Ludwig Topf, and (iii) was afraid of losing his position and of being arrested, he was not ethically responsible for what was done with the ventilation systems he installed. Is that a reasonable claim? Why or why not?
2. From an ethics point of view, is there anything problematic about Fritz Sander's memorandum to the Topf Brothers on September 14, 1942? Why or why not?
3. In an attempt to show he was not guilty of ethically irresponsible conduct, engineer Fritz Sander argued as follows: "I was a German engineer and key member of the Topf works and I saw it as my duty to apply my specialized knowledge in this way in order to help Germany win the war, just as an aircraft construction engineer builds airplanes in wartime which are also connected with the destruction of human beings." Is this a compelling analogical argument? Why or why not?
4. Co-managing director Ernst Wolfgang Topf argued, "You can't hold the builders of the ovens responsible for the deaths of the people who burned in them." Is that a compelling argument? Why or why not?
5. Asked why he continued to build cremation ovens even after he knew that "innocent human beings were being liquidated in Auschwitz gas chambers and that these corpses were subsequently incinerated in the crematoriums," senior engineer Karl Prüfer argued, "I had my contract with the Topf firm and I was aware of the fact that my work was of great importance for the national socialist

[353]Knigge (2005), p. 7.

state. I knew that if I refused to continue with this work, I would be liquidated by the Gestapo." Does that argument show that he was not ethically responsible for what he did in constructing additional cremation furnaces? Why or why not?

6. How is the conduct of the top Topf engineers discussed related to the claim that it is potentially dangerous to isolate or decontextualize engineering work from the social setting in which it is commissioned and in which its fruits will be used? Explain.

CASE 13: TRW AND THE U.S. BALLISTIC MISSILE DEFENSE SYSTEM

Background

In March 1983, the U.S. government announced an R&D initiative aimed at enabling the construction of a national missile defense system based on a network of satellites equipped with multiple X-ray lasers. The lasers were to be powered by onboard nuclear explosions. The resultant high-energy X-ray beams would supposedly destroy ballistic missiles that might be launched by the former Soviet Union. The U.S. government spent tens of billions of dollars on this initiative over the next decade. However, in 1993, the Clinton administration changed course and elected to pursue a region-specific or "theater" ballistic missile defense system, one intended to defend against missiles launched by "rogue," "terrorist," and "renegade" states.[354] The Strategic Defense Initiative Organization (SDIO) was renamed the Ballistic Missile Defense Organization (BMDO). BMDO awarded contracts to Rockwell International,[355] TRW,[356] and other companies to develop parts of the planned defense system that met government specifications.

As envisioned in the mid-1990s, the national ballistic missile defense system had three main elements:

infrared early-warning satellites, ground-based radars to precisely track warheads and decoys from thousands of kilometers away, and multistage, rocket-powered homing interceptor missiles launched from underground silos. The most critical element of this defense is the roughly 1.5-meter-long

[354] Raikow (1993).
[355] Boeing acquired Rockwell's defense and aerospace divisions in 1996.
[356] Northrop Grumman acquired TRW in 2002.

"exoatmospheric kill vehicle" that the homing interceptor deploys after being launched to high speed by its rocket stages.[357]

Based in underground silos, and working in concert with the aforementioned satellites and radars, these "ground-based interceptors" were intended to "strike high-speed ballistic missile warheads in the mid-course of the exoatmospheric phase of their trajectories and destroy them by force of impact."[358] The ground-based interceptor consisted of an "exoatmospheric kill vehicle" (EKV), containing an infrared light sensor, a navigational system, a computer brain and memory, and a booster rocket that would carry the EKV into space. Once outside the atmosphere, at an appropriate moment the EKV would be released to seek out, intercept, and destroy an enemy missile's nuclear-warhead-bearing reentry vehicle by colliding with it. Reporting to Rockwell—later to Boeing—one of TRW's main roles was to develop technology for the EKV.

TRW's hope was that its EKV, including purpose-designed technology for distinguishing warheads from warhead decoys, would be chosen by the Department of Defense over those of its competitors. To that end, in September 1995, TRW hired Dr. Nira Schwartz, a native-born Israeli and naturalized U.S. citizen, with a PhD in engineering from Tel Aviv University. She held a number of U.S. patents, "including ones involving computerized image analysis and pattern recognition."[359] TRW believed that Schwartz would be able to help develop software that would enable an EKV to accurately identify a warhead amidst decoys, then hone in on and destroy it by physical collision. After TRW hired Schwartz as a Senior Staff Engineer,[360] her boss wrote to personnel officials, "She is almost uniquely qualified to strengthen the interplay" among TRW specialists pursuing image analysis and object recognition.[361]

One of Schwartz's jobs at TRW was to help assess various computerized algorithms designed for application to the data picked up by the kill vehicle's sensors. These programs were supposed to work by matching selected incoming sensor data—infrared light emitted from objects in space—with images of threatening warheads stored in the computer's memory. One such program was called "the Kalman Feature Extractor,"

---

[357] Postol (2002).

[358] U.S. District Court, Central District of California (2001), Schwartz v. TRW, Plaintiff's Fourth Amended Complaint, §23, p. 9. Henceforth, this document will be referred to as *Schwartz v. TRW* (2001).

[359] Broad (2000a), p. 19.

[360] *Schwartz v. TRW*, §14, p. 7.

[361] Broad (2000a).

another the "Baseline Algorithms." Regarding the former, from incoming data gathered by the EKV's sensors, the extractor was supposed to "find critical characteristics of scanned objects, teasing out familiar 'signatures' that could separate decoys from warheads."[362] Schwartz tested the Kalman extractor with "nearly 200 types of enemy decoys and warheads in computer simulations, using secret intelligence data."[363]

The minimum acceptable performance level specified by the government for successful kill-vehicle discrimination technology was roughly 85%.[364] However, Schwartz claimed, "actual test results showed that the probability of a correct identification by TRW's Kalman Filter program was significantly less than fifty percent."[365] In the case of the Baseline Algorithm, she claimed to have determined that the program failed to identify the warhead about 90% of the time.[366]

Schwartz claimed she told her boss and colleagues of her conclusions about the disappointing probabilistic success levels of the discrimination technologies under development.[367] " 'They said not to worry,' she recalled. Flight tests of mock objects would always be arranged so discrimination would be relatively easy."[368] Apparently believing that doing so was inappropriate in a test of such a critical system component, Schwartz allegedly "pressed her superiors to tell industrial partners and the military of the shortcomings, especially of the extractor, the purported star of the show."[369] Schwartz claimed that her boss, Robert D. Hughes, "instructed her not to tell anyone and specifically directed her not to inform or contact the Government" about the discrimination program's capability.[370] She further alleged that at a meeting on February 26, 1996, "TRW decided not to inform the Government about the fundamental flaws in the Baseline Algorithms and Kalman Filter or the system's failure to meet the standards established in the Technical Requirements Document incorporated into the TRW Contract."[371] Schwartz claimed that the next day, February 27, 1996, TRW "summarily suspended [her] from employment and prevented [her] from accessing the tests she and other

---

[362] Ibid.
[363] Ibid.
[364] *Schwartz v. TRW*, §44, p. 15.
[365] Ibid., §11a, p. 4.
[366] Broad (2000a).
[367] *Schwartz v. TRW*, §95, p. 26, and Broad (2000a).
[368] Broad (2000a).
[369] Ibid.
[370] *Schwartz v. TRW*, §99, p. 27.
[371] Ibid.

TRW employees had conducted on the Kalman Filter and Baseline Algorithms."[372] She further claimed that when she made clear her intention to disclose those shortcomings to the government, she was fired. Shortly afterward, Schwartz allegedly wrote a letter to TRW stating, "If you will not notify the U.S. government, then I will."[373]

Schwartz contacted federal investigators and, on April 29, 1996, filed suit against TRW and Boeing in Federal District Court in Los Angeles. In Plaintiff's Fourth Amended complaint, filed in 2001, Schwartz alleged that TRW had "knowingly and falsely" certified to the government discrimination technology that, in the case of the Kalman filter, TRW "tests showed that over 85% of the time the technology failed to perform."[374] She claimed that TRW had wrongfully terminated her employment for threatening to bring her conclusions about the reliability of the discriminators to the attention of the government. Schwartz sought to recover "substantial sums paid [to TRW and Boeing] by the Government over the period 1992–1999,"[375] money she believed had been wasted because of what she viewed as TRW's misconduct. She also sought to get reinstated in her job with back pay and interest, and to receive a portion of whatever money was recovered by the government, the percentage depending on whether the U.S. government joined her suit against TRW and Boeing.

Ethical Analysis

Among the important ethical issues involving Schwartz that arose in this case are two previously encountered (for example, in Case 2): *whistleblowing* and *conflict of interest*. However, since these issues played out quite differently in this case than in Case 2, they deserve reconsideration.

Given what Schwartz claimed to have observed at TRW, and the conclusions she reached from the computer simulations that she and others had carried out, what, ethically speaking, should she have done? Did Schwartz have an ethical responsibility to notify the government about her concerns and conclusions in the name of heeding FERE2 as applied to this case? Or did she have an ethical responsibility to do the best she could to heed FERE4 as applied to this case, that is, to work on the

---

[372] Ibid., §100, p. 27.
[373] Broad (2000a).
[374] *Schwartz v. TRW*, §131, p. 36.
[375] Ibid., §127, p. 35. One source reported that the amount Schwartz sought to recover for the government was "more than a half-billion dollars." Broad (2000a).

discriminator to the best of her ability while setting aside her concerns about the efficacy of the software-based extractor and about what TRW was allegedly telling the government concerning its performance, as she claimed her employer instructed her to do?

From her studies of the computer-simulated discrimination tests, and possibly from reading or hearing written and/or oral TRW communications with the government about the performance of its discrimination technologies,[376] Schwartz allegedly came to believe that TRW was deliberately overestimating to the government the ability of the Kalman extractor and the Baseline Algorithms to distinguish warheads from decoys.

One fundamental problem that factored into her assessments of the discriminators' respective levels of reliability was that the identifying signatures of warheads and decoys were not constants. Rather, they differed "wildly, depending on variables like spin, attitude, temperature, wobble, deployment angle and time of day and year."[377] Comparing such dynamic signatures with fixed images of threatening warheads in the EKV's computer memory was not a recipe for a high likelihood of success in picking out which were the actual warheads with which the kill vehicles were supposed to collide. Another fundamental problem that any discrimination effort faced was that in real battle, the extractor would likely be faced with countermeasures. For example, the signatures of decoys could easily be deliberately made to closely resemble those of warheads to confuse the discriminators.

Why did Schwartz want to disclose to the government the serious problems she believed she had uncovered with the discrimination technologies? It does not appear that concern about national security was her primary motivation, as it was for David Parnas in his efforts against SDI in 1985 (Case 2). Schwartz appears to have been more concerned about the large amount of money the U.S. government was paying TRW for what she viewed as ineffective work on discrimination technology and the prospect of more of the same happening in the future. "It's not a defense of the United States," she said, referring to what TRW was doing with its contract work. "It's a conspiracy to allow them to milk the government. They are creating for themselves a job for life."[378] If an engineer's government is effectively wasting a large amount of public money on a contract with the engineer's employer because of allegedly

[376] *Schwartz v. TRW*, §127 and §§129–32, pp. 35–37.
[377] Broad (2000a).
[378] Ibid.

fraudulent actions on the part of that employer, it is possible that under certain conditions the engineer could have an ethical responsibility to try to prevent that financial harm from continuing and perhaps to try to reverse it if possible. I shall return to this issue below.

When the discriminator technology was tested on a mock enemy missile, in June 1997, the Pentagon hailed the test a "success." However, Schwartz and other experts stated that the Baseline Algorithms had actually selected a decoy to destroy.[379] A retired TRW senior engineer, Roy Danchick, claimed in an affidavit filed in connection with Schwartz's suit, that he had firsthand knowledge of TRW's " 'impermissibly manipulating' a study of the antimissile technology and 'censoring the test data' so that it appeared more successful than it was."[380] He told Pentagon investigators that in a postflight analysis of the data sent back from the kill vehicle, TRW had "select[ed] a more favorable stream of data for the baseline algorithm to digest" and "then the program zeroed in on the warhead."[381] TRW "categorically denie[d] that it misrepresented or improperly manipulated any flight test data."[382]

TRW's defense against Schwartz's charge of underperformance of the discrimination technology and misrepresentation of its efficacy to the government had several aspects. Robert D. Hughes, Schwartz's former boss, argued that Schwartz was completely wrong, misread her own tests, and failed to understand TRW's process of product development.[383] He argued that the extractor was not a finished product; it was still under development. In effect, he argued that Schwartz was being premature in demanding that TRW tell the government what she believed to be the true rate of discrimination success in the computer simulations. In a March 1, 1996, letter to his own boss, Hughes wrote, "Stating that there is a 'defect' that we should immediately report to the customer makes no sense. . . . We continually improve and verify system performance." According to Hughes, the extractor appeared to "perform properly under all conditions that have been specified to us" by the government and failed only when flight environments were "highly improbable" and "appear[ed] not to be within our requirements" (as laid out in the government performance specifications).[384]

[379] Ibid.
[380] Ibid.
[381] Ibid.
[382] Ibid.
[383] Ibid.
[384] Ibid.

Consider the argument that since the performance of a system like the extractor would gradually improve over time, immediate notification of the customer about alleged underperformance would be premature. This argument may strike some as a convincing refutation of Schwartz's charge that the firm's alleged nondisclosure to the government of what she regarded as the discriminator technologies' fundamental defects was deceptive and dishonest. However, while that argument may be based on a plausible generalization, the critical issue is *what TRW wrote or implied in its status/progress reports, applications for renewed funding, or other written or oral communications to the government about the* current *capabilities of its discrimination technologies.*

If TRW, in its interactions with its government "customer," told or intentionally misled the government to believe that one or both of its discrimination technologies was *already* able to successfully pick out a warhead from accompanying decoys at least 85% of the time, and if Schwartz's estimates of TRW's current discrimination technology success levels as being below 50% were correct, then her charge about deception and inflated discriminator efficacy would seem to have been confirmed.

If, however, TRW told the government that, say, one or both of its discrimination technologies was currently "on track to reach" or "moving in the direction of reaching" the specified roughly 85% probability-of-success level in the future, then it is unclear whether such a claim would constitute misrepresentation or deception. Whether it did would depend on the strength of the evidence available to support that claim. If there was strong, uncontrived empirical evidence that supported that claim, then TRW would arguably not be guilty of misrepresentation or deception. If there was no such evidence, it would arguably be guilty of same. In that context, it is important to note that the plausible general proposition that the performance of complex systems under development typically improves over time *would not by itself count as evidence that the discriminator was on track to meet the roughly 85% specification.*

Not being privy to TRW's written reports or proposals to, or to its oral communications with, the government, the author is unable to say with confidence which of the two scenarios more closely represents what happened between the parties. What is known is that TRW eventually dropped the Kalman extractor, relied on the Baseline Algorithms, and, in 1998, lost the competition for the contract to build the kill vehicle to Raytheon.[385]

[385] Ibid.

1. *Whistleblowing.* Did Schwartz have an FERE2-based derivative ethical responsibility to inform the government about the conclusions she had reached about TRW's discrimination technology? To address this question, let us revisit the author's revised versions of DeGeorge's five necessary conditions (C1–C5) for public whistleblowing to be an engineer's presumptive ethical responsibility, as laid out in the Ford Pinto case discussion (Case 11).[386]

> C1*. The amount of public money involved in the TRW contract to develop discrimination technology for the EKV was easily enough to give rise to a serious (financial) harm if all or a significant part of it was wasted through contractor misconduct. Granted, deploying a discriminator that was not very good at distinguishing warheads from decoys under realistic conditions could have jeopardized the security of the United States. However, as noted, concern over this risk did not seem to be the primary factor underlying Schwartz's suit.
>
> C2*. Schwartz stated that she informed her boss and colleagues of her concerns about the performance of the discrimination technologies.[387] Given the earlier-quoted memo her boss wrote to his boss, it is virtually certain that he shared Schwartz's concerns with that individual. Hence, condition C2* was in all likelihood satisfied.
>
> C3*. Schwartz does not seem to have received any promise or indication of forthcoming constructive action from her superiors to address the claimed problems to which she had allegedly called attention. Thus C3* seems to have been satisfied.
>
> C4*. Schwartz may have initially believed she would have access to documentary evidence that would convince a reasonable, impartial observer, at least one technically literate about computer algorithm–based discrimination technology, that she was correct and that company claims about the reliability of its discriminator technology were false or misleading. However, as noted previously, Schwartz

---

[386] DeGeorge believed his five necessary conditions for an engineer to have an ethical obligation or responsibility to publicly blow the whistle applied to situations in which individual safety was apparently at risk because of some party's actions. However, since putting someone's safety at risk puts her or him at risk of incurring *harm*, I submit that my revisions of DeGeorge's five conditions also apply to situations in which significant harms other than risks to individual safety are involved, such as the public harm involved when a government wastes or loses a large amount of public money because of alleged corporate wrongdoing. The reader is invited to consider for her- or himself whether a different set of necessary conditions (for public whistleblowing to be an engineer's presumptive ethical responsibility) applies to disclosing covert prospective harms unrelated to safety.

[387] *Schwartz v. TRW*, §95, p. 26.

alleged that on February 27, 1996, TRW prevented her from accessing the data from her computer tests with the algorithms. But Schwartz may have felt that even without this key documentary evidence in hand, her special expertise in the optical recognition area and her insider knowledge about what had been going on with TRW's discrimination work would suffice to have her concerns taken seriously by fair-minded technically literate observers.

But was that belief, if she subscribed to it, borne out in practice? In fact, some notable nontechnical outside parties did take her views seriously, including a U.S. senator and a member of the U.S. House of Representatives.[388] However, all things considered, whether condition C4* was satisfied is debatable, for it is not clear how persuasive Schwartz would have been to impartial observers technically literate in her field when, because of allegedly being blocked from access to her test data before she was fired, she may have lacked critical documentary evidence to support her claims about the efficacy of TRW's discrimination technologies.

C5*. It may have seemed to Schwartz that her insider expert testimony would be strong enough that giving it to the government would suffice to prevent—or at least have a decent chance of preventing—serious harm. But it is not clear whether she considered the possibility that pertinent governmental units might, for their own reasons, not be interested in hearing and acting on what she had to say. Such a stance could make the chances of her being able to prevent the harm in question (or its repetition) by engaging in public whistleblowing less than decent or reasonable, in which case C5* would not have been satisfied.

If and to the extent that the Pentagon's ballistic missile defense unit did not wish to hear and vigorously act on Schwartz's claimed findings and allegations, such a posture could have stemmed from any of several factors. For example, that unit might have wished to avoid being perceived as incompetent and unreliable by those controlling its budget and programs, as might have happened if the unit was shown to have spent major public money on a contract with a firm of its choice that yielded little progress on warhead discrimination technology. Alternatively, the unit might have believed that not compromising U.S. national security interest by pursuing Schwartz's charges vigorously in a public legal setting

[388] Broad (2003a).

was far more important than trying to recover allegedly wasted contract money from her former employer and preventing the continuation of such alleged waste in the future.

Schwartz's action of disclosing her findings and views to the federal government may have been motivated by a desire to prevent what she saw as (financial) harm from being caused through a government contract with her employer for antiballistic missile discrimination technology she viewed as fundamentally flawed. However, under the particular circumstances that prevailed, it is not clear that all five conditions (C1*–C5*) for publicly blowing the whistle to be an engineer's *ethical responsibility* were unambiguously satisfied, especially the last two.

2. *Conflict of interest.* In June 1996, several months after Schwartz's firing and the filing of her lawsuit against TRW and Boeing, the Pentagon's Defense Criminal Investigative Service entered the case and interviewed all the principals. Its lead investigator was Samuel W. Reed Jr. After the first mock test with an enemy missile, and in light of what retired TRW senior engineer Ray Danchick had told Pentagon investigators about the posttest selection of data for the discriminator to analyze, Reed concluded that Schwartz's charges warranted "further review."[389] He noted that TRW was "the antimissile program's overall systems engineer, with responsibilities to 'analyze and validate results' of all flight tests, [including those of the EKV discriminator technology], and thus had a potential conflict of interest."[390]

Pursuant to Reed's conclusion about "further review" of Schwartz's charges being desirable, in 1997 and 1998 the Pentagon commissioned two studies of the credibility of Schwartz's charges about TRW. One was done by Nichols Research of Huntsville, Alabama, "a regular advisor to the antimissile program."[391] It found that while TRW's discrimination work was "no Nobel Prize winner," it did meet contract requirements. The other was done by a group of five scientists "drawn from a high-level antimissile advisory board with Pentagon ties dating back to the Reagan 'Star Wars' era."[392] That group expressed doubts about the viability of the Kalman extractor but found the Baseline Algorithms "basically sound." It also gave its approval to TRW's manipulating data from the first mock EKV interception test, saying that a computer program was being developed that, if perfected, would do the same in space automatically."[393]

[389] Broad (2000a).
[390] Ibid.
[391] Ibid.
[392] Ibid.
[393] Ibid.

Nichols Research and the members of the five-person committee had histories of doing consulting work for the Pentagon on the antimissile program.[394] Hence, it is reasonable to assume that neither of those groups, that is, the firm or the five scientists, had an interest in doing anything that would estrange it from those in charge of the national missile defense program, including those who controlled its R&D grant- and contract-awarding functions.

> A federal official unconnected to the Pentagon but familiar with the case, who spoke on condition of anonymity, faulted the military's two scientific studies of Dr. Schwartz's charges as biased. "They don't go biting the hand that feeds them," he said of the antimissile advisory groups.[395]

Whether the Pentagon's choice of groups of researchers to evaluate Schwartz's charges reflected a conflict of interest on its part is debatable. However, had the Pentagon picked reputable independent groups without prior relationships to it, that would probably have allayed suspicions of possible conflicts of interest. It also would probably have enhanced the perceived credibility of the solicited evaluations. Whether either group of technical professionals enlisted to evaluate the credibility of Schwartz's claims about TRW's discrimination technology had an actual conflict of interest that could have influenced the findings of its report to the Pentagon about the validity of Schwartz's claims, appears, at this point, to be indeterminable.

Lessons

David Parnas, the main protagonist in the SDI case (Case 2), was a tenured academic with considerable autonomy when he worked on the SDIO committee. In contrast, Schwartz was employed "at will" by a private, for-profit firm and completely lacked such autonomy. Whereas Parnas seems not to have been at risk of losing his job for his public efforts against SDI, Schwartz was allegedly quickly fired for threatening to divulge to the government her claimed findings about the efficacy of TRW's discrimination technologies and her beliefs about the veracity of her employer's communications with the pertinent Pentagon unit.

It can be admirable and laudable for an engineer to attempt to prevent harm, an unreasonable risk of harm, or the recurrence of harm. Doing

[394] Ibid.
[395] Ibid.

so would be in accord with FERE2. However, acting with laudable intent does not by itself ensure that all necessary conditions for publicly blowing the whistle to be an ethical responsibility of the engineer in question are satisfied. Invoking Richard DeGeorge's distinction, it might be *ethically permissible* for an engineer to try to prevent what he or she regards as harm to the public interest by engaging in public whistleblowing; yet, for reasons beyond the engineer's control, it might be that, given prevailing circumstances, trying to prevent harm by engaging in public whistleblowing may not rise to the level of being that engineer's FERE2-based derivative *ethical responsibility*.

It is interesting to note the outcome of Schwartz's suit against TRW and Boeing N. America. TRW subpoenaed 38 military documents, including ones listing the technical requirements for its work on the antimissile system. TRW argued that it needed those documents to defend itself.[396] The U.S. federal government, through the U.S. Departments of Defense and Justice, took the position that it could not afford to have those documents become public, so it asserted the so-called state secrets privilege.[397] On February 24, 2003, U.S. District Court Judge Ronald Lew dismissed the case, stating there was "a reasonable danger that compulsion of the evidence will expose matters which in the interest of national security should not be divulged. . . . The only way to eliminate the security risk posed by these lawsuits is to dismiss them."[398] This quashed Schwartz's lawsuit against TRW and Boeing. As a result, interested members of the public were denied an opportunity to assess the credibility of her claims about the efficacy of TRW's discriminator technologies, the circumstances surrounding her firing, and the accuracy of TRW's communications with the Pentagon about what it had accomplished in the area of warhead discrimination technology.

Discussion Questions

1. Given the author's revised criteria for public whistleblowing discussed in Chapter 11, did Schwartz have an ethical responsibility to publicly blow the whistle on TRW? Why or why not?

---

[396] Broad (2003b).
[397] Center for Constitutional Rights (2007). See also http://en.wikipedia.org/wiki/State_secrets_privilege.
[398] Broad (2003b).

2. Schwartz's boss argued it was wrong for her to publicly blow the whistle on TRW regarding the discriminators' alleged underperformance when the technologies in question were still works in progress that would improve over time. Is this argument sufficient to show it was premature of Schwartz to contact the government when she did and charge TRW with misrepresenting the performance of the discriminator technologies on which she and the company were working?

3. The Pentagon selected Nichols Research and a committee of five technical professionals to evaluate Schwartz's charges about TRW's claims regarding the efficacy of its discrimination technologies. Did the fact that both had previously done consulting work for the Pentagon on the national antimissile program create a real and/or apparent conflict of interest for both groups in their work evaluating Schwartz's charges? Why or why not?

4. Suppose one of the groups retained by the Pentagon to evaluate Schwartz's claims and charges had a real conflict of interest, for example, a financial interest in not displeasing Pentagon management, that was in tension with a professional interest in working to the best of its ability to serve the legitimate interests of the client which engaged its services to provide a high-caliber assessment of credibility of Schwartz's claims. Would the engineers working for that group have an ethical responsibility to recuse themselves from the assessment effort? Why or why not?

CASE 14: THE HYATT REGENCY KANSAS CITY HOTEL

Background

Hyatt hotels are known for their innovative architectural designs. In 1978, construction began on a new Hyatt hotel, the 40-story Hyatt Regency Kansas City, in Kansas City, Missouri.[399] The hotel opened officially on July 1, 1980.[400] One of its most distinctive features was an atrium that rose from the lobby to the ceiling above the fourth floor. The

---

[399] The hotel was renamed the Hyatt Regency Crown Center in 1987. It changed owners and was renamed the Sheraton Kansas City Hotel at Crown Center in December 2011. See "Sheraton Kansas City Hotel at Crown Center," http://en.wikipedia.org/wiki/Sheraton _Kansas_City_Hotel_at_Crown_Center.

[400] Ibid.

atrium was spanned by three steel, concrete, and glass walkways or "sky-walks," each 120 feet long and weighing about 64,000 pounds.[401] Each walkway was to be held up not by supportive columns but by three pairs of threaded "hanger rods," with attached nuts and washers, that were anchored in the ceiling and went through the tops and bottoms of the three box beams on which each walkway rested. The walkway on level 4 (L4) was directly over the walkway on level 2 (L2). The walkway on level 3 (L3) was parallel to those on L2 and L4 but off to the side, with no walkway above or below it.

On Friday evening, July 17, 1981, a band was playing for a weekly "tea dance" in the atrium lobby. An estimated 1500 people were at the dance and/or in the atrium lobby, with tens of observers watching the proceedings from the overhead walkways.[402] At 7:05 p.m., the L4 walk-way collapsed and fell onto the L2 walkway. Both crashed onto the lobby floor (Figure 7), hitting many people and pinning some under concrete and steel debris. The collapse caused the death of 114 people and injured at least 239 others.[403] Prior to the destruction of the World Trade Center's Twin Towers in New York City on September 11, 2001, the Hyatt Regency walkway collapse had been the most lethal structural engineering episode in U.S. history.[404] It remains the most lethal structural engineering *accident* in U.S. history.

The National Bureau of Standards conducted an extensive investigation into the cause of the collapse.[405] The original design of the walkways showed three steel box beams perpendicular to and under each concrete walkway. Three pairs of hanger rods were to go from the ceiling through the three transverse L4 box beams and continue down to and through the three L2 walkway box beams, while three pairs of hanger rods were to run from the ceiling down to and through the three box beams on which the L3 walkway rested. Investigation revealed that *each* of the six long continuous rods intended to hold up the L4 and L2 walkways had been replaced with a *pair* of partly threaded shorter rods. As shown in

[401] Haskins (1983) and Robbins (1985).

[402] "45 Killed at Hotel in Kansas City, Mo., as Walkways Fall," *New York Times*, July 18, 1981, http://www.nytimes.com/1981/07/18/us/45-killed-at-hotel-in-kansas-city-mo-as-walkways-fall.html.

[403] Robbins (1985) and *New York Times* (1983), "Jurors Award $15 Million to Victim of Hyatt Disaster," September 23, http://www.nytimes.com/1983/09/23/us/jurors-award-15-million-to-victim-of-hyatt-disaster.html.

[404] Ibid.

[405] R. D. Marshall et al. (1982). The National Bureau of Standards (NBS) is now called the National Institute of Standards and Technology (NIST).

Figure 8, one rod in each pair of shorter rods went from the atrium ceiling through the L4 walkway's box beam and into a nut (with washer) *underneath* that box beam to hold it up. The other rod in each pair went from the *top* of the same L4 box beam (held in place by a threaded nut with washer screwed onto the threaded rod protruding through the top of that box beam) to the *bottom* of the L2 box beam (held in place by a threaded nut with washer screwed onto the threaded rod that protruded through the bottom of the L2 box beam).

In the original design, six rod-and-box-beam connections were to hold up the L4 walkway from below, and six different rod-and-box-beam connections *on the same rods* were to hold up the L2 walkway from below. In contrast, in the changed, as-built design, the six box-beam connections on L4 (whose other ends were connected to the atrium ceiling) had to support *both* the L4 walkway above them *and* the L2 walkway that was hanging below them from six other L4 rod-and-box-beam connections. Thus, the connections on the L4 rods connected to the ceiling had about *twice as much stress* on them as in the original design. The locations of the

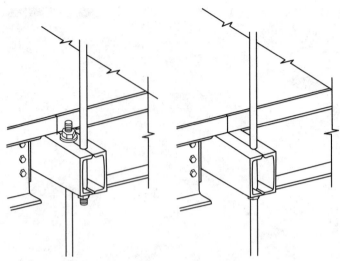

Fig. 8. Original single-rod (R) and revised double-rod (L) box-beam connections, pedestrian skywalks, Hyatt Regency Kansas City hotel.

bolts (and washers), right at the points where the box beams were welded together from two C-channels, made those vulnerable stress points. The L4 box beams simply could not support the L4 walkway, the L2 walkway, and the weight of the people standing on those walkways while watching the dancers on the lobby floor below. The nuts were pulled right through the places on the box beams where the C-channels were welded.

The NBS/NIST investigating team reached the following conclusion:

> Two factors contributed to the collapse: inadequacy of the original design for the box beam-hanger rod connection, which was identical for all three walkways, and a change in hanger rod arrangement during construction that essentially doubled the load on the box beam-hanger rod connections at the fourth floor walkway. . . . Based on measured weights of damaged walkway spans and on a videotape showing occupancy of the second floor walkway just before the collapse, it is concluded that the maximum load on a fourth floor box beam-hanger rod connection at the time of collapse was only 31 percent of the ultimate capacity expected of a connection designed under the Kansas City Building Code. It is also concluded that had the original hanger rod arrangement not been changed, the ultimate capacity would have been approximately 60 percent of that expected under the Kansas City Building Code. With this change in hanger rod arrangement, the ultimate capacity of the walkways was so significantly reduced

that, from the day of construction, they had only minimal capacity to resist their own weight and had virtually no capacity to resist additional loads imposed by people.[406]

Ethical Analysis

As with many other structural engineering cases, for example, the Citicorp Center case, the overarching general ethical issue in this case is that of safety. A more specific question is this: in light of the FEREs, did any engineers or engineering-related groups involved directly or indirectly in the Hyatt Regency Kansas City hotel project do anything arguably at odds with the FEREs and that contributed to causing the tragic outcome?

To address this question, I shall focus on three aspects of the case: (1) the prechange structural engineering drawings; (2) the change from single-rod to double-rod box-beam connections; and (3) the role of the building industry and related professional engineering societies in this episode.

1. *The prechange structural engineering drawings.* Consider the original single-rod box-beam connections that were to link the ceiling with the L2 and L4 walkways. The original drawings of these connections sent to the fabricators, Havens Steel Company, were not detailed. They were only conceptual in nature.[407] However, even if the project's architectural firm did not specify in detail the nature of the single-rod box-beam connections, it would presumably be the FERE1-based derivative ethical responsibility of the structural engineer of record for the Hyatt project, Jack D. Gillum of G.C.E. International, to ensure two things: that all connections in the structure had been checked for strength and structural integrity, and that the drawings of the connections sent to Havens Steel were sufficiently precise and adequately detailed to make it unambiguously clear exactly what the fabricator was to do. I shall return to this responsibility below.

2. *The change from single-rod to double-rod box-beam connections.* The structural drawings of the original box-beam connections sent to Havens Steel showed single continuous hanger rods. However, this design

---

[406] Ibid., p. iii.

[407] "The connection that failed was never designed. The original detail of the connection . . . , as shown on the contract structural drawings, was only conceptual in nature. Neither the original detail nor the double rod connection, as built . . . , was designed." "The most obvious of all the items that should have occurred are the following: Designed detail of the connection should have been shown on the engineer's drawing." Gillum (2000), p. 69.

elicited concern. According to one author, the single continuous hanger rod design "seemed to be highly impractical" to "the construction contractor" (Eldridge Construction Company).[408] But for another, it was "the fabricators" (Havens) who "found that design to be impracticable."[409] Regardless of who first voiced that concern, by early 1979 Havens had come to believe that the original single continuous hanger rod/box-beam connection design was impracticable.

A strong difference of opinion emerged about what happened next. "The fabricator, in sworn testimony for the administrative judicial hearings after the accident, claimed that his company (Havens) telephoned the engineering firm (G.C.E.) for change approval [from single-rod to double-rod connections] and received it over the phone. G.C.E. denied ever receiving such a call from Havens."[410]

Nevertheless,

[the fabricator] made the change, and the contractor's Shop Drawing 30 and Erection Drawing E-3 were changed. . . . On February 16, 1979, G.C.E. received 42 shop drawings (including the revised Shop Drawing 30 and Erection Drawing E-3). On February 26, 1979, G.C.E. returned the drawings to Havens, stamped with Gillum's engineering review seal, authorizing construction. The fabricator (Havens) built the walkways in compliance with the directions contained in the structural drawings, as interpreted by the shop drawings, with regard to the hangers. In addition, Havens followed the American Institute of Steel Construction (AISC) guidelines and standards for the actual design of steel-to-steel connections by steel fabricators.[411]

While G.C.E. denied having received a telephone call from Havens requesting a change from a one-rod to a two-rod hanger-and-beam connection, the drawings that Havens received back from G.C.E. in late February 1979 had the engineer of record's review stamp on them. This may have led Havens to believe that Gillum had reviewed the proposed new connections and found them structurally sound. Yet, to the best of the author's knowledge, no evidence was ever presented during the 1984 administrative trial of Gillum and Daniel Duncan[412] that Gillum or any

---

[408] Moncarz and Taylor (2000), pp. 47–48.

[409] American Society of Civil Engineers (2007).

[410] Texas A&M University (1993a). See also http://ethics.tamu.edu/Portals/3/Case%20Studies/HyattRegency.pdf.

[411] Ibid.

[412] Daniel M. Duncan was a structural engineer employed by G.C.E. International. Gillum had designated him as project manager for the Hyatt Regency hotel building project.

other G.C.E. structural engineer(s) had carried out a detailed technical analysis to demonstrate the adequacy and safety of the proposed new double-rod connections.

If in fact no careful technical analysis had been done by G.C.E. engineers of the proposed double-rod connection, then whoever placed the engineer of record's review stamp of approval on the shop drawings with the revised L4 connector designs returned to Havens, or allowed that stamp to be placed on those drawings, arguably contributed to creating a significant risk of harm—one that later materialized. Affixing the engineer of record's review stamp of approval to the shop drawings (or allowing it to be affixed by another party), and G.C.E. engineers allegedly failing to carry out a rigorous detailed technical analysis of the proposed double-rod connection heightened risk, both seem to the author to be actions at odds with FERE1. The apparent failure to carry out a careful detailed technical analysis engendered an unreasonable risk of harm, while affixing the engineering review stamp effectively sustained that risk by obscuring its presence.

In 2000, reflecting on the Hyatt disaster, Gillum published an essay about the role and responsibility of the engineer of record.[413] In a section titled "Factors Contributing to Collapse," Gillum listed 17 factors he believed contributed to the walkway collapse.[414] Two merit special notice here: (1) "The engineer relied on the fabricator to design the connection as well as all other steel-to-steel connections in the atrium. (This was the industry custom and practice at the time.)"; and (2) "The engineer, while reviewing the shop drawings, checked the piece size but not the connection, as it was not drawn or shown on the shop drawing, i.e., *there was no detail on the shop drawings of the assembled connection to be checked* (emphasis in original)."[415]

These factors prompt several comments. First, the expression "the engineer," used in both factors, is vague. It is not clear whether the individual being referred to is the structural engineer of record (Gillum) or the structural engineer of record's designated project manager for the Hyatt building (Duncan).[416] Second, citing the first of these two alleged contributory factors could be viewed as an attempt by the engineer of record to deny he behaved irresponsibly, since letting fabricators design steel-to-steel connections was "industry custom and practice at the time."

---

[413]Gillum (2000), pp. 67–70.

[414]Ibid., pp. 68–69.

[415]Ibid., p. 69.

[416]Presumably the former is what was meant. But it is possible that the reference was deliberately left vague.

Third, citing the second factor could be viewed as an attempt by the engineer of record to deny he had acted irresponsibly in not checking the technical details of the new connection on the revised shop drawings. He did not do so because there was no detailed new connection depicted on the shop drawings that could be checked.

Under FERE1, the structural engineer of record arguably has a derivative ethical responsibility to conscientiously check any shop drawings sent back to her or him by the fabricators that incorporate nontrivial changes from what the structural engineers first sent to the fabricators. Further, if there was no detailed shop drawing of the revised connection to be checked, then, in my view, the structural engineers, who are the presumptive experts on the structural integrity of connections, should have *requested* that they be given a detailed shop drawing for them to check. Moreover, the structural engineer of record's engineering review stamp should not have been put on the revised shop drawings unless and until the engineer of record had checked the adequacy and integrity of the revised connection. Finally, the fact that the connection was now a two-rod connection, rather than the one-rod connection originally sent to the fabricators, arguably should have triggered a precautionary impulse in the structural engineer of record and impelled him to scrutinize the structural adequacy of the newly configured connection.

The organizational structure of Gillum's firm during the design of the Hyatt Regency is of interest in this connection. According to an article on the ASCE website, "roughly" 10 associate engineers reported to Gillum. Each supervised 6–7 projects.[417] Since Gillum could not personally oversee every aspect of the structural design work on 60–70 projects, he entrusted oversight responsibility for a given project to whichever associate engineer he had designated as its project engineer. In the case of the Hyatt Regency Kansas City hotel, Gillum assigned oversight responsibility to "an associate structural engineer" whom he designated as its project engineer: Duncan.

According to the same article, "the fabricators" allegedly "requested approval of the double-rod system by telephone." "The structural engineer"—Gillum or Duncan?—"verbally approved the change, with the understanding that a written request for the change would be submitted for formal approval. This follow-up request was never honored."[418] Hence, formal approval of the change was not given. One reason why

---

[417] American Society of Civil Engineers (2007).
[418] Ibid.

Havens did not submit a written request for formal approval could be an unexpected development at the company:

> [T]he fabricators had just begun work on the shop drawings when a sudden increase in workload required them to subcontract the [Hyatt] work to an outside detailer. That detailer, in turn, mistakenly assumed that the double-rod connection on the shop drawings had already been designed and therefore performed no calculations on the connection himself.[419]

This account does not indicate who allegedly gave verbal approval to the request the fabricators allegedly made by telephone. Nor does it note that giving informal approval without doing careful detailed technical analysis opens the door to increased risk. Moreover, this account seems to place the blame for what eventually happened on the fact that the supposedly promised *written* request for formal approval was never submitted. It does not consider the view that "the structural engineer," who allegedly informally approved the proposed change without technically analyzing it in detail, should have *followed up* when the fabricator did not submit the allegedly promised written request for formal change approval.

According to the same article, the shop drawings, completed by the "outside detailer" to whom the fabricators had outsourced their design work, "were returned to the engineer of record with a request for expedited approval." The ASCE essay states that Gillum assigned "a technician" on his staff to do the review.[420] However,

> the connections were not detailed on the drawings and the technician did not perform calculations on the connections. The structural engineer performed "spot checks" on portions of the shop drawings, and the engineer of record affixed his seal to the documents. The latter stated that he had not personally checked all calculations and had relied on the work of his project engineer and design team.[421]

If this account accurately reflects what occurred regarding the calculations made on the revised connections, what might account for it? In his 2000 essay, Gillum wrote the following:

> Proper engineering takes time to provide proper oversight, and overloaded engineers, draftsman [sic], and project managers are susceptible and prone

[419] Ibid.
[420] Ibid.
[421] Ibid.

to making mistakes. Dan Duncan had many projects to manage, and much of my time was spent traveling and out of the office.[422]

However, if designated Hyatt Regency Kansas City project engineer Duncan was "overloaded" and had too many projects to permit him to focus on each one in appropriate detail, then, under FERE2, it would arguably have been his derivative ethical responsibility to let his supervisor, Gillum, know that, so that either more project engineers could have been hired, or the supervisor/owner could have accepted less new work, or Duncan's work on his assigned projects could have been prioritized or reprioritized on the basis of criticality.

Moreover, also under FERE2, the engineer-owner of G.C.E., who was also the engineer of record, arguably had a derivative ethical responsibility to take whatever steps would enable each project engineer to give adequate attention to each project for which he or she was responsible. Not taking those steps would be imprudent, for it could easily engender an increased risk of harm from any project given inadequate attention.

As indicated in the preceding quote from his essay, Gillum traveled and was away from his office a lot during the erection of the Hyatt. While this could help *explain* why he didn't pay closer and more detailed attention to the Hyatt project than he did, it does not, by itself, *justify* his not having done so. Gillum presumably regarded his absences as ethically acceptable because he had assigned responsibility for the Hyatt project to engineer Duncan. However, when dealing with a safety-related element of a structure that if it fails, could cause death or serious injury, a structural engineering firm's principal arguably has two additional derivative ethical responsibilities. First, she or he should accept new work for the firm only if it can be done in a careful, fastidious manner without compromising the quality of work already underway on other projects. Second, she or he also has an ethical responsibility not to *adapt to* an increasing load of recently accepted new work but rather to take whatever steps are needed to assure a balance is maintained—or restored—between the firm's workload and the number of technically qualified engineering employees available to do all work assigned to them with competence, care, and due attention to detail.

Interestingly, on the question of who or what was to blame for the accident, the authors of another article on the case did not focus on the deeds or nondeeds of Gillum and Duncan, or on those of the fabricators and their subcontractors. Instead, they focused on "a flaw in the design

---

[422] Gillum (2000), p. 70.

process control" "within the A&E [architecture and engineering] group as well as the construction industry."[423]

> The collapse of the Hyatt Regency walkways was the result of a flaw in the design process control. However, the post-failure investigation efforts concentrated on design procedures and not on the process. It can well be argued that no clear industry-wide accepted definition of responsibilities existed at the time the hotel was designed and built. This argument might yet be applied to the next catastrophic failure. However, it is to be hoped that as with the other major engineering failures mentioned in the introduction of this paper, the Hyatt collapse has also contributed to improvement in public safety and in the quality of built facilities by promoting a more advanced set of rules governing the design/construction process.[424]

This account seems to absolve the individual structural engineers involved of any responsibility for what happened. The authors blame a "flaw" in the design process, something that was not the fault of any individual. However, consider two general scenarios: (i) an engineer of record puts his or her review approval stamp on shop drawings of a proposed safety-related connection change submitted by someone else without first personally subjecting the proposed change to technical scrutiny; and (ii) an engineer of record does not prevent another engineer in his or her firm from putting the former's review stamp of approval on shop drawings without the engineer of record's having scrutinized the other engineer's work. The author finds it difficult to reconcile either scenario with FERE1, and perhaps also with FERE4.

3. *The role of the building industry–related professional associations in this episode.* The professional societies of architects, structural engineers, and construction engineers may have played a background role in this episode. Allowing—by not clearly and forcefully prohibiting and penalizing—the making and approving by phone of requests for changes in safety-related structural components to become an acceptable professional practice in the building field seems difficult to reconcile with FERE1. Perhaps the economic savings from employing a "fast-track" construction method, in which construction of a building starts before its design is finalized, had made building professionals tolerant of the timesaving but risky practice of approving structural changes by phone.

[423] Moncarz and Taylor (2000), p. 46.
[424] Ibid., p. 49.

The societies of building professionals presumably had (and still have) the ability to delegitimize, penalize, or even prohibit such a practice. However, to the best of the author's knowledge, this had not been done. Allowing that practice to be culturally acceptable arguably helped set the stage for what occurred in this case: a less-than-transparent and fastidious structural-change approval process that was not conducive to full accountability and led to strong disagreement over whether a particular telephone conversation took place, one in which a safety-related structural change was allegedly requested and informally approved, conditional upon subsequent submission of a formal printed request for that change to be made. Time and financial pressures for moving projects ahead notwithstanding, resolving such change requests should be acceptable only when done on paper; where there is no doubt about what is being requested, by whom; what is being approved, by whom; when; on what grounds; and subject to what conditions. Telephone communication is apt to be rushed and nonreflective, hence conducive to misunderstanding and the serious consequences that can flow therefrom. Building-industry professional societies that purport to care about ethically responsible engineering practice should explicitly oppose, stigmatize, and penalize—if not prohibit outright—any practice that contributes to making the structural-change approval process less transparent, less fastidious, and less conducive to full accountability.

Lessons

What lessons can be learned from the Hyatt Regency disaster? First, under FERE1, a structural engineer of record has a derivative ethical responsibility to analyze and confirm the technical viability of all safety-related features of the structure on which she or he is working. She or he also has derivative ethical responsibilities to provide detailed structural drawings for all structural connections to be sent to fabricators, and to scrutinize any changes proposed in safety-related features of the structures for which she or he is serving as structural engineer of record, including those proposed by others.

She or he also has derivative ethical responsibilities to not put her or his engineering review stamp of approval on any structural or shop drawings, *and to prevent it from being placed thereupon by someone else*, until she or he has carefully checked all safety-related features of the structure in question and found them satisfactory. These responsibilities are even weightier when the revised drawings involve a safety-related element significantly different from the originally approved one.

Under FERE1, a structural engineer who is the head of a structural engineering firm has an ongoing derivative ethical responsibility to ensure that the relationship between the number of (competent) structural engineer employees he or she has engaged to work for the firm, and the amount of work that the firm has taken on, is such that all projects can be handled with due care, attentiveness to detail, and technical competence by the firm's current employees. If that relationship would become a mismatch if new work is taken on, then the structural engineer–owner has an FERE1-based derivative ethical responsibility to not take on new work until a balance is restored between work load and number of qualified engineering employees, either by hiring additional competent engineering employees and/or by establishing the order in which projects will be professionally addressed.

The structural engineer who is the principal of his or her firm has a derivative ethical responsibility to monitor and supervise subordinate employees in a substantive way. A head structural engineer who turns over responsibility for a particular project to a subordinate structural engineer must then substantively monitor, supervise, and check his or her work. In the author's view, it is an ethically irresponsible engineering practice for a head structural engineer to allow his or her review stamp to be used by a subordinate without first carefully checking on the accuracy and reliability of the engineering work the subordinate has done on the project in question.

The Citicorp Center case and the Hyatt Regency Kansas City case differ in at least one respect. In the Citicorp case, the top structural engineer apparently did not learn until *after* the building was finished that, at the behest of a potential building erector, his subordinates had approved changing the butt-welded connections he had originally specified to bolted connections. In contrast, in the Hyatt case, the owner/engineer of record knew *before* the structure was erected that a proposal had been made by the fabricators to change from one-rod to two-rod box-beam connections on the L4 walkway.

That difference notwithstanding, both of the following general scenarios seem to me to be at odds with the clause of FERE1 concerning creation of unreasonable risk of harm: (a) the owner/head structural engineer of a structural engineering firm finds out only after a building she or he had worked on as engineer of record had been erected that a safety-related structural change had been approved by her or his engineering subordinates without first being run by her or him for technical scrutiny and approval; and (b) the owner/head structural engineer of a structural engineering firm knows before a building he or she was working on as engineer of record is erected that a safety-related structural change has been proposed by the fabricator, incorporated into revised shop drawings, and sent to him or her for approval, but, for whatever reason, the engineer of record does not personally carry out a careful detailed technical analysis of the proposed change (or carefully check the detailed technical analysis he or she ordered be done on the proposed change by a qualified subordinate) before the revised shop drawings are stamped with his or her engineering review stamp of approval and returned to the fabricator.

In short, the engineer-owner of a firm who is the structural engineer of record for a building project can be remiss either by (a) allowing conditions to exist in her or his firm such that a safety-related structural change proposed by others might not be brought to her or his attention for scrutiny and approval before erection, or (b) not carrying out a thorough technical analysis, or not carefully reviewing a subordinate's thorough technical analysis, of a safety-related structural change proposed by others that is brought to her or his attention before erection, but nevertheless affixing her or his review stamp of approval (or not preventing it from being affixed by another) to shop drawings sent back to the fabricator for implementation.

Discussion Questions

1. Jack Gillum wrote: "Proper engineering takes time to provide proper oversight, and overloaded engineers, draftsman [sic], and project managers are susceptible and prone to making mistakes. Dan Duncan had many projects to manage, and much of my time was spent traveling and out of the office." If true, does the last sentence absolve Gillum of ethical responsibility for the collapse of the pedestrian walkways? Why or why not?

2. Did Gillum have an ethical responsibility to see to it that, contrary to fact, his engineering review seal was *not* placed on the shop drawings that his firm returned to Havens Steel on February 26, 1979, until he had done or validated the quantitative analysis needed to determine whether the strength and structural integrity of the new two-rod connection was adequate to support the L4 and L2 walkways?

3. Is it ethically acceptable for a structural engineer–owner of a firm to take on new work without knowing that the distribution of the current workload is such that each of the firm's structural engineers will have enough time and resources to work on each project for which he or she is responsible with due care and attentiveness, even after the new work is taken on?

CASE 15: THE MANHATTAN WESTWAY PROJECT

Background

"Westway" was the name given in 1974 to a controversial, ultimately abandoned New York City roadway-and-urban-development project.[425] The roadway was to be 4.2 miles long, six lanes wide, and to cost $2+ billion.[426] Its planned location was the southwest side of Manhattan, partly tunneled through and partly resting atop 169 acres of landfill to be placed in the Hudson River.[427] The project's urban-development component was intended to reclaim the waterfront of southwestern Manhattan, marked by decaying piers remaining from past maritime commerce. Project promoters envisioned public access to the new waterfront, a large park, and numbers of commercial and residential buildings on the new land. As originally conceived, Westway was an enormous project, extending Manhattan an average of 600 feet into the Hudson River along the project's length. It was slated to use 1.5 million cubic yards of concrete, require the dredging of more than 3 million cubic yards of "fluff" from the river bottom, and deposit into the river more than 8 million cubic yards of landfill, "enough to raise the level of all 840 acres of Central Park by six feet."[428]

HISTORICAL BACKGROUND

In December 1971, then New York City Mayor John Lindsay and New York State Governor Nelson Rockefeller signed an agreement establishing the "Westside Highway Project" as part of the New York State Department of Transportation (NYSDOT). A former Federal Highway Administration (FHWA) official, Lloyd Bridwell, principal of SYDEC,

---

[425] For an overview of the history of the Westway project, see Roberts (1984a,1984b).
[426] Roberts (1983, 1984c).
[427] Roberts (1984a), p. 17.
[428] Roberts (1984b), p. 13.

a private consulting firm, was appointed executive director of the project. His company was responsible for, among other things, obtaining all environmental permits for the project.[429] Thus, a powerful state agency (NYSDOT) became the main organizational engine behind Westway, but it outsourced to private consultants the technical work needed to get the required environmental permits.

In 1973, after further interaction between state and city officials, the urban-development component of the project was scaled down, to 243 acres and 20,000 apartments. That same year, the existing Westside Elevated Highway collapsed at 14th Street, ironically under the weight of a truck hauling asphalt to repair potholes on its deck.[430] As a result, the project picked up new momentum. However, in what was to prove a significant boost for project opponents, the U.S. Congress passed legislation enabling roads initially designated as future parts of the federally funded Interstate Highway System, as Westway was, to be "traded in" for federal money to be used to fund mass transit and alternative transport projects. However, this option was available only for qualifying roads withdrawn from the Interstate Highway System by September 1, 1985.[431]

In 1974, the project administrator issued a draft original environmental impact statement (DOEIS) for the first scaled-down project. In response to extensive public hearings about and mixed reactions to the first scaled-down version from about 400 special interest groups, the project was further downsized—to 181 acres and 7,100 apartments. The second scaled-down version was dubbed "Westway," in hopes of distinguishing it from the earlier, more vociferously opposed versions of the project.[432]

In January 1977, the final original environmental impact statement (FOEIS) was issued, signed by NYSDOT and the FHWA. In April 1977, per The River and Harbors Act of 1899, NYSDOT applied to the Army Corps of Engineers (ACE) for a landfill permit. However, the federal Environmental Protection Agency (EPA) questioned the impact of Westway on Hudson River fisheries, as described in the FOEIS, and requested that NYSDOT do a study of this issue. In response, NYSDOT reluctantly engaged Lawler, Matusky, and Skelly Engineers (LMS), a New York State consulting firm specializing in environmental engineering

[429] Schanberg (1984).
[430] Perlmutter (1973). See also "West Side Highway," http://en.wikipedia.org/wiki/West_Side_Highway.
[431] Roberts (1984a), p. 19.
[432] Ibid., pp. 18–19.

and science, to do a fishery study. The study lasted from April 1979 to April 1980.[433]

In March 1981, with the LMS study results in hand, ACE issued a landfill permit. Two months later, the Sierra Club and several other organizations brought suit against ACE for issuing that permit. After a two-month trial, in March 1982, U.S. District Court Judge Thomas P. Griesa found that ACE's landfill permit process was deficient in its treatment of the projected impact of Westway on fishery resources. He issued a temporary injunction against proceeding with Westway landfill work until compliance with the National Environmental Policy Act (NEPA) was demonstrated via preparation of a supplementary environmental impact statement. The judge stipulated that this was to be done on the basis of further evaluation of existing data and possible additional fishery study, to be determined after consultation with EPA, the Fish and Wildlife Service (FWS), and the National Marine Fishery Service (NMFS).[434]

In August 1982, Malcolm Pirnie Inc. (MPI), an environmental consulting firm retained by ACE, recommended that additional fish sampling be done to determine the "relative abundance" of striped bass in the projected Westway landfill area compared with the river channel area. The firm also recommended conducting a "habitat survey" to assess the quality of the Westway area in relation to other relevant river areas, that is, to determine the likely impact on the habitat of the striped bass of the loss to landfill of the areas between the decaying piers on Manhattan's lower western side.[435]

In October 1982, MPI convened a workshop of experts to determine the parameters of the fishery study. It recommended that ACE do a 17-month, two-winter fishery study. However, in December 1982, ACE Col. Walter Smith decided that ACE would *not* perform further studies. In March 1983, Col. Fletcher Griffis replaced Smith as New York District Engineer of ACE responsible for Westway and reviewed Smith's decision. In July 1983, after convening a new workshop of experts, Griffis decided that ACE should do a 17-month, two-winter fishery study starting in December 1983, plus a striped bass habitat study. However, in December

[433] Sierra Club v. U.S. Army Corps of Engineers, 536 F. Supp. at 1242 (S.D.N.Y. 1982).

[434] Sierra Club v. U.S. Army Corps of Engineers, 541 F. Supp. 1367 at 1367–83 (S.D.N.Y. 1982). See also Sierra Club v. U.S. Army Corps of Engineers, 614 F. Supp. 1475 at 1485 (S.D.N.Y. 1985), https://www.courtlistener.com/opinion/2142198/sierra-club-v-us-army -corps-of-engineers/.

[435] Ibid.

1983, the chief of ACE overruled Griffis and decided that ACE would do only a four-month, one-winter fishery study.[436]

That study was carried out from January through April of 1984. In May 1984, a draft supplementary environmental impact study (DSEIS) was issued by ACE and FHWA. In late June 1984, ACE held public hearings on the DSEIS. After going through six drafts, the last of which differed substantially from the first five, a final supplementary environmental impact statement (FSEIS) was issued by ACE and FHWA. In January 1985, Col. Griffis issued his decision that the Westway landfill permit should be granted.[437] The next month, ACE issued a landfill permit to NYSDOT[438], which then moved to vacate Judge Griesa's 1982 injunction. Not surprisingly, the Sierra Club and other organizations filed suit to oppose that motion.

In August 1985, Judge Griesa again found ACE's FSEIS deficient, for several reasons. He did not believe that the evidence presented supported ACE's key stated conclusion: that Westway posed only a minor or insignificant risk to the striped bass fishery. Nor did he find in the FSEIS any reasons that could explain why the risk Westway allegedly posed to the striped bass fishery had changed from "significant" to "insignificant" between the fifth and sixth drafts of that document. He also found ACE's four-month one-winter fish study inadequate. Having concluded that the FSEIS did not meet NEPA EIS requirements, Judge Griesa permanently enjoined the awarding of a landfill permit for Westway, thereby effectively killing the project.[439]

Griesa's decision was upheld on appeal to the U.S. Court of Appeals for the Second Circuit in September 1985.[440] In late 1985, New York State received $1.725 billion to use on other highway and mass transit projects in lieu of the money previously designated for the now-abandoned Westway Project.[441] In 1989, the West Side Highway was demolished between 43rd and 57th Streets, and in 1996 construction began on the more modest "Route 9A," a west-side, ground-level boulevard with traffic lights, between the northern end of Battery Park, at the southern tip of

[436] Sierra Club v. U.S. Army Corps of Engineers, 614 F. Supp. 1475 at 1488 (S.D.N.Y. 1985).

[437] Roberts (1985).

[438] "Permit Is Granted for Westway; Opponents Appeal," *New York Times*, February 26, 1985.

[439] Lubasch (1985a).

[440] Lubasch (1985b).

[441] Oreskes (1985).

Manhattan, and 59th Street.[442] This roadway was basically completed in 2001, but the segment in lower Manhattan required additional work because of damage caused by the destruction of the World Trade Center's Twin Towers on September 11, 2001.[443]

Ethical Analysis

Several interesting ethical issues faced engineers involved in various ways with the Westway Project. Let us explore the situations and issues that faced three of them.

1. *Brian Ketcham.* Some engineers who faced Westway-related ethical issues were not directly involved in the project. Brian T. Ketcham earned a BS in mechanical engineering from Case Institute of Technology in 1962 and completed all coursework for a master's degree in mechanical engineering and urban and transportation planning at MIT.[444] A transportation engineer, he specialized in traffic engineering, air quality, and pollution control. In 1969, Ketcham became a Licensed Professional Engineer in New York State. For three decades he was coprincipal and executive vice president of Konheim & Ketcham,[445] a New York City environmental consulting firm founded in 1981.[446]

Brian Ketcham worked for many years to protect the public interest as he saw it in various engineering contexts. In 1974, when the DOEIS for the Westway project was presented, Ketcham and attorney William Hoppen initiated legal action to block proceeding with the Westway project. They did so on the grounds that allowing the project to proceed would have effects that violated the State Implementation Plan, required of New York by the U.S. Clean Air Act, in particular, the part of the plan that pertained to mobile air pollution sources in New York City.[447] In 1977, when the FOEIS for Westway was published, Ketcham was vice president and

---

[442] Levine (1989) and Pierre-Pierre (1996).

[443] See http://www.renewnyc.com/displaynews.aspx?newsid=d12b0142-f47b-4784–81ca -2af0683b5ca1.

[444] Curriculum vitae, http://digitalcollections.library.cmu.edu/awweb/awarchive?type= file&item=54097 and http://hudsonhighlandsenviro.com/personnel/brian-t-ketcham-p-e.

[445] http://www.konheimketcham.com/.

[446] http://digitalcollections.library.cmu.edu/awweb/awarchive?type=file&item=54097 and http://www.konheimketcham.com/KKWeb/btk_resume.html.

[447] Ketcham had prepared this plan in 1972–73 while working for the New York City Department of Air Resources; Ketcham, email to the author, January 6, 2017. Also Ketcham, letter to Jonathan Glazer, October 14, 2014 (copy shared with author).

chief engineer of a nonprofit public interest group, Citizens for Clean Air (CCA).[448] In that role, he challenged what he viewed as optimistic claims in the FOEIS about the expected impact of new Westway traffic on New York City air quality. His challenge helped prompt evidentiary hearings before the New York State Department of Environmental Conservation (NYSDEC), the unit responsible for deciding whether to issue an air quality permit to the Westway Project, just as the ACE was for the landfill permit. Hearings began in May 1977, took 76 days over more than three years, and generated 15,372 pages of testimony. Initially, Peter A. A. Berle, commissioner of NYSDEC, denied the air quality permit on the grounds that the official traffic figures were underestimated. However, the governor of New York, Hugh Carey, a strong Westway supporter, "ousted" Berle and appointed a successor who approved the Westway air quality permit in October 1980.[449]

That same year, the U.S. EPA approved an application by Brian Ketcham and CCA for a grant to help New York City and New York State meet the State Implementation Plan he had devised to meet the Clean Air Act's air quality standards. New York Senator Daniel Moynihan, an ardent Westway supporter, criticized EPA for awarding a grant to a Westway opponent and instigated a formal investigation of EPA grant making.[450] EPA subsequently withdrew the grant to Ketcham.

In 1981, after Ketcham had moved to Konheim & Ketcham, the firm was selected to conduct

> a study of taxi traffic in midtown Manhattan, under a contract with the Tri-State Regional Planning Commission. . . . The study was to be funded by FHWA with the New York City Department of Transportation as a sponsor and NYSDOT having "sign off" authority. When, by April 1982, the final awarding of the contract had still not occurred, Ketcham and his partner, Carolyn Konheim, scheduled a meeting with [then] Commissioner [of NYSDOT] William Hennessy.[451]

Ketcham claimed he had "been informed indirectly" that Hennessy, a strong Westway supporter, "was responsible for the lack of progress on the contract."[452]

---

[448] Ketcham (1977).

[449] Roberts (1984a).

[450] "Pat Seeks Probe of Possible Link Between EPA and Westway Foe, *New York Daily News*, August 20, 1980.

[451] State of New York Commission of Investigation (1984), p. 95.

[452] Ibid.

According to Ketcham and Konheim's memo to their files about their meeting with Hennessy, Ketcham told Hennessy he believed his firm's contract was being held up because of his previous opposition to Westway.[453] According to the same memo, Hennessy replied thus:

Well, I'm a very narrow-minded guy. I work with people I trust. Westway is an all-consuming issue with me. I would find it very disturbing to do anything which seemed to give a stamp of approval to you and then at some future date you were in a position to turn it against me. My kids say, "Dad, why can't you just trust us." Well, I can't. I need proof. Now I realize I have no official power over this contract except . . . to pass it along to Federal Highway, but I can assure you that they won't act on it without an endorsement from me. I need something more, something my lawyers could look over and assure me that if anything came up on Westway you wouldn't be able to do something without coming to me first. I'm not trying to neutralize you. I don't expect you to come out in favor of Westway—that would be too much to ask. . . . [Why] don't you draft me an agreement—something I can have my lawyers look over that I can hold you to and if they say I can, then I'll I think about it. I'm not saying I'll okay it. It would be very difficult for me to do.[454]

Ketcham agreed to draft an agreement for Hennessy. He sent a letter to Hennessy "concluding that"[455]

[w]hile I cannot envision any circumstances which would cause me to alter these facts [about traffic and air pollution] in regard to Westway, I can assure you that if unforeseen extraordinary conditions forced some change, I would not engage in any activity inconsistent with the present facts [about traffic and air pollution] without prior consultation with you.[456]

---

[453] Ketcham's opposition stemmed from two factors: concern over the health effects of worsened air pollution he believed would result from new, Westway-induced traffic, and what he believed would prove to be an unfair distribution of Westway's benefits and costs. "[T]he crucial urban issue of our time is the collapse of the region's transit lifelines which daily serve 3 million [people], not a distant super highway to service 30,000 affluent, largely white, male suburban motorists." Citizens For Clean Air, press release, August 21, 1980, p. 2. Ketcham was concerned not only with whether the benefits of Westway were likely to exceed its costs but also with the likely distribution of those benefits and costs over various socioeconomically better-off and worse-off groups of affected parties. This concern echoed the spirit of Rawls's Difference Principle, discussed in Chapter 3.

[454] Konheim and Ketcham, "Memo to Files," April 14, 1982, p. 2.

[455] State of New York Commission of Investigation (1984), p. 96.

[456] Ibid., p. 97.

Hennessy replied by letter, stating that, based on their conversation and Ketcham's letter,

> I am confident that any potential problems that would affect a professional relationship between your firm and this Department [NYSDOT] . . . have been allayed.[457]

Upon reviewing this episode in June 1984, the State of New York Commission of Investigation focused not on the fact that an engineer was being punished for his past harm- and distributive justice–based opposition to Westway but on the following public-interest-related concern:

> [T]he effect of Hennessy's actions and the message they conveyed to Ketcham were unmistakable: future opposition to Westway would come only at the risk of foregoing governmental employment. In a field where that employment is a significant source of revenue, such a sanction is heavy indeed. Were it to be imposed generally, it would ensure that engineering consultants, the individuals most capable of critiquing a project, would be intimidated from doing so.[458]

What can be said, from an ethics perspective, about Ketcham's interaction with Hennessy over his grant money? If after receiving the delayed grant money for the proposed taxi study and completing the study Ketcham had reached a conclusion that had negative implications for Westway because of its likely environmental impact, his promissory letter to Hennessy might seem to have given the latter a basis for blocking dissemination of his findings.[459] Although understandable, given his vital interest in earning a livelihood, Ketcham's assurance to Hennessy might therefore seem at odds with the FERE1-based derivative ethical responsibility to not create or contribute to creating, directly or indirectly, an unreasonable risk of harm to the public interest or welfare.

Moreover, since the taxi study that Konheim & Ketcham was selected to conduct would be funded with public money, *the public*, not the entity that awarded or had sign-off authority over the grant—FHA or NYSDOT—was Ketcham's client. Hence, he arguably had an FERE4-based derivative ethical responsibility to do the best he could to serve that client's legitimate interest in protecting its environmental resources, something his letter to Hennessy might seem to have jeopardized.

---

[457] Ibid.

[458] Ibid., pp. 97–98.

[459] One purpose of Hennessy's, "as he described it, was to . . . withhold approval of a study 'which would be difficult for Westway.'." Ibid., p. 97.

However, if the account of the Ketcham-Hennessy encounter and the ensuing exchange of letters given in the State of New York Commission of Investigation report is accurate, the assurance Ketcham gave Hennessy in the conclusion of his letter to him was fully compatible with FERE1 and FERE4. Ketcham's carefully crafted assurance left him sufficient room to maneuver so that he could choose to act in an ethically responsible way if the research on taxi traffic in midtown Manhattan he proposed to carry out revealed information about congestion or air pollution that, if disseminated, could pose problems for Westway. Ketcham's use in his letter to Hennessy of "consultation" instead of "permission" was deft. That shrewd choice of words did not give Hennessy a veto over the dissemination of his future research findings, something that might arguably have been ethically irresponsible. It simply assured Hennessy that Ketcham would *consult* with him before disseminating those findings.

This strategy clearly differentiates Ketcham from a university researcher who in order to obtain research funding reaches an agreement with a company, under which it funds her or his work if she or he allows the company to determine, mindful of its private financial interests, what findings she or he can publish, and when and possibly where they can be published. In contrast, through adroit and precise use of language, Ketcham overcame a barrier to his firm's getting the grant, for which it had been selected several times. He did so while preserving for himself the option of acting in a way that was fully in accord with the FEREs, something his long record of fighting to protect the public interest in engineering settings strongly suggested he would continue to do.

2. *Walter Smith*. Col. Walter M. Smith Jr. was New York District Engineer of ACE when, in March 1982, Judge Griesa remanded to that organization the landfill permit it had issued to Westway a year earlier. To ensure compliance with NEPA, Griesa ordered that before he would lift his injunction on ACE's landfill permit to Westway, ACE would have to prepare a supplementary EIS. This document was to reflect further evaluation of existing data and possible additional fishery study, the latter to be determined after consultation with EPA, the U.S. FWS, the NMFS, and a panel of experts.

Upon remand, Smith retained the environmental consulting firm Malcolm-Pirnie Inc. (MPI) to analyze existing fish data and recommend whether additional fishery study was necessary. MPI recommended a fish-sampling program of at least three years and a striped bass "habitat survey" of at least one year to assess the quality and importance of the striped bass habitat projected to be displaced by Westway compared with other striped bass habitats. Since Smith resisted its recommendation, MPI convened a workshop in October 1982 of experts on striped bass ecology, sampling design, statistics, and hydroacoustics. In light of the workshop, MPI recommended a 17-month, two-winter study.[460]

On December 1, 1982, Smith announced his decision: the MPI workshop-shaped recommendation notwithstanding, the Corps would not commission any further fishery studies.[461] On February 1, 1983, the Sierra Club and other plaintiffs filed a contempt of court motion alleging misconduct by the Corps and NYSDOT for deciding not to commission any more fish studies when they had been ordered by the court to consult experts and cognizant federal agencies, all of whom had recommended that additional fish studies be done.[462]

In preparing for that trial, the U. S. Department of Justice took Col. Smith's deposition, from which an interesting development emerged. Smith planned to retire from the army in June 1983, after 26 years of service. On October 13, 1982, six weeks before his December 1, 1982, announcement, Smith had written a letter to Seymour S. Greenfield, top executive at Parsons, Brinkerhoff, Quade, and Douglas (PBQD), the principal engineering consultants for the Westway Project. Smith included a résumé with his letter and wrote that he was seeking a post-army-retirement executive position "in the private sector." He wrote the following to Greenfield:

> Obviously when considering opportunity potential, the name Parsons Brinkerhoff is near the top of any list. I would look forward to an opportunity to discuss my objectives with you and would deeply appreciate your advice and counsel.[463]

Smith met with Greenfield and two other PBQD executives on January 5, 1983. When this communication came to light from the deposition, Smith denied he had an actual conflict of interest and that his interest in securing future employment had influenced his fishery decision. However, probably because of concern about the *appearance* of a conflict of interest, Smith's no-more-fishery-study decision was vacated by the U.S.

---

[460] *Sierra Club v. U.S. Army Corps of Engineers*, 614 F. Supp.1475 at 1485–86 (S.D.N.Y. 1985).
[461] Ibid.
[462] Ibid.
[463] Lubasch (1983).

Attorney, and Smith was relieved of his decision-making authority over the Westway landfill permit application.[464]

The revelation that Smith had sent PBQD a letter inquiring about his employment prospects with the firm shortly before he announced his fishery-study decision engendered suspicions of a real conflict of interest. His presumptive public and professional interest in making an objective decision about whether to authorize additional fishery studies to ensure the preservation of the striped bass fishery could be seen as being in conflict with the fact that his personal interest in securing future employment with Parsons Brinkerhoff would be advanced by deciding against commissioning any more fishery studies, a decision likely to lead to further profitable engineering work on the project for the engineering consulting firm to which he had made employment overtures.

Whether the conflict of interest involved in making his decision was real or only apparent, Col. Smith's decision to reject MPI's recommendation and to not authorize any more fishery studies was arguably at odds with FERE1, since it brought the Westway project a major step closer to proceeding without solid knowledge of how it would affect the striped bass fishery. This would put the preservation of that fishery, and hence the protection of an important component of the public interest, at significant risk, something that would be at odds with FERE1.

3. *Fletcher Griffis*. Col. Fletcher H. Griffis replaced Col. Walter Smith as ACE's New York District Engineer in March 1983. Griffis attended the U. S. Military Academy, earned a PhD in civil engineering at Oklahoma State University, and served as adjunct professor of civil engineering at Columbia University. After he replaced Smith in March 1983, Griffis revisited the question of additional study of the striped bass fishery. After convening a new workshop of experts, he decided to recommend a 17-month, two-winter study, plus a striped bass habitat survey. He sent a letter to then New York State Governor Mario Cuomo explaining why he believed such a study was necessary.

Cuomo proceeded to send a letter to the Secretary of the Army objecting to the delay in issuing the Westway landfill permit the additional fishery study would entail. Cuomo asked that the ACE be stripped of the authority to make the landfill permit decision. The influence of this letter filtered down to the top ranks of the ACE. In spite of Griffis's proposal of a two-winter study as the minimum necessary for putting together a defensible supplementary environmental impact study, ACE's chief of

[464]Schanberg (1983).

engineers at the time, Lieutenant Gen. Joseph K. Bratton, decided to permit only a four-month, one-winter study. He rejected outright the proposal of a striped bass habitat study.[465]

Griffis acquiesced in Bratton's decision, *thereby putting himself in the position of having to defend downstream the conclusions of what he realized upstream would be scientifically indefensible findings of the permitted four-month, one-winter study.* He testified at trial in 1985 that he believed that a two-winter study was the minimum necessary to do the job in a scientifically sound way.[466]

The truncated, four-month study approved by Bratton and implemented by Griffis took place between January 1 and April 30, 1984.[467] Arguably, the key finding in the *draft* supplementary environmental impact statement (DSEIS) of May 14, 1984, was the following:

> The project site represents a significant portion of [the striped bass] habitat; its loss would have a significant adverse impact on the Hudson River stock of the species. Though such an impact would not be a critical blow to the species, it is likely to cause some long-term repercussions that could result in depressed population levels in the foreseeable future.[468]

Under the guidelines for implementing the Clean Water Act, such a finding would have required denial of the landfill permit NYSDOT sought, because that act requires denial of a landfill permit if a project will cause "significant degradation" of waters of the United States, where one factor in determining if a project will do so is whether it will have "significantly adverse impacts" on fish, fish habitat, and ecosystem productivity.[469]

Yet, six months later, the FSEIS contained two curious claims. First, whereas the supposedly empirically derived "relative abundance" estimate for the percentage of juvenile striped bass that used the Westway interpier project area had been reported orally to Griffis's office in July 1984 as 44%, the FSEIS used a "reasonable range" of 26%–33%.[470] Second, instead of referring to Westway as likely to have "a significant adverse impact" on the striped bass fishery, as the DSEIS had done, the FSEIS stated that the expected impact of removing the existing habitat

---

[465] *Sierra Club v. U.S. Army Corps of Engineers*, 614 F. Supp. 1475 at 1488 (S.D.N.Y. 1985).

[466] Ibid.

[467] Ibid.

[468] Ibid., pp. 1494–1495.

[469] Ibid., pp. 1495–1496.

[470] Ibid., pp. 1496 and 1506.

to make way for Westway would be "minor," "inconsequential," "difficult to notice," "insufficient to significantly impact," "difficult to discern," "too small to noticeably affect commercial or recreational fishing," and "not a critical (or even minor) threat" to the well-being of the striped bass stock.[471] Col. Griffis and the ACE technical professional who drafted the DSEIS and FSEIS took the position at trial that there were no substantive differences in the conclusions of the two documents.[472] They argued that the conclusions in the FSEIS were just *clarifications* of what was already found in the DSEIS; they were merely bringing out what was really meant in the earlier document.[473]

At trial before Judge Griesa in early 1985, it emerged that the FSEIS had gone through six draft versions before assuming its final form. Shortly before significant changes first appeared in the draft versions, Griffis had held a briefing meeting with the new ACE chief of engineers and had taken a river trip with the assistant secretary of the army. For these circumstantial reasons, the Sierra Club argued an inference could be drawn that Griffis had been subjected to improper political influence during the interval and that was what was behind the striking differences in the estimates of Westway's impact on the striped bass fishery between the early drafts and the final version of the FSEIS.

Whether Griffis was subjected to improper political influence will probably never be known with certainty. However, remarks he made in an interview shed light on one factor that might have influenced his testimony and landfill permit decision. A *New York Times* article about Griffis, published in June 1984, while hearings were being held to get public reaction to the recently published DSEIS, contains the following passage, with the quotations in it being from Col. Griffis:

> While he has gained new respect for the bass, he has maintained scorn for biologists. "If we dealt with the data the way biologists do, not one of these buildings would be standing," he said, pointing at the skyline of lower Manhattan Beach beyond his office window. "These guys study and study and study and get all the data and it means nothing." He recalled gathering two dozen striped-bass experts at a conference a year ago and asking them what the impact of the Westway would be on the fish. "All I got from these PhD's was I don't know," he said. "They studied this fish

---

[471] Ibid., p. 1496.
[472] Ibid.
[473] Ibid., p. 1500.

CHAPTER 4

their entire lives and they can't even predict behavior; engineers would never tolerate that."[474]

Perhaps civil engineer Griffis, renowned for completing transportation projects in Vietnam and Israel in much less time than expected, decided that further delays, for more analysis of fishery data, were intolerable. Whether his decision to sign the FSEIS in its final revised form, one considerably more optimistic about Westway's impact on the striped bass fishery than all earlier drafts, reflected political influence brought to bear, frustration with delays, a sense of military-organizational duty, or other factors, ACE issued a landfill permit for the Westway Project on February 25, 1985, and Griffis defended the FSEIS's conclusions in court.[475]

Griffis's conduct in this case reflects the ethically difficult situations in which many contemporary engineers who work in large-scale organizations can find themselves. He had argued in-house for a fishery study sufficient to provide the evidence needed to make a rational decision on the landfill permit application, but was overruled. He testified at trial that he was ordered to prepare a final supplementary EIS without the studies he deemed the minimum necessary for this purpose. Thus, Griffis had to implement a fishery study and defend in court an EIS he knew was deficient. As ACE decision-maker for the critical Westway landfill permit application, Col. Griffis was subjected to a complex web of scientific, organizational, cultural, and political-economic influences that would challenge the best intentions of any would-be ethically responsible engineer. As he said in the same interview, six months before issuing the landfill permit, "[I]t's lonely in the middle. . . . I see good things and bad about the highway. I really worry about making the wrong decision. But if it were easy, it would already be made."[476]

Lessons

One important moral of this case has to do with cultural conflict. Conflict can and sometimes does exist between a culture that promotes the "build it, ship it!" mentality characteristic of a results-oriented engineering organization that has not internalized the value of environmental protection,

[474] Dowd (1984).
[475] *Sierra Club v. U.S. Army Corps of Engineers*, 614 F. Supp. 1475 at 1513 (S.D.N.Y. 1985).
[476] Dowd (1984).

and a culture that promotes the value of carrying out systematically the time-consuming scientific research needed to do a thorough job on environmentally complex and sensitive problems. Such cultural conflict can lead to frustration over increased project costs and delayed schedules. This can tempt engineers to resort to questionable conduct, especially when political pressures are brought to bear.

The pattern of ethical conflict exemplified in the Westway case is likely to recur. As large-scale civil engineering projects are increasingly perceived as possibly impinging on complex ecological systems, extensive scientific studies of likely ecological consequences of proceeding with the projects will be expected or mandated. This will engender the aforementioned cultural conflict and tempt some engineers to engage or acquiesce in questionable conduct. Carrying out solid scientific studies of the ecological impacts of large proposed civil engineering projects is likely to lengthen total project times and elicit clever but sometimes arguably irresponsible ways of trying to circumvent such delays.

In recent decades, *alta aqua* (high water) has invaded the historic city of Venice, Italy, with increasing frequency, threatening its sustainability. In the late 1980s, the Italian authorities finally responded by deciding to build an innovative system of hinged underwater barrier gates to protect the city from flooding. However, the identical pattern of cultural conflict seen in the Westway case impeded these plans. The momentum of the floodgate project was slowed by disagreement between the civil engineers charged with building the floodgate system and a group of scientifically knowledgeable, ecologically oriented opponents. The latter argued that a detailed scientific model of the hydrodynamics of the Venetian lagoon should be generated before the floodgates are installed, so that in erecting the flood-control system the zealous engineers would not inadvertently turn the lagoon into an algae- and insect-infested swamp. The tension has continued, between civil engineers eager to get on with the time- and money-sensitive task of building the system of floodgates, and scientifically oriented ecologists who contend that failure to do sound scientific studies first—here, complete an accurate hydrodynamic model of the lagoon—could result in greater harm to a public good (an ecologically healthy Venetian lagoon) than the harm that would be averted (episodic flooding of Venice) by quickly building the floodgate system before the scientific studies are completed.[477]

[477] On the Venice floodgates project, see http://en.wikipedia.org/wiki/MOSE_Project and http://www.salve.it/uk/.

Discussion Questions

1. Was it ethically acceptable for Col. Smith to write a letter of inquiry about the possibility of being offered a position at Parsons Brinkerhoff before he made his decision on whether to approve more fish studies? Why or why not?
2. Was it ethically responsible of Col. Griffis to defend in court the Army Corps of Engineers' final supplementary draft of the Westway environmental impact statement when he did not believe that the data gathering period was sufficient to enable a rational decision to be made on whether to grant Westway a landfill permit to proceed? Why or why not?
3. Was it ethically acceptable for engineer Brian Ketcham to write ardent Westway supporter William Hennessy a letter assuring him that if he released funding for the grant Ketcham had been awarded, and if Ketcham's research yielded a finding that was "unfavorable to Westway," Ketcham would first consult with Hennessy before disseminating the finding? Why or why not?

CASE 16: INNOVATIONS FOR RURAL KENYAN FARMERS

Unlike a number of preceding cases, in this case there is no misconduct by any engineer. In fact, the actions of the engineer at the center of this case are clearly in accord with the pertinent FEREs and he appears to have acted with laudable intent. After reading this case, the reader will most likely agree with the author that the individual in question is an impressive example of an ethically responsible engineer.

Background

Martin Fisher earned a BS in mechanical engineering from Cornell University and MS and PhD degrees in mechanical engineering from Stanford University. After obtaining his doctorate, Fisher went to South America, where he encountered poverty in developing countries. This led him to wonder "if engineering could . . . play a critical role in solving global poverty."[478] Fisher subsequently received a Fulbright Scholarship "to

---

[478] http://www.kickstart.org/about-us/people/martin-fisher.php.

travel to Kenya to study the appropriate technology movement and the relationship between technology and poverty."[479] He wound up staying in Kenya for 17 years. In 1991, with his British friend Nick Moon, Fisher cofounded a nonprofit organization initially called "ApproTEC."[480] In 2005, ApproTEC became "KickStart International," of which Fisher is cofounder and CEO.[481]

Fisher came to believe the best way to help a developing country grow economically was to help its families substantially increase their wealth. As he put it, "income is development."[482] He decided to try to "[bring] together the entrepreneurial spirit of the poor, innovative tools and technologies, and the power of the marketplace."[483] Fisher and Moon undertook to design "affordable moneymaking tools," then mass-marketed them to poor people with entrepreneurial aspirations. The entrepreneurial poor used these items of technology to "establish highly profitable family businesses." Fisher's hope was—and remains—to help develop substantial and sustainable middle classes in the countries where Kick-Start is active.

Ethical Analysis

This case raises two interesting and important ethical issues. The first revolves around how engineers should design consumer products intended for use in less techno-economically developed countries. The second concerns whether engineers in an affluent developed country like the United States have an ethical responsibility to devote some of their professional effort to benefitting people in great need.

DESIGN FOR LOCAL CULTURAL COMPATIBILITY

Martin Fisher's innovations show how he sparked economic growth in the areas where KickStart was active. He chose to work on low-tech devices, especially for poor rural farmers. Fisher was determined to avoid the mistake of many aid organizations that promote the adoption of "advanced" time-and-labor-saving technological items. Because rural Africa has ample reservoirs of worker time and labor, the introduction of

---

[479] Ibid.
[480] "ApproTEC" was short for "Appropriate Technologies for Enterprise Creation."
[481] http://www.kickstart.org/about-us/.
[482] Fisher (2006), pp. 9–30.
[483] http://www.kickstart.org/about-us/people/martin-fisher.php.

such "advanced" devices can decrease work opportunities and provoke societal unrest.

Fisher's engineering innovations include the following devices: (1) a building-block press, (2) an oil-seed press, and (3) a micro-irrigation pump.[484] Let us look briefly at each.

1. *Building-block press.* KickStart's building-block press does not sound radically innovative, but in one sense it is. It uses only soil, a local natural resource to which rural farmers have ample access, and "a bit of cement," something that costs only a little money to obtain. Using the manually operated "Money Maker Block Press," four workers can produce 500 rock-hard blocks a day, "compacting a soil/cement mixture under high mechanical pressure." The press can be adjusted to work with various types of soil, and one bag of cement yields more than 100 bricks. The maker can sell the blocks at a profit, and buyers can build walls at half the cost of ones made with concrete blocks or stone.[485]

2. *Oil-seed press.* Fisher developed an oil-seed press for several reasons. Kenya had been importing 80% of its cooking oil, a practice that was sending precious local currency abroad.[486] Then, in 1992, the Kenyan government removed price controls on essential commodities, and the price of cooking oil "almost tripled in a few weeks." Fisher saw that if the right technology could be made available to poor entrepreneurial Kenyans, they could process locally available sunflower seeds to make and sell cooking oil locally at a profit. KickStart's oil-seed press, dubbed "Mafuta Mali," meaning "oil wealth" in Kiswahili, was "based on an original 'ram press' design by Carl Bielenberg." Fisher adapted it to be "more efficient, durable and profitable to use."[487]

Like the building-block press, the improved oil-seed press is manually operated. It can extract oil from sunflower, sesame, and other oil seeds. KickStart also designed an accompanying "Gravity Bucket Filter" that yielded clear oil ready for domestic use. In the spirit of an ethically responsible process of international technology transfer (ERPITT), Kickstart "developed a complete set of tooling for local mass production of

[484] Stevens (2002). http://www.sfgate.com/magazine/article/Martin-Makes-a-Middle -Class-Stanford-grad-2747565.php.

[485] https://webbeta.archive.org/web/20140403141903/http://www.kickstart.org /products/building-block-maker/.

[486] Stevens (2002).

[487] https://webbeta.archive.org/web/20140403151727/http://kickstart.org/products /cooking-oil-press.

both the press and filter." In 1994, his firm "trained four local engineering firms to manufacture the new presses."[488]

3. *Micro-irrigation pump.* By far the most important and influential KickStart product to date, one that has generated up to 90% of its business, is the micro-irrigation pump called the "Super Money-Maker." The word "micro" indicates that the pump is small and light enough to make it mobile. Its design illustrates sensitivity to existing local conditions in four ways.

First, since the pump is portable, users can move it inside at night, rather than leaving it outside, in the field. This feature appeals to prospective buyers because, given local social conditions, Kenyan farmers are concerned about theft. Second, the pump produces a *spray* of water, not a stream. This is critical because a developed network of irrigation ditches does not exist in Kenya. Third, the treadles on the pumps were designed for comfortable barefoot operation, since rural people who farm do not typically wear shoes. Fourth, the pumps were designed to have a short, low treadle stroke, so that "when women use them, they do not display provocative hip movements at eye level."[489] The women wear long garments and do not want to appear to be dancing suggestively when they are working (Figure 9). If they did, local men might get angry with them and restrict their economic activity, something that has happened in other less developed countries.[490] The micro-irrigation pump's design made the movements of its female users socially acceptable.

In short, Fisher designed affordable technologies that addressed local needs, used available local natural resources, and were good matches with rural Kenyan social, material, and cultural conditions: the prevalence of theft, the absence of ditch irrigation, the relative rarity of footwear, and men's sensitivity to women's physical movements in public. The design changes embedded in the KickStart irrigation pump show that Fisher acted in the spirit of FERE1. Those design changes spared purchaser-users risks of harm from theft, from wasting money (by buying something designed for a nonexistent ditch irrigation system), and from risks of ostracism or worse.

---

[488] As an added bonus, the seedcake that is left when the oil is extracted serves as "a high-protein animal feed supplement." Ibid.

[489] Fisher (2006), p. 23.

[490] For example, women of "the impoverished Wapishana and Macushi tribes of Guyana" started selling "their intricate hand-woven hammocks over the Web at $1,000 each." The money they made angered the status-sensitive tribal males who proceeded to drive out the young woman who ran the business website. Romero (2000).

Fig. 9. KickStart irrigation pump in use. Image provided by KickStart International.

Unlike Union Carbide's methyl isocyanate plant in Bhopal, India, KickStart's micro-irrigation pump fit harmoniously into the existing local Kenyan cultural-environmental system. The culturally mindful engineering design that is a KickStart trademark has enabled those who have purchased and used its products to enhance their economic situations, provide employment for others, and avoid economic and sociocultural harms that would have resulted from inadvertent cultural mismatches in the design of its products. The remarkable impact of KickStart's irrigation pumps can be seen in the statistics that as of October 8, 2017, it had sold approximately 308,225 pumps in Africa, created 230,000 businesses, and moved 1,200,000 people out of poverty.[491]

Many engineering students are familiar with the concept of "design for manufacturability" (DFM), explained in Case 5. In discussing that case, I introduced the concept of "design for ethical compatibility" (DFEC) as applied to engineering products, whether software or hardware. The case of Martin Fisher and KickStart in Kenya demonstrates that there is another kind of "design for" that engineers have an ethical responsibility to pursue. I shall call it "design for local cultural compatibility" (DFLCC). Even if a device does not directly cause harm when used, if it is not designed for local cultural compatibility it may indirectly engender harm when used in the existing local cultural-environmental system. Thus, it makes sense to say that engineers have an FERE1-based derivative ethical responsibility to ensure that their product designs embody DFLCC. Failure to do so might well amount to negligence in contexts like those in which Fisher and KickStart are active.

The Bhopal case study presented the idea of an ethically responsible process of international technology transfer, or ERPITT. I argued that for an ERPITT to qualify as ethically responsible, the engineers involved must not complete or sign off on commencing or resuming use of the (relatively advanced and risky) technology being transferred until they are sure *that all conditions for its safe and effective operation are satisfied locally*. It seems highly unlikely that this condition was met in the Bhopal case. In contrast, DFLCC involves adapting a relatively low-tech product so that it can be introduced into the local milieu without conflicting with aspects of the local cultural-environmental system, something Fisher did with his adapted-for-Kenya micro-irrigation pump. Whereas processes of international technology transfer often assume that changes will be made *in the social order* into which the technology is to be transferred so as to

[491] http://kickstart.org/impact/#by-the-numbers. Accessed October 8, 2017.

ensure its safe and effective operation and use, DFLCC focuses on making changes *in the technology* to be introduced so it will fit harmoniously into the existing local social order. It is telling that those changes are decided upon locally. Most of KickStart's design work is done in Nairobi, Kenya, at its Technology Development Centre.

> [The design team members] research raw material properties and ergonomics, use CAD and stress analysis to develop the designs, incorporate design for manufacturability from the start, and do many hours of building and testing of prototypes to ensure the performance and wear characteristics, cultural acceptability and durability. As a result it takes many months to invent, design and produce each new technology.[492]

Designing for local cultural compatibility may well be a derivative ethical responsibility of engineers when FERE1 and FERE2 are applied to the local circumstances of the technology's use in a target less developed society. Even if an engineer is determined to make a positive difference in the lives of rural poor people in developing countries such as Kenya, her or his laudable motives must be coupled with resolve to avoid the indirect harms—and product rejection—that can result from failure to design for local cultural compatibility.

An Engineer's Ethical Responsibility to Devote Some Effort to Trying to Benefit the Needy

A recurrent theme in this book has been engineers' fundamental ethical responsibility to combat harm—by not causing harm or creating an unreasonable risk of harm, by trying to prevent harm and unreasonable risks of harm, and by trying to alert and inform those at unreasonable risk of harm.[493] One can even recast FERE4 and speak of engineers' ethical responsibility to not harm or put at unreasonable risk of harm their firm's or client's legitimate business interests, for example, by doing shoddy or mediocre work, by stealing a competitor's intellectual property, or by being careless with the employer's or client's intellectual property. However, the Martin Fisher case is not the story of an engineer striving to combat harm. It is the story of an engineer working diligently to benefit people in great need.

---

[492] Fisher (2006), p. 20.
[493] By acts of either commission or omission.

This brings us to a second ethical issue raised by this case: whether engineers in a country like the United States have, in contemporary society, an ethical obligation or responsibility to devote at least a small part of their professional effort to trying to benefit people in great need. I shall approach this issue by relating a personal experience.

In December 2008, I was one of a group of eight faculty members from five U.S. universities, organized and led by Dr. Sandip Tiwari, professor of electrical engineering at Martin Fisher's alma mater, Cornell University. The faculty group traveled to India with 11 graduate students in engineering drawn from eight U.S. universities. At the Indian Institute of Technology in Kanpur, Uttar Pradesh, the faculty taught the U.S. engineering students and a similar-sized group of Indian counterparts an intense one-week course on nanotechnology, focusing on organic electronics and optoelectronics. Besides state-of-the-art technical subject matter, the course included discussion of social and ethical issues related to nanotechnology.[494]

But the short course per se was not the only focus of the trip. Once the formal course had concluded, most of the faculty-student group took a long trip by rail and bus to a remote and impoverished part of eastern India, in and around Paralakhemundi, at the Orissa-Andhra Pradesh border. This area, normally off-limits to visitors, is accessible only with government permission. The focus of the experience was on tribal living and rural needs. The idea was to spend time with the local residents, learn about their problems and needs, and see what the students could do to improve the lives of the people by using their engineering skills.

Several months after the trip, I read some reflections by the U.S. students about their experience. Many were moved by what they had seen. Since their advanced engineering studies were not readily relatable to the conditions under which the poor rural people were living, some struggled to reconcile pursuing them with what they had experienced. A couple of the students indicated they hoped to devote at least a part of their future professional effort to activities that would benefit people like those whom they had visited, rather than spending all their professional effort on esoteric endeavors that would likely benefit only those in the most developed and materially affluent parts of the world.

In their post-trip reflections, the engineering students did not use the phrase "ethical responsibility." However, reading their reflections

[494] http://www.nnin.org/education-training/graduate-professionals/international-winter-school/prior-years/2008.

conveyed the impression that several of them were wondering whether they did have some kind of obligation to devote at least a small part of their professional energy to engineering endeavors aimed at benefiting those in great need.

The visit to the rural area was designed to awaken and touch the students by bringing them face to face with extremely needy people, individuals whom, they recognized, their basic engineering knowledge and skills equipped them to help. The face-to-face encounters seemed to spark empathy in some of the students. Insofar as it did, the trip was an effective transformative experience.

Of course, such resolve could be promoted in other ways. Society could formally recognize and publicly laud engineers who devote some of their time, energy, and resources to trying to benefit the most needy by designating them as "engineering heroes" or some such honorific. Some engineers might, when alone and looking in their personal mirrors, want to see someone who devotes a part of her/his professional effort to benefiting others in great need. That, too, could move them to modify their personal portfolios of engineering endeavors.

Cultivating empathy directly by effecting exposure to people in great need, or indirectly by deploying societal honorifics or inducing engineers to "look at themselves in the mirror" with the hope that conscience will prompt them to undertake ameliorative social action, are approaches to raising consciousness in engineers about what they could and perhaps should do as would-be responsible professionals. But is there an ethics-related *argument* that might help persuade engineers they have an ethical obligation or responsibility to do something to try to benefit those in great need, rather than viewing doing so as laudable but "above and beyond the call of duty"?

Although somewhat abstract, one relevant line of reasoning I call "the antiparasite argument." The individual engineer who pursues only his or her own private interest and makes no effort to benefit public welfare, particularly as regards the most needy, can be viewed as a kind of parasite. He or she takes full advantage of the enabling resources and opportunities society has made available to engineers solely to serve his or her own private interests but does nothing intended to benefit the commonweal or help the neediest. Although society does not hold nonhuman parasites ethically accountable for their behavior, perhaps it should hold professionals ethically responsible for their parasitic behavior, because when aggregated, parasitic behavior by many individual members of a given profession can yield two kinds of harmful consequences: harm to the reputation of the profession in question[495] and harm to important consensual social goods.

Social goods can be *institutional*, such as a legal system that treats all parties fairly, independent of their means, or *ideational*, such as the idea that all people have certain fundamental interests ("human rights") that deserve protection independent of whether they have the financial means to do so on their own. Such *social goods* or, put differently, *items of social capital*, are ones in whose protection each of us has a significant interest. Going beyond the pursuit of self-interest and contributing to benefitting society, normally deemed laudable or heroic but not an individual duty or responsibility, can be plausibly viewed through the lens of a classical ethical principle: that it is wrong for an agent to unjustifiably harm or impose an unreasonable risk of harm on another human being or on society.

In this instance, however, that principle would be applied not to the actions of single *individuals* that directly harm or create an unreasonable risk of harming others but rather to the behavior of *a large group of professionals*, whose self-interested actions, *taken in the aggregate*, undermine some key social good or item of social capital in whose preservation everyone has a long-term interest.

In discussing Case 1, I suggested that the Cadillac design engineers arguably had an ethical responsibility to not design a new engine-control chip for the stalling cars that although relatively harmless at the level of the individual car, in the aggregate substantially harmed the item of social capital we call clean air.[496] Similarly, each engineer fortunate enough to be educated, live, and work in a materially affluent developed society in the current era might have an ethical responsibility not to devote all of her or his professional activity solely to furthering her or his own economic self-interest, for example, by doing engineering work prescribed only by firms that cater to well-off individuals in materially affluent countries. Although the harm done to society by each individual engineer who directs all his or her professional work effort toward private gain would be negligible, in the aggregate such effort might do significant harm to some as-yet-unidentified social good or item of social capital.

---

[495] The image of the legal profession has been tainted by the notion of lawyers as hired guns.

[496] That is, installed in almost half a million cars over five years, the new chip they designed contributed to causing a public harm of aggregation.

An important engineering-related social good or item of social capital is the idea that all people should have access to quality basic engineering services—hydraulic, sanitation, energy, shelter, and the like—regardless of whether they live in rich or poor countries and can afford to pay the going market prices for those services.[497]

In short, mass parasitism by engineering professionals would cause a significant public harm of aggregation by undermining a vitally important ideational social good or item of social capital: *the ideal of universal access to basic engineering services.* Conversely, if individual engineers in large numbers from the most materially well-off countries devote at least a small fraction of their professional effort to trying to benefit people in great need, as Martin Fisher has done so impressively, that would contribute to both *not causing* and *preventing* a socially harmful outcome. I contend that most contemporary engineers living in the most well-off countries have an FERE1- and FERE2-based derivative ethical responsibility to devote at least a small fraction of their professional effort to trying to benefit or improve the lot of the most needy, either in less developed or in their own more developed countries.

That said, it is unclear which approach would be most effective in inducing engineering students and young engineers to devote some of their professional effort to trying to benefit those in great need: in-person exposure to people in great need whom they see they could help with their engineering skills; lauding and recognizing doing so as ethically admirable or heroic; or making the case that doing so is an FERE-based derivative ethical responsibility incumbent on contemporary engineers in the most materially affluent countries.[498]

---

[497] An important reason why universal access to these basic engineering services deserves to be regarded as a nonarbitrary social good is that each of the services can be plausibly linked to one or more basic human needs, ones attributable to all humans in all cultures at all times.

[498] In elaborating "the antiparasite argument," the author did not ground the claimed ethical responsibility of some engineers (to devote some of their professional effort to activities intended to benefit the needy) in the principle of *beneficence.* This principle, often invoked in biomedical ethics, states that one is morally obligated to act so as to benefit others. Rather, the author attempted to derive this ethical responsibility from the FEREs, which are grounded in a variation of the principle of *nonmaleficence,* under which one is morally obligated to "combat harm" to others from one's work or work about which one is technically knowledgeable. For further discussion of beneficence, see Beauchamp (2013).

Lessons

It may be that whether an individual engineer, lawyer, or doctor is moved to try to incorporate efforts to benefit the lives of the needy into her or his professional practice will depend most on whether or not she or he is content to perceive herself or himself in the mirror as a competent hired gun who is well remunerated for working productively at some advanced technical frontier. My impression is that many engineering students and practicing engineers do not relish thinking of themselves as mere hired guns whose work profits only their employers and helps the materially most well-off become more so. Indeed, many engineers would recoil at any suggestion that they were behaving in a parasitic manner. I also believe that many engineering students and practicing engineers would like to think of themselves as concerned not just with combating harm but also with benefiting society—and not just its currently most well-off members. I suspect that such individuals would find furthering the ideal of universal access to quality basic engineering services appealing.

The work of engineers like Martin Fisher, and of engineering students and practicing engineers who belong to organizations like Engineers Without Borders[499] and Engineers for a Sustainable World,[500] suggests there is a chance that doing some pro bono engineering work aimed at benefiting the most needy of society, whether at home or abroad, could yet become an element of being an ethically responsible engineer in the most advanced sectors of contemporary global society. That would be a welcome development.

Discussion Questions

1. Would it have been ethically responsible for Fisher or other Kick-Start engineers to design and sell items for everyday use in rural Kenyan society that were highly efficient but also widely viewed by Kenyans as incompatible with or subversive of their local culture? Why or why not?

[499] http://ewb-usa.org/.
[500] http://www.eswusa.org/.

2. Do engineers from a more developed society who are charged with designing products for everyday use in a less developed society have a derivative ethical responsibility to do prior field work (or to commission such field work) in that society to become aware of possible relevant features of the local culture of the society in question that could affect how their planned products are received? Why or why not?

3. Is it ethically acceptable for an engineer to devote her or his entire career to work that is foreseeably likely only to serve the wants of well-off people in the most affluent societies? Why or why not?

4. Do engineers who obtain a world-class engineering education in more developed societies have an ethical responsibility to devote at least a part of their professional effort to improving the lot of the needy, either in a less developed society, if they are citizens of such a society, or in the more developed society of which they are citizens? Why or why not?

CASE 17: THE GOOGLE STREET VIEW PROJECT

Background

Google Maps is a Web-based service that provides online maps and travel directions for user-selected areas and destinations. Google launched the service in February 2005. In May 2007, the firm announced the launch of a new component of Google Maps:

> Street View is a new feature of Google Maps that enables users to view and navigate within 360 degree street level imagery of various cities in the U.S. Street View provides users with a rich, immersive browsing experience directly in Google Maps, enabling greater understanding of a specific location or area. Street View imagery will initially be available for maps of the San Francisco Bay Area, New York, Las Vegas, Denver and Miami, and will soon expand to other metropolitan areas. By clicking on the "Street View" button in Google Maps, users can navigate street level, panoramic imagery. With Street View users can virtually walk the streets of a city, check out a restaurant before arriving, and even zoom in on bus stops and street signs to make travel plans.[501]

[501] http://googlepress.blogspot.com/2007/05/google-announces-new-mapping_29.html.

Google subsequently expanded Street View (SV) to include street-level imagery for thousands of cities and towns in many countries.

Within days of its introduction, privacy advocates expressed concern that some of the images SV made available in Google Maps violated the privacy of certain individuals, for example, those whose faces could be clearly identified or whose license plates or home addresses were legible.[502] Others objected that certain facilities, such as shelters for female victims of domestic abuse and the Lincoln Tunnel under the Hudson River, were too sensitive for their addresses or other critical characteristics to be determinable from SV images. Google denied that SV violated privacy interests, arguing that

> Street View only features imagery taken on public property. . . . This imagery is no different from what any person can readily capture or see walking down the street.[503]

Google also noted that SV "allows users to request that a photo be removed for privacy reasons."[504]

In September 2007, Canada's Privacy Commissioner wrote to Google "to seek further information [regarding Street View] and assurances that Canadians' privacy rights [would] be safeguarded if [Street View were] deployed in Canada."[505] Google's global privacy counsel, Peter Fleischer, responded by promising that Google would "try not to have identifiable faces and identifiable license plate numbers in any Street View images in Canada."[506] Perhaps in response to the expression of such concerns, in May 2008 Google announced that it had begun testing a new technology that would automatically recognize and blur faces (and license plates) in GSV images.[507]

In April 2009, Google reached agreement with German data protection officials that people "whose faces or license plates showed up in Street View pictures" could "register complaints with Google Germany to have those images blurred."[508] Johannes Caspar, the German data protection regulator for Hamburg, threatened Google with "unspecified sanctions" if it did not change the SV project to conform with "German

---

[502] Helft (2007).
[503] Ibid.
[504] Ibid.
[505] https://epic.org/privacy/streetview/#timeline.
[506] Ibid.
[507] https://maps.googleblog.com/2008/05/street-view-revisits-manhattan.html.
[508] Ibid., in the "Germany" entry.

privacy laws, which prohibit the dissemination of photos of people or their property without their consent."[509] Eventually, when Google gave German citizens the option of removing pictures of their houses or apartments from SV, about 244,000 Germans exercised it.[510]

During a subsequent inquiry into SV, Caspar discovered that "all vehicles used for the Internet service of Google Street View are equipped with technological devices for mapping" wireless networks.[511] The German Federal Commissioner for Data Protection and Freedom of Information disclosed this discovery publicly on April 23, 2010.[512] This discovery and revelation marked the beginning of a new phase in the conflict between Google and many national data protection authorities over the ethical acceptability of SV as it existed at the time.

It eventually became clear that the Google vehicles collecting digital photographic images for SV as they drove through the streets of cities and towns in many countries were also equipped with antennas and computers that enabled them to gather nonphotographic information. "Between May 2007 and May 2010, as part of its Street View project, Google Inc. . . . collected data from Wi-Fi networks throughout the United States and around the world."[513] The purpose of Google's Wi-Fi data collection effort "was to capture information about Wi-Fi networks that the Company could use to help establish users' locations and provide location-based services," such as showing nearby restaurants, movie theaters, or friends.[514]

Prompted by the German data authority's findings and by the findings in a report by Stroz Friedberg, a consulting firm engaged by Google "to evaluate the source code used in Google's global Wi-Fi data collection effort,"[515] in November 2010 the U.S. Federal Communications Commission's (FCC's) Enforcement Bureau launched an investigation into Google's "Wi-Fi data collection activities to assess whether those activities violated Section 705(a)" of the Communications Act of 1934, legislation that prohibits the "unauthorized interception and publication of radio communication and the unauthorized reception and use of interstate radio communications."[516]

---

[509] O'Brien (2009).

[510] K. O'Brien (2010).

[511] https://epic.org/privacy/streetview/#timeline.

[512] Ibid.

[513] U.S. Federal Communications Commission (2012), p. 1, sec.1; referred to hereafter as FCC.

[514] Ibid., pp. 1 and 10.

[515] Ibid., p. 5, sec.10.

[516] Ibid., p. 3.

In July 2010, shortly after the German data protection authority disclosed publicly that Google was collecting Wi-Fi network identification data, and before the FCC launched its investigation, a post appeared on Google's official blog:

> The Wi-Fi data collection equipment has been removed from our cars in each country and the independent security experts Stroz Friedberg have approved a protocol to ensure that any Wi-Fi related software is also removed from the cars before they start driving again.[517]

The FCC published the results of its investigation of Google's Wi-Fi data collection activity in April 2012. It fined Google $25,000 for "noncompliance with Bureau information and document requests."[518] However, the FCC declined to take enforcement action under Section 705 (a) against the company for its Wi-Fi data collection activity. One reason it declined to do so is that, through counsel, a Google engineer centrally involved with the Wi-Fi data-collection effort—called "Engineer Doe" in the FCC report—invoked his Fifth Amendment right against self-incrimination and declined to be deposed by the FCC.[519]

SV continues to operate in the U.S. and many other countries in Europe and Asia, although sometimes on a limited territorial basis and subject to certain conditions and constraints. SV has proven to be a frequently used feature of Google Maps.

Ethical Analysis

As with many cases involving computers and networks, arguably the main ethical issue raised by the SV project involved individual privacy. An important engineering-related ethical issue raised by this case can be framed thus: whether it is ever ethically acceptable for an engineer-employee of a firm, or a group of same, to devise software that when implemented will covertly collect personal data with the potential to serve the employer's economic interests but whose collection and retention violates the privacy of affected individuals.

The ethical analysis that follows is divided into two parts, corresponding to the two kinds of data collected by Google. The first kind took the

[517] Ibid., pp. 6–7.
[518] Ibid., p. 2, sec. 4.
[519] Ibid., p. 2, sec. 3.

form of digital photographic images captured by Google vehicles. When that imagery, after appropriate stitching, became accessible through Google Maps, certain potentially delicate information about some individuals became publicly available without their consent. Such information included their face and automobile license plate letters/numbers, and the address or other characteristics of certain special locations related to them, such as special human shelters, prisons, and homes. In some cases, making such information available in SV arguably contributed to creating an incremental risk of harm for certain individuals. For example, someone planning to rob a residence might find it helpful in selecting a target to carefully scan images of candidate residences and to determine which ones are located on streets with quick-getaway potential.

As noted, Google initially justified its photographic image–gathering effort by arguing that "Street View only features imagery taken on public property. . . . This imagery is no different from what any person can readily capture or see walking down the street."[520] Although superficially plausible, this analogical argument is at bottom unconvincing. It is not so much *what* imagery SV vehicles captured or even the fact that the imagery was *captured and retained*. Rather it is *what was done with* the imagery in question—being put on the Internet for anyone to inspect at any time for any purpose, innocuous or problematic—that differentiates what Google did from what "any person" can do "walking down the street." Putting that collected imagery online arguably helped engender incremental risks of harm for certain identifiable individuals.

What consideration, if any, did Google engineers who designed or were otherwise involved in the SV project give not only to the technical question of how they could most efficiently gather the desired imagery but to the ethics question of whether it might be inappropriate or irresponsible to put some of the collected imagery online? More specifically, did cognizant Google managers, whether engineers or nonengineers, and/or the engineers working on the SV project take meaningful *precautionary* steps to avoid creating and imposing any nontrivial incremental risks of harm on individuals rendered vulnerable by having their likeness, license plate numbers, and/or home or other sensitive addresses put online? If they did not, did any of the cognizant managers and involved engineers take the position that, or acquiesce in the position being taken that, the photographic image collection project should move ahead full speed without serious precautionary steps first being taken, and that any adjustments

---

[520]Helft (2007).

would be made only downstream once it became clear what concerns, objections, or political-economic risks surfaced after SV was launched? The author knows of no evidence that permits definitive answers to be given to these ethics-relevant questions.

The second kind of information that Google's SV vehicles gathered, between May 2007 and May 2010, was data collected as part of Google's concurrent Wi-Fi data-collection effort. In turn, this second kind of data fell into two categories. The first consisted of technical identification data about the Wi-Fi networks that Google vehicles drove by.[521] The second consisted of data, including information about users of *unencrypted* Wi-Fi networks and their online activities, collected by Google vehicles driving by residences and commercial buildings in the United States and other countries.[522] The FCC report and Google refer to this category of data as "payload data."[523]

This second category of Wi-Fi network data revealed sensitive information about many individual computer users, including the substance of many of their communications transactions and social relationships, things that most people would not want to be accessible to others without their permission. The capture and retention of some such information would reasonably be regarded by some users as putting them at incremental risk of incurring harm. The amount of Wi-Fi private data captured by Google between 2008 and 2010 was substantial. In the United States alone, Google cars collected "approximately 200 gigabytes" of payload data from private Wi-Fi network access points and related LANs.[524] For Australian regulator Stephen Conroy, Google's Wi-Fi private data collection effort was "probably the single greatest breach in the history of privacy."[525]

How did Google engineers come to collect the first category of this second kind of information? According to the FCC report,

> As Street View testing progressed, Google engineers decided that the Company should also use the SV cars for "wardriving," which is the practice of driving streets and using equipment to locate wireless LANs using Wi-Fi, such as wireless hotspots at coffee shops and home wireless networks. By

---

[521]FCC (2012), p. 10, sec. 21.
[522]Ibid. and p. 22, sec. 51.
[523]Ibid., pp. 1 and 4. The FCC report characterizes "payload data" as "the content of Internet communications—that was not needed for its location database project. This payload data included email and text messages, passwords, Internet usage history, and other highly sensitive personal information." FCC, p. 1.
[524]Ibid., p. 11, sec. 24.
[525]Streitfeld (2013).

collecting information about Wi-Fi networks (such as the MAC address, SSID, and strength of signal received from the wireless access point) and associating it with global positioning system (GPS) information, companies can develop maps of wireless access points for use in location-based services.[526]

The FCC report asserts that "[t]o design the Company's [wireless data–collection] program," Google "tapped" one of its engineers with expertise in wireless networking. The FCC report refers to that Google engineer/programmer as "Engineer Doe."[527] Doe authored at least six versions[528] of a "design document . . . describing the hardware, software, and processes he proposed the Company should use in its Wi-Fi data collection program."[529]

Based on his expertise in Wi-Fi networks, Doe apparently proposed to take Google's SV Wi-Fi data-collection effort to another level, one that went beyond gathering locational and identifying information about residential and business Wi-Fi networks. Noting that wardriving "can be used in a number of ways," including "to observe typical Wi-Fi usage snapshots," Doe proposed that besides collecting data useful for mapping the location of wireless access points, Google should also "collect, store, and analyze payload data from unencrypted Wi-Fi networks."[530]

Did the engineer or engineers involved in the Google Wi-Fi data-collection effort behave in an ethically responsible manner? Regarding the collection of the *first* category of Wi-Fi data, namely, technical identification data about the Wi-Fi networks being mapped, it is not clear whether gathering that data constitutes a violation of individual privacy. Mapping Wi-Fi access points is useful for more efficiently providing location-based services to users. However, an argument could be made that such data, even if it does not refer to any specific computer user by name, would allow parties who collect or gain access to it to know

[526]FCC (2012), p. 10, sec. 21.

[527]In April 2012, it was reported that a former state investigator involved in another inquiry into SV had revealed the name of the Google engineer/programmer whom the FCC called "Engineer Doe." See Lohr and Streitfelt (2012). However, in this case discussion, the Google engineer in question will be referred to as "Engineer Doe" or "Doe."

[528]That Doe wrote at least six drafts of his design document suggests, but does not by itself demonstrate, that he was getting feedback from engineers and/or managers on earlier drafts, in response to which he would revise his most recent draft. One draft was completed on October 26, 2006, another on August 23, 2007. See FCC, p. 10, n. 76.

[529]FCC (2012), p. 10, sec. 22.

[530]Ibid., p. 11, sec. 22.

potentially interesting things about the network owner, for example, that she or he is economically well-off enough to have a computer, access to the Internet, and a LAN, suggesting that such a party might make an inviting target, for example, for theft or hacking.

Regarding the *second* category of Wi-Fi network data, collected by Google from January 2008 to April 2010,[531] collecting user payload data is arguably a clear violation of user (informational) privacy interests. Several things merit comment about Engineer Doe in this connection.

First, Doe *did* recognize that privacy was a concern raised by the collection of "payload data." In one version of his design document, he wrote:

> A typical concern might be that *we are logging user traffic* along with sufficient data to precisely triangulate their position at a given time, *along with information about what they were doing.*[532] (italics in original)

Second, Doe seems not to have regarded privacy violation as a "serious concern," because the Google cars would not be "in proximity to any given user for an extended period of time" and "[n]one of the data gathered . . . [would be presented to end users of [Google's] services in raw form."[533] However, the Dutch Data Protection Agency (DDPA), the organization that reviewed payload data collected by Google in the Netherlands over a two-year period, concluded that

> [t]he recorded data are not meaningless fragments. It is factually possible to capture 1 to over 2,500 packets per individual in 0.2 second. Moreover, the car may have captured a signal from a single Wi-Fi router several times.[534]

Nevertheless, it is possible that Doe did not *believe* that the Google cars could capture any substantive data about individual private users that would jeopardize or violate their privacy. Hence he may have thought that the collection effort was *not* violating individual privacy, or that if it was, it was not a noteworthy privacy violation.

Third, Doe's design document includes, "as a 'to do' item," "*[D]iscuss privacy considerations with Product Counsel.'*" However, the FCC report stated, "That never occurred."[535] Whether Doe ever *attempted* to have that discussion with a cognizant Google lawyer is unclear. If he did not, it could be argued that he was negligent in going ahead with the

[531] Ibid., p. 11, sec. 24.
[532] Ibid., p. 11, sec. 22.
[533] Ibid.
[534] Ibid., p. 13, sec. 25.
[535] Ibid., p. 11, sec. 22, and p. 22, sec. 51.

payload-data collection part of the Wi-Fi data-collection project without having done so. That view might be rebutted by noting that since Doe apparently believed that any payload data about an individual user that might be collected would be fragmentary and not substantively revealing, he did not think that a meeting with a "product counselor" was imperative. But then, if he believed that, why did he list doing so as a to-do item?

Regarding the possible applicability of FERE1—the engineer's fundamental ethical responsibility not to cause harm or create or contribute to creating an unreasonable risk of harm (by acts of commission or omission)—a question arises. If Engineer Doe did *not* check to see whether his assumption—that only fragmented payload data about individual Wi-Fi network users would be collected—was correct, was *not checking* a reasonable thing for him to do because he had good reason for making that assumption, or, if he did *not* have good reason for making that assumption, was not checking on its validity negligent of Doe and an unreasonable thing for him to do because of the privacy violation that would likely result from acting on that assumption?

It may be relevant to answering this question to note that collecting substantive payload data about private Internet users' personal online activities without their consent is arguably a far more sensitive matter than collecting technical identification data about user Wi-Fi networks. Hence, Doe had an FERE1-based derivative ethical responsibility to be extra vigilant that the protection of individual privacy he presumably believed was assured regarding the collection of Wi-Fi network location and identification data was also valid in the case of collecting more revealing payload data about specific individuals. It would arguably have been negligent to assume uncritically that what was the case for the collection of technical network identification data was also valid for the collection of personal payload data.

Was Engineer Doe the only Google engineer involved with the SV project who knew about the full Wi-Fi data collection effort? A comment on an official Google blog by the firm's vice president for engineering and research seems to suggest that Doe was essentially the only Google engineer involved with the company's full Wi-Fi data-collection effort:

> In 2006 an engineer working on an experimental Wi-Fi project wrote a piece of code that sampled all categories of publicly broadcast Wi-Fi data. A year later, when our mobile team started a project to collect basic Wi-Fi network data like SSID information and MAC addresses using Google's

Street View cars, they included that code in their software [by mistake]—although the project leaders did not want, and had no intention of using, payload data.[536]

However, from emails Google produced in response to FCC letters of inquiry, the FCC discovered that

> on October 31, 2006, Engineer Doe sent an e-mail with links to his draft design document and draft software code to the Street View team, . . . [and] on at least two occasions Engineer Doe specifically informed colleagues that Street View cars were collecting payload data.[537]

In light of this and other Google emails, the FCC reached a different conclusion about the involvement of other Google engineers than the then–vice president of engineering and research had suggested:

> As early as 2007 and 2008 . . . SV team members had wide access to Engineer Doe's Wi-Fi data collection design document and code, which revealed his plan to collect payload data. One Google engineer reviewed [Doe's] code line by line to remove syntax errors and bugs, and another modified the code. Five engineers pushed the code into Street View cars, and another drafted code to extract information from the Wi-Fi data those cars were collecting. Engineer Doe specifically told two engineers working on the [SV] project, including a senior manager, about collecting payload data. Nevertheless, managers of the Street View project and other Google employees who worked on SV have uniformly asserted in declarations and interviews that they did not learn the SV cars were collecting payload data until April or May 2010.[538]

If Doe actually shared his design plans regarding collecting personal payload data with those SV engineers, before payload data started to be collected, then they may have failed to act in accord with FERE2. Neither the FCC Enforcement Bureau report, nor any other document with which the author is familiar, contains any evidence that other Google engineers tried to prevent the risk of harm that gathering and retaining payload data about specific individuals would clearly engender.

Perhaps trying to reconcile its view with that of Google's then vice president for engineering and research, the FCC concluded that

---

[536] https://googleblog.blogspot.com/2010/05/wifi-data-collection-update.html. Note the poster's use of the question-begging phrase "*publicly broadcast* WiFi data." (emphasis added).

[537] FCC (2012), p. 15, sec. 30.

[538] Ibid., pp. 17–18, sec. 39.

[t]he record . . . shows that Google's supervision of the Wi-Fi data collection project was minimal. . . . [I]t appears that no one at the Company carefully reviewed the substance of Engineer Doe's software code or the design document.[539]

Whether Google's Wi-Fi network data gathering effort violated the Communications Act of 1934 may never be known with certainty, for, as noted, Engineer Doe invoked his Fifth Amendment right against self-incrimination and refused to be deposed by FCC staff. However, if the FCC's conclusion in the following quotation *is* correct, it might reflect something noteworthy about the *culture* that prevailed in the Google engineering workplace at the time:

> During interviews with Bureau staff, Google employees stated that any full-time software engineer working on the Street View project was permitted not only to access and review the code, but also to modify it *without prior approval from project managers* if the engineer believed he or she could improve it.[540] (emphasis added)

Might Google's entrepreneurial engineering workplace culture have helped make it legitimate and acceptable for a Google engineer doing work related to SV to propose adding a qualitatively new dimension to an already-approved project; for such an addition to not be scrutinized upstream by senior engineers or managers to see if there was anything technically, ethically, or legally problematic about it; and/or for the added dimension to be implemented by its proposer/designer without explicit managerial permission?

If the FCC conclusion about minimal managerial supervision of Engineer Doe's endeavor is correct, then any senior managers or engineers who were responsible at the time for establishing or sustaining the norms and practices of Google engineering workplace culture were arguably negligent under FERE1, because their lack of oversight arguably allowed Doe to add the payload-data component to the Wi-Fi data-collection effort and have it executed. It may never be known whether Google engineering workplace culture played a contributory causal role here or whether Doe simply violated or neglected recognized norms of Google engineering workplace culture and did what he did on his own, without any Google senior engineers or managers feeling obliged to know

[539] Ibid., sec. 51.
[540] Ibid. pp. 16–17, sec. 36.

and evaluate what he was proposing or to apprise themselves of what happened when his proposal was first implemented. Interestingly, and possibly revealingly, in an interview with the FCC, one Google "senior manager of Street View said he 'pre-approved' [Doe's] design document before it was written."[541]

To conclude this case discussion, let us explore the outcomes and implications of two important court cases involving SV. The first was brought in 2009 by Switzerland's federal data protection agency. It was prompted by privacy concerns about the original SV project of collecting and putting online digital images of street views in Swiss cities. The second was initiated in 2010 by the attorneys general of 38 U.S. states. It was prompted by the revelation in early 2010 that since 2008 Google had also been covertly collecting payload data about individual private unencrypted-Wi-Fi-network users.

The Swiss court case was "the last one pending in Europe" that challenged "the basic legality of Street View's photographing methods.[542] In 2010, the Swiss federal data protection and information commissioner demanded that for SV to operate in Switzerland, Google's pixilation technology, which blurred images selectively, must work 100 percent of the time.[543] However, in June 2012, the Swiss Federal Supreme Court, the *Bundesgericht*, held that Google

> did not have to guarantee 100 percent blurring of the faces of pedestrians, auto license plates and other identifying markers captured by Google Street View cars; 99 percent would be acceptable.[544]

However, the Swiss court upheld four other privacy-related conditions demanded by the Swiss federal data regulator: that Google (i) lower the height of its SV cameras so that they would not "peer over garden walls and hedges," (ii) "completely blur out sensitive facilities like women's shelters, prisons, retirement homes and schools," (iii) provide communities with at least a week's notice in advance of upcoming SV shootings, and (iv) "[allow] people to opt out of the photo archive through traditional mail services as well as online."[545]

The Swiss court ruling suggests a position of considerable importance when dealing with controversial new technologies. Rather than SV being

[541] Ibid., p. 16, sec. 35.
[542] O'Brien and Streitfeld (2012).
[543] Ibid.
[544] Ibid. Google claimed that its pixilation technology was 99 percent accurate.
[545] O'Brien and Streitfeld (2012).

found to be either intrinsically ethically unacceptable or ethically accept-able without restrictions, SV was found **conditionally ethically accept-able**. While SV has been prohibited in a few countries,[546] in most coun-tries in which it was introduced (and elicited protests over its privacy implications), it has been conditionally permitted, with the conditions imposed on its operation varying from country to country.

The second noteworthy court case began in June 2010 when a number of U.S. state attorneys general launched an inquiry into what Connecti-cut's then attorney general called "Google's deeply disturbing invasion of personal privacy."[547] He was referring to Google use of SV vehicles equipped with antennas, GPS equipment, laptops, and open-source software to collect private payload data from individual Internet users using unencrypted Wi-Fi local area networks without their consent.[548] In March 2013, Google settled the case by

> acknowledg[ing] to state officials that it had violated people's privacy dur-ing its Street View mapping project when it casually scooped up passwords, e-mail and other personal information from unsuspecting computer users.[549]

As part of the settlement, Google paid a fine of $7 million to the states. The settlement agreement also required Google to

> engage in a comprehensive employee education program about the pri-vacy or confidentiality of user data; to sponsor a nationwide public service campaign to help educate consumers about securing their wireless net-works and protecting personal information; and to continue to secure, and eventually destroy, the data collected and stored by its Street View vehicles nationwide between 2008 and March 2010.[550]

In 2013, Connecticut's attorney general stressed the nonfinancial impor-tance of the agreement:

> While the $7 million is significant, the importance of this agreement goes beyond financial terms. Consumers have a reasonable expectation of

---

[546] For example, in 2016, the Indian government refused Google permission to carry out a Street View mapping of the country, on grounds of national security. Indian authorities claimed that SV imagery could be useful to terrorists determined to carry out attacks on important Indian sites. See Singh (2016).

[547] Streitfelt (2013).

[548] As noted earlier, Google asserted that it had collected private payload data "mis-takenly" and "in error." See https://googleblog.blogspot.com/2010/05/wifi-data-collection-update.html.

[549] Ibid.

[550] State of Connecticut, Office of the Attorney General (2013).

privacy. This agreement recognizes those rights and ensures that Google will not use similar tactics in the future to collect personal information without permission from unsuspecting consumers.[551]

In light of aforementioned remarks about Google's engineering workplace culture and its possible bearing on ethically responsible engineering practice, the following statement from the settlement announcement is noteworthy:

> Under terms of the agreement, Google agreed to corporate culture changes, including a corporate privacy program that requires in part direct notification of senior management, supervisors and legal advisors about the terms of the agreement; enhanced employee training about the importance of user privacy and the confidentiality of user data and the development and maintenance of policies and procedures for responding to identified events involving the unauthorized collection, use or disclosure of user data.[552]

This passage does not indicate whether the "corporate cultural changes" that Google agreed to make will target engineers along with managers, although some Google managers are engineers by training. However, it is encouraging to read that in the attorney general's view, "Google deserves credit for working in good faith with my office to develop policies and best practices to protect consumer privacy going forward."[553] Whether adherence to "best practices" for protecting consumer privacy becomes an integral operative element of Google's engineering workplace culture remains to be seen.

Lessons

What useful lessons can be learned from this case? First, although a course of action may be legally permissible, that does not imply or guarantee that carrying it out is ethically acceptable or responsible. Discussing the "to-do" "privacy considerations" raised by a company project with a company "product counsel" might result in this lawyer's finding the proposed project legally permissible. However, that by itself would not resolve the question of whether the proposed project or practice was ethically acceptable or responsible. That issue would need to be addressed separately. Legality is not a stand-in for ethical acceptability. Whether the

---

[551] Guynn (2013).
[552] State of Connecticut, Office the Attorney General (2013).
[553] Ibid.

question of ethical acceptability is actually addressed in an engineering workplace is likely to depend critically on the norms of the dominant engineering workplace culture of the company in question.

Second, it may be that an IT firm to which private data about individual computer user activity is a potentially important economic resource and which aspires to a reputation for ethically responsible engineering practice will have to continuously socialize its managers and engineers to view user privacy protection *not* as an ethereal *cost* that is greatly outweighed by the potential economic benefit that appears derivable from a trove of covertly obtainable payload data but rather as an important ethics-related *constraint* that must be satisfied in the design and implementation of all data-collection projects. Put differently, engineers and managers may need to be socialized into regarding privacy protection as *a trumping factor* that when violated on a substantial scale calls for denying permission to launch a payload-data-collection effort, even if its projected economic benefit is believed to greatly outweigh its aggregate cost to individual privacy.

The SV case stands in an interesting relationship to Case 5, about Diane the software engineer. A client requested that Diane design a database management system that, by virtue of its low-level security features, she believed would put sensitive employee information at serious risk of being accessed without authorization, possibly putting those employees at unreasonable risk of harm. The question for Diane was whether it would be ethically responsible for her to accept the design job from her potential client, given the low level of informational privacy protection the latter specified for the system. In the Google SV case, the question for Engineer Doe was not whether he should acquiesce in ethically problematic design constraints imposed on him by senior management or top SV engineers but whether it was ethically responsible for him to push toward implementation the personal payload-data effort he designed and wrote the code for without first taking reasonable steps to ensure that collecting such, without permission, was ethically as well as legally acceptable. Diane had to decide what would be an ethically appropriate response to her potential client's troubling specification, while Doe seems to have conceived and enabled a broad expansion of Google's Wi-Fi data–gathering effort, possibly without first having carefully explored the ethical acceptability of the expanded initiative that, once known, provoked broad public opposition or without having undertaken to verify his assumption that the information gathered would be fragmentary and not linkable to specific individuals.

In one respect the SV case is reminiscent of the Citicorp case, namely, as regards the importance of the corporate culture factor. LeMessurier effectively tolerated or did not oppose or seek to change a culture in his firm that allowed safety-related structural changes to be made by subordinates without first being run by him, the most qualified relevant technical expert. Similarly, senior engineers and/or managers at Google effectively seem to have tolerated or not opposed an engineering workplace culture that allowed significant new data-gathering efforts to be designed and launched by lower-level engineers without their first having to get informed and considered judgments that the proposed undertakings were legally permissible and ethically acceptable. A relatively hands-off culture in an engineering workplace may foster or expedite fruitful initiatives, but, without appropriate checks in place, it can also allow certain problematic activities to be launched that might otherwise have been caught and prevented through responsible managerial oversight.

Discussion Questions

1. If an engineer sends his or her engineering manager the design for a new project that he or she is proposing be carried out, and the superior does not read or cursorily scans it, and/or does not respond to it, does the superior bear any ethical responsibility if when the plan is implemented as designed, it is widely and reasonably judged ethically objectionable? Why or why not?

2. Suppose an engineer realizes that her or his proposed data-collection project may violate the privacy of certain individuals targeted in the process of implementing it. If she or he neither attempts on her or his own to determine before proceeding whether, on balance, the project is ethically acceptable, nor proactively solicits the judgment of her or his engineer or nonengineer superiors or peers about its ethical acceptability, can such an engineer plausibly claim to have behaved in an ethically responsible manner? Why or why not?

3. If an engineer believes that the innovative payload-data project that he or she is proposing has major economic potential for his or her company, can it be ethically responsible for that engineer to proceed to implement the project if he or she knows that the privacy of a substantial number of individuals whose personal payload data is to be covertly collected will be violated? Why or why not?

4. If an IT company's engineering workplace culture strongly encourages its engineers to initiate technically innovative data-collection projects if they believe they have significant economic potential, is the existence of that culture enough to exonerate a company engineer from a charge of being ethically irresponsible if the implementation of her or his project involves covertly collecting and retaining personal information about individual computer users without their consent? Why or why not?

5. Does the precautionary derivative ethical responsibility to be extra vigilant when dealing with paradigm-departing engineering product designs, as discussed in the Citicorp Center case, apply to the Wi-Fi data collection effort discussed in Google Street View case? Explain.

CASE 18: OPIOID BIOSYNTHESIS
AND NEURAL ENHANCEMENT

Background

The final cases to be discussed pertain to a burgeoning multidisciplinary engineering field: *bioengineering*, also known as *biological engineering*.

Bioengineering lies at the intersection of engineering, biology, and medicine. It brings a range of engineering methods to bear on the rapidly growing body of knowledge about the basic physical properties and principles of living matter. Some bioengineers are interested in diagnostic and therapeutic applications, such as medical imaging, prosthetics, and tissue repair and regeneration. Others are interested in manufacturing applications or in developing biomaterials for energy- and environment-related applications. While areas of research priority vary from one bioengineering department to another, Brey's characterization of the field's terrain is helpful:

> As a field, [biomedical engineering] is very broad, with applications ranging from molecular imaging to the construction of artificial hearts. Biomedical engineering is however narrower in scope than *bioengineering*, or *biological engineering*, with which it is sometimes equivocated. Bioengineering focuses on the engineering of biological processes and systems in general, and includes not only biomedical engineering but also agricultural engineering, food engineering, and biotechnology.[554] (italics in original)

[554]Brey (2009).

Given the variety of subfields of bioengineering, no single case can illustrate the diverse ethical issues that bioengineers face and are likely to face in the future. Hence, in what follows I shall explore a mini-case from each of two subareas of bioengineering, ones in which ethical issues arise that differ from those characteristic of long-established engineering fields. Some activities referenced in the mini-cases are current and seem ethically benign and straightforward. Others are embryonic or projected to occur in the foreseeable future and raise controversial ethical issues.

In the mini-cases, the focus will not be on resolving the ethical issues raised. Rather, it will be on highlighting that bioengineering is undergoing rapid development in many of its areas, that some of its current and projected future developments raise challenging ethical issues that will be hard for bioengineers to ignore, and that bioengineers need to think critically and comprehensively about complex ethical issues posed by many advances in their field to be effective and responsible practitioners. Doing so is especially important in this field, since law, policy, and codes of engineering ethics are finding themselves hard pressed to keep up with the ethically challenging innovations being generated by work being done at the interface of engineering and organismic life.

Mini-Case 1: Opioid Biosynthesis

BACKGROUND

Opioids are an important class of painkilling medicines. They include the natural opiates—morphine, thebaine, and codeine—all derived from opium, and synthetic opioids, such as hydrocodone and oxycodone. The sole source of opioids has long been—and remains—the opium poppy plant. It can take a year to go from harvesting plants on licensed poppy farms, to the extraction of active drug molecules, to their refinement into medicines in pharmaceutical factories.

In 2015, working in a laboratory "permitted and secured for controlled substances,"[555] a team of researchers led by Christina Smolke, a Stanford bioengineer/synthetic biologist with a PhD in chemical engineering and a postdoctoral fellowship in cell biology, demonstrated "proof of principle"[556] biosynthesis—starting from sugar—of the opioids

[555]Galanie et al. (2015).
[556]Ibid., p. 1095.

thebaine and hydrocodone. The team first identified the "enzymes that help convert glucose into a morphinan scaffold."[557] These enzymes were taken from three species of poppy, goldthread, rat, bacteria, and yeast.[558] Then the team "made the DNA that gives the yeast the instructions to produce those enzymes" and inserted the engineered DNA "snippets" into the genome of baker's yeast cells. The inserted genes enabled the yeast to express the 21 enzymes needed to synthesize thebaine, a precursor to other opioids, including morphine, and the 23 enzymes needed to synthesize hydrocodone.[559]

This "major milestone"[560] has the potential to yield important benefits. It may lead to a radical shortening of the opioid production process (to 3–5 days), ensure that sufficient supplies of opioid painkillers are regularly available to meet demand (rather than being at the mercy of variable environmental factors), and lower production costs, a factor especially relevant to pain sufferers of modest means who live in less developed countries.[561]

An equally if not more important result of the team's work may be the methods and techniques it used for the proof-of-concept production of the two opioid compounds. According to Smolke,

> This is only the beginning. . . . The techniques we developed and demonstrate for opioid pain relievers can be adapted to produce many plant-derived compounds to fight cancers, infectious diseases and chronic conditions, such as high blood pressure and arthritis.[562]

The medical benefits derivable from such compounds make the development of those techniques appear to be an unmixed blessing.

ETHICAL ANALYSIS

However, the bioengineering of opioids in yeast also has a problematic side. Lower opioid production costs could lead to lower prices and increased demand, a development that could result in increased opioid

---

[557] Peplow (2016). A "morphinan scaffold" is a chemical compound that can serve as the basis for the production of the morphine-like alkaloids (morphine, codeine, and thebaine) and semisynthetic morphine derivatives, such as oxycodone and hydrocodone.

[558] Galanie et al. (2015).

[559] As of mid-2015, the developmental path followed by the genetically modified yeast strains had not been optimized to substantially increase opioid yield. Yields remain extremely small. Given these yields, scaling up the lab production process in industrial fermenters is not yet economically viable. See ibid., pp. 1095 and 1100.

[560] Jens Nielsen, quoted in Service (2015).

[561] Peplow (2016).

[562] Abate (2015).

abuse, opioid addiction, and opioid overdose–related death, phenomena that have plagued the United States in recent years.[563] Aware that the stakes are high, Smolke has expressed the following general view:

> We want there to be an open deliberative process to bring researchers and policymakers together. . . . We need options to help ensure that the bio-based production of medicinal compounds is developed in the most responsible way.[564]

The biosynthesis of opioids has a related worrisome aspect. Suppose that the same bioengineering techniques the team developed and demonstrated in making small amounts of two opioids from genetically engineered yeast and sugar could also be used to biosynthesize a dangerous opioid like heroin, and do so more efficiently, cheaply, and locally.[565]

That possibility is more than theoretical. According to U.C. Berkeley bioengineer/synthetic biologist John Dueber, it will not be long before illegal drug researchers "could come up with the same technology":

> We're looking at a timeline of a couple of years, not a decade of more, when sugar-fed yeast could reliably produce a controlled substance. . . . The field is moving surprisingly fast, and we need to be out front so we can mitigate the potential for abuse.[566]

Whereas Dueber once believed that the biosynthesis of a morphine-making yeast was a decade away, by mid-2015 he had come to believe that a low-yielding strain could be made in two or three years.[567] According to Peter Facchini, once the mechanism by which a plant produces a particular substance naturally is understood, there is no reason why yeast cannot be engineered to produce it biosynthetically. Thus, for example, "someone could potentially produce cocaine in yeast."[568]

This possibility raises a disturbing question: could the Smolke team's methods be adapted to "home-brew" opioids for illicit use? Her view is that home-brewing is "less of a concern,"[569] presumably meaning that

[563] "In 2015, United States drug overdose deaths exceeded 50,000; 30,000 involved opioids. There were more deaths from opioid overdose than not only from motor vehicle accidents, but also than from HIV/AIDS at the peak of the epidemic in 1995." Gawande (2017), p. 693.
[564] Abate (2015).
[565] "Poppy fields are not readily available to someone in Chicago, whereas yeast can be made available to anyone." Vincent Martin, quoted in Service (2015).
[566] Perlman (2015).
[567] LePage (2015).
[568] Ibid.
[569] Peplow (2016).

she doesn't find her team's research results especially worrisome. Smolke holds that view for several reasons.

First, the opioid strains her team biosynthesized have extremely low yields. For example, "a single dose of hydrocodone, as used in Vicodin® (5 mg) would require thousands of liters of fermentation broth, which no home-brewer would reasonable pursue."[570] Moreover, previous work on converting thebaine to morphine—a substance closely related to heroin—realized only a 1.5% yield.

Smolke, Drew Endy, and Stephanie Galanie claim to have shown experimentally that "the first example of yeast engineered to produce opioids from sugar under laboratory conditions does not produce detectable amounts of natural opiate or semi-synthetic opioid drug molecules in simple home-brew fermentations."[571]

However, obtaining extremely low yields of heroin precursors is not a preordained, immutable outcome of genetically engineering yeast strains. Indeed, Smolke writes, "substantially improved production of opioids via yeast should be expected in the next several years."[572] That anticipated development should arguably be cause for serious concern. Smolke has founded a company, Antheia, in Palo Alto, to greatly increase the outputs of the opiates her team was able to synthesize.[573]

Second, Smolke believes "the [opioid producing yeast] strains would also be tightly regulated, both for intellectual property reasons and because of drug enforcement."[574] However, the theft of yeast strains from bioengineering labs or legal fermentation factories, as well as the rise of do-it-yourself genetic engineering of yeast strains by curious or economically motivated members of the public cannot be ruled out a priori and may not easily be prevented.[575]

Third, Smolke argues that

> it's not clear that the way we make the drugs determines how much gets into the black market. Other things also impact this—policies toward prescription drugs and how we help addicts, for example—so the conversation has to go a lot further than the technology.[576]

---

[570] Galanie et al. (2015).
[571] Endy et al. (2015).
[572] Ibid.
[573] Service (2015).
[574] Peplow (2016).
[575] Le Page (2015).
[576] Peplow (2016).

Although it may well be true that factors other than the way opioids are produced affect how much of a given opioid reaches the black market, this does not negate the fact that, barring effective countermeasures, public disclosure of the methods and processes used in biosynthesizing thebaine and hydrocodone in yeast could be a contributory factor to greater quantities of dangerous opioids eventually reaching the black market. Indeed, Dueber has been paraphrased as follows: "If all the instructions are out there, re-creating the strain [of a desired opioid] from scratch starting with baker's yeast would not be that difficult, Dueber says."[577]

The preceding considerations notwithstanding, if the methods used by Smolke's team (and others) do make the biosynthesis of an opioid like heroin more feasible, what can be said, from an ethics perspective, about the team's decision to publish the details of their bioengineering achievement in *Science*?

As noted, the research team leader commented, "We need options to help ensure that the bio-based production of medicinal compounds is developed in the most responsible way."[578] A key word in this sentence is "ensure." But, from an ethics perspective, the *timing* of the team's publication is a no-less-important consideration. The question posed in the last paragraph needs revision: what can be said from an ethics perspective about the team's (and the journal's) decision to publish its manuscript in September 2015, well *before* the societal "options" needed to "ensure" that the biosynthesis of medicinal compounds is done in "the most responsible way" have been developed, tested, and implemented?

This situation raises an even more general ethics question: under what circumstances, if any, do bioengineering researchers (and counterparts in other fields of engineering) have an FERE1-based derivative ethical responsibility to temporarily withhold from publication or significantly redact a manuscript reporting research results expected to yield significant medical benefits but that could also facilitate harmful outcomes if individuals with malign intent gained access to or adapted them?

This scenario is not unprecedented. For example, in 2002, in the wake of 9/11, the U.S. government decided that technical papers about biological agents with the potential to be weaponized should be subject to censorship, to prevent misuse of the potent knowledge they contained by individuals with malevolent intent.[579] There are, however, two differences.

[577] Le Page (2015).
[578] Abate (2015).
[579] Broad (2002).

In the opioid case, the issue, at this point at least, does not involve not censorship by others but rather authorial self-censorship. Moreover, in the opioid case, there is no fear of bioterrorism. Rather, the major concern is the possibility of efficient, localized, difficult-to-detect home-brew production of illegal addictive drugs that could lead to increased use, with all its consequences.

In 1998, two Australian researchers who were trying to develop a genetically engineered virus to combat common pests inserted the gene for interleukin-4 (IL-4) into mousepox virus.[580] The result was a genetically engineered mousepox virus of unprecedented lethality. The researchers quickly realized that their method could be used to turn smallpox into a potent biological weapon. After considering the situation, in 2000 the researchers submitted a manuscript to the *Journal of Virology* describing what they had done but made no reference to possible applications of their technique to bioterrorism.[581] They presumably did so even though aware that what the manuscript revealed could possibly be seized upon and adapted by individuals intent on developing a bioweapon of mass human destruction.[582]

The authors apparently concluded it was better to diffuse their findings than to suppress them, in hopes that doing so would engender fruitful efforts to develop antidotes for any genetically modified human-targeting virus that might be developed by adapting what they had disclosed. That decision effectively placed a large wager on a time-sensitive outcome that was by no means assured.

The publication of the bioengineers' paper on the biosynthesis of opioids from genetically engineered yeast raises the same delicate ethical challenge as the Australian virus researchers faced. Publication of the methods used may help pave the way for the development of safer, less addictive painkillers and other therapeutic drug products.[583] However, publication of the team's manuscript by no means guarantees that societal "options" for managing the enhanced opioid production in "the most responsible way"—presumably a way that would effectively preclude illicit biosynthesis of, say, heroin and cocaine—will have become operational *before* the revealed production methods are appropriated and applied by individuals bent on inflicting harm on selected social groups or societies.

---

[580] Cohen (2002).
[581] Jackson et al. (2001).
[582] Cohen (2002).
[583] Le Page (2015).

Apart from wanting to secure recognition as the first team to completely synthesize opioids in yeast, what was the dominant mindset of the members of Smolke's team about the relationship between publishing the paper in *Science* when they did and ensuring that opioids would be biosynthesized in the most responsible way? It is difficult to know, but several possibilities come to mind.[584]

The first possibility (P1) is that most team members, after doing due diligence, *believed* that a practical way—combining technical, educational, legal, regulatory, and policy measures—exists or could be found and successfully implemented in time to ensure that the biosynthesis of opioids in yeast would proceed in the most responsible way, for example, such that the risk of exploiting to harmful effect what is revealed in their paper was extremely low. A second possibility (P2) is that most team members *assumed*, without doing due diligence, that such a way could be found and implemented in time. A third possibility (P3) is that most team members simply *hoped* such a way could be found and implemented in time but had no solid reasons for believing that was possible.

If the team submitted its manuscript for publication under P1, then doing so would seem to have been ethically responsible precautionary conduct, fully in accord with FERE1. But if, however unlikely, they submitted the manuscript under P2 or P3, then, given the "unreasonable risk of harm" clause of FERE1, it is unclear whether that submission would qualify as ethically responsible.

There is, however, a fourth possibility (P4) that could reflect what actually transpired: the team submitted the manuscript for publication even though most or all members were *unsure* of whether a practical way could be found of ensuring that opioids would be biosynthesized in yeast in the most responsible way, and did so with the intention of *provoking serious discussion* about how to structure a governance scheme that would make the illicit biosynthesis of opioids virtually impossible. If P4 was the dominant mindset about the relationship between methods disclosure and a preventive governance scheme informing the team's submission of its manuscript to *Science* and its subsequent publication in September 2015, that mindset might qualify as ethically praiseworthy in intent and be in accord with FERE2. However, consequentially speaking, an impartial ethics "jury" might request more time to discuss whether

<hr/>

[584] All include the premise that the team members believed the projected downstream benefits of biosynthesis of opioids are substantial and highly likely to result from their achievement.

publishing the paper at that point was prudentially responsible or negligently premature.

Some engineers and scientists deny they have an ethical responsibility to refrain from publishing papers reporting research findings with significant dual-use potential, that is, with the potential to be used to realize not only the benefits that the researchers had in mind in doing the work but also the potential to be "misused by state and non-state actors for nefarious purposes."[585] However, others believe that under certain social conditions, withholding publication of certain findings, albeit temporarily, would be the appropriate, ethically responsible thing to do.[586]

All things considered, was the Smolke team's decision to publish its paper in *Science* in September 2015 in accord with FERE1? That might depend on whether in contemporary U.S. society it *is* possible to devise a governance scheme capable of ensuring that the biosynthesis of medicinal compounds proceeds in the most responsible way.[587] If the Smolke team's methods are not, or not easily, adaptable by nonprofessionals with ill intent, and/or if putting such a governance scheme in place in fairly short order was plausibly possible, then it is arguable that publication of their research in September 2015 was in accord with FERE1. However, to the extent that (a) the Smolke team's methods *are* or will shortly be adaptable, with modest additional intellectual effort and affordable infrastructure cost, to permit the biosynthesis of a harmful opioid like heroin, and (b) it is *unrealistic* to think that a governance scheme could be devised and implemented in time to effectively prevent the illicit biosynthesis of opioids, it is arguable that publishing the paper in question at that juncture was not in alignment with FERE1.

---

[585] Selgelid and Weit (2010).

[586] During the controversy over recombinant DNA research in the 1970s, biologist Robert Sinsheimer posed a question: "Can there be 'forbidden,' or, as I prefer, 'inopportune' knowledge?" See Sinsheimer (1978), p. 23. He held that under certain conditions there could be, e.g., if a compelling case was made that freedom to pursue that knowledge at that point in time would undermine other valued liberties. Sinsheimer spoke of implementing a moratorium on the pursuit of such potentially disruptive knowledge until doing so was no longer 'inopportune,' i.e., until society was adequately prepared to grapple with it.

[587] See Oye et al. (2015). The authors make four "recommendations" for preventing the home-brewing of opiates: (i) develop yeast strains that are "less appealing to criminals"; (ii) require that "commercial organizations that make stretches of DNA to order" screen for "sequences [used] for opiate-producing yeast strains"; (iii) "keep opiate-producing yeast strains in controlled environments licensed by regulators"; and (iv) extend current laws covering opiates "to cover opiate-producing yeast strains" and "make their release and distribution illegal."

As with the 2001 publication of the lethal mousepox paper in the *Journal of Virology*, the publication of the opioid biosynthesis paper in *Science* in September 2015 seems tantamount to placing a major wager, here on the proposition that an effective mode of governance, including laws, education, regulations, and policies, could be devised and made operational *before* any adaptive applications of the published methods were implemented by unauthorized parties with malign intent.

Whether or not the reader believes that, all things considered, including Smolke's promotion of widespread discussion of the societal risks of their research methods and of how to minimize them,[588] it was ethically appropriate for the team to publish its research results when and where it did, the conclusion of the Smolke team's *Science* article clearly evidences genuine concern about the responsible conduct of bioengineering research. After noting "the complexity and diversity of both the potential concerns and possible benefits," the lead synthetic biologist and her team added in the article's final paragraph that they

> strongly endorse an open deliberative process that develops options for the governance of medicinal compound biosynthesis before economically competitive processes are realized.[589]

Thus, in the very paper in which Smolke and her team reported their results, Smolke urged, in the spirit of FERE2, that "researchers, policy experts, regulatory and enforcement officials, health and medical professionals, and representatives of communities in which essential medicines are either unavailable or abused,"[590] convene and identify the best preventive governance options.

Dueber took a different precautionary step. Before publishing the results of their recent research on opiate biosynthesis, Dueber and another bioengineer/synthetic biologist approached biotechnology policy specialists at MIT and "requested advice on how they might maximize the benefits of their research while mitigating the risks."[591] Although Dueber's view is that substantial technical work lies "between the aspiring drug lord and his closet bioreactor," he believes that "[e]ventually [the illicit

---

[588] "We've been working on this for a decade, and we're definitely aware of the issues involved, so we are seeing to it that they are discussed widely." Smolke, quoted in Perlman (2015).

[589] Galanie et al. (2015).

[590] Endy et al. (2015).

[591] Oye et al. (2015).

biosynthesis of dangerous opioids] will happen."[592] However, he holds that the potential benefit that the new biosynthesized strains of genetically modified yeast offer as a platform for developing new drugs is so substantial that it makes doing such work and publishing its results worth the risk doing so carries—*as long as appropriate precautions are taken*.[593] However, Dueber is concerned that what he sees as a conditional positive benefit-risk analysis may not be obvious to others:

> It's easy to point to heroin; that's a concrete problem. . . . The benefits are less visible. They are going to greatly outweigh the negative, but it's hard to describe them.[594]

The present author's view about this striking example of what some call "the dual-use dilemma"[595] is that it is extremely important to protect researchers' ethical right to publish the results of their inquiries. A strong presumption in favor of respecting that right should be adopted and maintained. However, depending on key factors, such as (i) the nature and implications of the research results, (ii) the society-specific characteristics of the social context into which those findings would be released, (iii) the financial and epistemological start-up costs of unauthorized parties' being able to adapt and exploit the research in question, and (iv) the magnitude of the harm that could be done through adapting and abusing the research results in question, it should not be ruled out a priori that a compelling case could be made in a particular instance that publishing research immediately would be ethically inappropriate and that important societal interests that would be protected by delaying, redacting, or not publishing it deserve to take precedence *for the time being* over a researcher's right to diffuse the results of her or his work. *However*, the burden of proof should be on those who want to override the researcher's right, not on the researchers who wish to invoke it.

Even if a bioengineer with research results with significant projected medical benefit and significant potential for being appropriated and turned to harmful social account concludes, after doing due diligence, that publishing his or her results is consonant with FERE1, his or her ethical responsibilities do not end there. It is arguable that in such cases bioengineers who undertake to publish such research have FERE2- and FERE3-based derivative ethical responsibilities to make explicit the risks

---

[592] Twilley (2015).
[593] Ibid.
[594] Ehrenberg (2015).
[595] See Miller and Selgelid (2007), pp. 524–527.

their publications carry, to take significant proactive steps to try to mitigate or eliminate them, and to try to alert and inform those at risk of incurring harm from the abuse of the bioengineers' work.

Mini-Case 2: Neural Enhancement

BACKGROUND

Interest in pharmaceutical approaches to cognitive and affective enhancement has grown in recent years. Some drugs, originally developed and prescribed to treat disabling brain-related conditions, have subsequently been used by healthy individuals for performance-enhancement purposes. Ritalin, initially used to treat attention-deficit disorder, and modafinil, initially used to treat narcolepsy and other sleep disorders, have been shown in some laboratory studies to help healthy individuals "sharpen focus and improve memory."[596]

Similarly, some technologies originally developed with therapeutic goals in mind could eventually lead to the ability to enhance mental processes in healthy human subjects. Consider the young but rapidly growing area of bioengineering called *neural engineering* or *neuroengineering*, "a field at the intersection of engineering and neuroscience."[597] It "uses engineering techniques to study and manipulate the central or peripheral nervous system," with goals that include "restoration and augmentation of human function."[598]

Several studies published in recent years have shown that, under laboratory conditions, the neural technology called "transcranial direct current stimulation" (tDCS), in which electrodes are attached to the subject's head and a mild current is run through the brain, enhances "learning, working memory, decision-making and language" in healthy research subjects.[599] "Research shows that tDCS . . . may increase brain plasticity, making it easier for neurons to fire. This in turn improves cognition,

---

[596] Masci (2016), p. 8. See also Fisher (2008) and J. O'Brien (2010).
[597] Brey (2009).
[598] Ibid.
[599] Farah (2011), p. 766. For example, according to Sparing et al. (2008), "Our finding of a transient improvement in a language task following the application of tDCS together with previous studies . . . suggest that tDCS applied to the left PPR [posterior perisylvian region] (including Wernicke's area [BA 22]) can be used to enhance language processing in healthy subjects." Dockery et al. (2009) found that "both anodal and cathodal tDCS can improve planning performance as quantified by the Tower of London test."

making it easier for test subjects to learn and retain things, from new languages to mathematics."[600]

Some believe that this external approach to augmenting brain function will eventually be complemented—or displaced—by an internal approach, in which a stimulative technology is embedded *in* the brain in hopes of enhancing cognitive, sensory, motor, or affective abilities, skills, feelings, or states of healthy humans.

A significant step in that direction was taken in April 2016, when Battelle Memorial Research Institute researchers used computer technology to enable a person with quadriplegia to regain a measure of control over his right hand and fingers.[601]

The technological system used includes a computer chip that had been implanted in the research subject's brain and connected to an external computer, in turn connected to a cuff on the patient's arm. By intently imagining making certain movements with his hand, fingers, and/or wrist, the subject transmitted signals from his brain to the chip, which was linked by cable to an external computer equipped with specially designed software. The latter decoded the pattern of brain signals sent to the computer, then sent instructions directly to the electronic sleeve that induced the fingers and/or hand and wrist to execute the intended actions. Thus, "limb reanimation"[602] was accomplished by having signals from the patient's brain bypass the injured spinal cord and, after computer mediation, go directly to the limb in question. After a long and difficult process of learning, the individual in question was able to "pour from a bottle and to pick up a straw and stir."[603]

The researchers used brain imaging to "isolate the part of [the subject's] brain that controls hand movements." Then, in surgery, a microchip the size of an eraser head was implanted in "the exact location" of the brain's motor cortex region that controls hand movements. "The chip holds 96 filament-like 'microelectrodes' that record the firing of [a few hundred] individual [motor] neurons." When the subject concentrates on/ imagines hand or finger movements he wants to make, "the firing patterns [of the neurons] [are] picked up by the chip and run through a cable . . . fixed to a port on the back of his skull and connected to a computer." The computer software contains an algorithm to "decode those [neuron] firing patterns" and send appropriate signals to 130 electrodes

[600] Masci (2016), p. 9.
[601] Carey (2016).
[602] Ibid.
[603] Ibid.

embedded in a sleeve or cuff the patient wears on his forearm.[604] The cuff then transmits impulses through the skin to the muscles in the patient's hand and wrist to trigger the contractions needed to execute the desired finger, hand, or wrist movements.

Whereas tDCS was applied to the brain as a whole, the Battelle achievement required knowing the exact location of the part of the brain that controls finger and hand movements and actions. Moreover, while tDCS has proven to augment certain cognitive skills in some studies, the Battelle achievement was therapeutic-restorative in intent and involved motor skills. It was not an attempt to demonstrate enhancement and did not target cognitive, sensory, or affective skills, feelings, or states.

However, let us consider an ethically provocative conceivable future scenario in neuroengineering. Suppose that in the next few decades brain researchers generate a detailed, accurate map of the whole brain showing which parts control the exercise of which cognitive, sensory, motor, and affective skills and feelings. In principle, such a map could make it possible for specialized microchips to be implanted in different localized parts of the brain. This could be done *not* to record and transmit patterns of firing neurons to help overcome a motor-skill disability but to stimulate the neurons in selected parts of the brain to *augment* specific cognitive abilities or skills connected with those parts of the brain.

ETHICAL ANALYSIS

What can be said from an ethics perspective about this speculative brain-enhancement possibility? Assuming that the informed consent of the patient has been secured, the dominant mainstream position among ethicists, doctors, and relevant technical professionals has been that noncosmetic, noninheritable technological interventions in the human body, such as genetic engineering and neural engineering activities, are ethically acceptable only if used for therapeutic purposes, and that such interventions are ethically unacceptable and irresponsible if undertaken for purposes of "enhancement."[605]

For example, in February 2017, a panel of the U.S. National Academy of Sciences and the U.S. National Academy of Medicine conditionally

---

[604] Ibid.
[605] See Harmon (2017).

approved "the modification of human embryos to create genetic traits that can be passed down to future generations." However, the panel endorsed only therapeutic "alterations designed to prevent babies from acquiring genes known to cause 'serious diseases and disability,' and only when there is no 'reasonable alternative.'" In contrast, the panel's report "called for prohibiting any alterations resembling 'enhancement,' including 'off-label' applications."

In 2016, the Pew Foundation surveyed the attitudes of the U.S. public about emerging and possible future technologies of enhancement.[606] One of the three scenarios presented to the respondents was as follows:

> New developments in understanding the brain are creating the possibility that doctors will be able to surgically implant a small computer chip in the brain. Right now, these implanted devices are being developed for people with some kind of illness or disability. But in the future, these implanted devices could potentially be available for use by healthy individuals, giving people a much-improved ability to concentrate and process information in everyday life.[607]

The survey's main finding about public reaction to this scenario was as follows:

> Seven-in-ten Americans say they are "very" (28%) or "somewhat" (41%) worried about the possibility of technology that would allow an implanted computer chip in the brain to give healthy people an improved ability to concentrate, but only about a third are "very" (9%) or "somewhat" (25%) enthusiastic about this prospect. A majority (64%) say they are "not too" or "not at all" enthusiastic about this potential technology.[608]

While about one-third (32%) of the respondents indicated they would want such a chip implanted in themselves, about two-thirds (66%) indicated they would not.[609]

In the discussion of Case 10, it was noted that a 2006 survey of more than a thousand nanotechnology researchers revealed that when asked how ethically acceptable they found the goal of "increasing human mental abilities" by using nanotechnology, 34.8% of the respondents replied "very" (18.4%) or "quite" (16.4%) morally acceptable, while 38.1% responded "slightly" (20.1%) or "not at all" (18%) morally acceptable.

---

[606] Masci (2016).
[607] Funk et al. (2016c).
[608] Ibid.
[609] Ibid.

This finding and the aforementioned findings of the Pew survey strongly suggest that even if its safety and efficacy are assured, any proposal or attempt to enhance a human mental skill or ability through advanced technology is likely to provoke serious ethical controversy, in both the research community and society at large.

To help clarify thinking about the controversial ethical issue of human cognitive enhancement, let us distinguish *four categories of technology/ science-based (T/S-based) brain interventions* that improve, increase, or otherwise enhance a cognitive ability or skill of a healthy human subject.

C1: those that *preserve* a targeted human cognitive skill by preventing its deterioration from its current level in the normal range to a significantly "lower" level, perhaps one below the normal range;

C2: those that *restore* a targeted human cognitive skill to a level it had formerly reached, or would have reached but for an illness, disease, or genetic, neurological, or physiological flaw or defect.

C3: those that effect a *minor quantifiable* improvement in some cognitive ability or skill of a healthy person, to a level greater than she or he had ever previously realized but that many other humans already exemplify without such interventions.

C4: those that effect a *major quantifiable* improvement in some cognitive ability or skill of a healthy person, to a level that he or she had not even come close to attaining previously and has rarely been realized by other humans.

What can be said about each of these categories of T/S-based brain interventions in relation to the entrenched ethics-relevant distinction between "therapy" and "enhancement"? C3- and C4-type interventions seem to qualify as unambiguous enhancements of the skills involved.

C1- and C2-type brain interventions appear to be obvious examples of therapeutic efforts. However, if a skill whose level has been preserved and maintained by a C1-type intervention would normally, over time, have fallen significantly below the preserved level, then that intervention could plausibly be called an "enhancement" of the skill involved. Similarly, a C2-type intervention would arguably qualify as an "enhancement" if the skill involved was restored to a level significantly greater than where the skill had naturally fallen over time.

C4-type interventions would probably be most likely to elicit the most ethical controversy. Given the major increases in skill levels they would bring about, they would probably evoke strong ethical concerns that those able to access and partake of such enhancements would gain unfair advantage over those unable or unwilling to do so. Since C3-type interventions result in skill levels already realized by many, they would probably evoke weaker opposition and would likely be regarded as less ethically objectionable than C4-type interventions.

In light of this analysis, for a neural engineer/neuroengineer (or anyone else) to make the same ethical judgment about the acceptability of all T/S-based brain interventions that enhance human cognitive skills would arguably muddy the ethical waters. Instead of painting all T/S-based enhancement interventions with the same broad ethics brush, it would arguably be preferable to situate each one (or each category) someplace along an ethical acceptability/unacceptability scale. Its location on the scale would depend on various criteria, such as (i) the extent to which allowing that type of intervention to be freely carried out, in the society in question, would foster increased human inequality and (ii) the extent to which a particular cognitive enhancement would, in the society in question, exert pressure on others to acquiesce in undergoing that intervention to keep up with those who had already done so. Interestingly, the Pew survey found that 73% of respondents "believe inequality will increase if brain chips become available because initially they will be obtainable only by the wealthy."[610]

This perspective notwithstanding, the ethical judgments of many individuals who oppose the idea of implanting brain chips to enhance human cognitive skills appear to be based not on beliefs about the *consequences* of such interventions but rather on the belief that they are *inherently* ethically unacceptable. Such interventions are seen by many highly or moderately religious U.S. adults as "changing God's plan," "meddling with nature," and as "crossing a line that should not be crossed."[611] But individuals whose ethical judgments about technological innovations are not primarily shaped by beliefs that they are violations of natural/divine orders, such as agnostics and atheists, are considerably more likely to judge different categories of T/S-based neural enhancement interventions as ethically acceptable or unacceptable to different degrees, depending on the acceptability or unacceptability of what they believe will be their consequences.

Given this background, it appears increasingly likely that a controversial ethical issue will confront neural engineers in the next half century: whether they should participate in R&D work about brain-chip

---

[610]Funk et al. (2016d).
[611]Funk et al. (2016c).

implants that, if successful, will enable human cognitive abilities or skills to be enhanced. More specifically, given (a) the projected individual benefits and societal costs of neural enhancement work in societies like the United States, (b) the ample resources likely to be available to engineers who pursue such work, and (c) the natures of FERE1 and FERE2, future neural engineers may eventually be confronted with a dual ethical challenge. If such an engineer has reason to believe that access to technology-enabled cognitive enhancement is likely to occur in society in such a way as to bestow significant economic advantage on those who access it, thereby exacerbating human inequality—*arguably an incremental societal harm*—then she or he will have to decide whether participating in such work is something that she or he has an FERE1-based derivative ethical responsibility to avoid. Such an engineer will also have to decide whether she or he has an FERE2-based derivative ethical responsibility to try to prevent the projected societal harm from materializing or at least to mitigate its magnitude if such work is consummated.

If neural engineers working on human cognitive enhancement, like the synthetic biologists discussed in the opioids mini-case, believe the benefit of their work will greatly exceed any harm it may contribute to causing, they arguably still need to determine in good conscience whether any trumping factors, such as the magnitude of the harm involved (even if outweighed by the aggregate benefit) or the violation of Rawls's Difference Principle, suggest it might be advisable to forego or temporarily postpone pursuit of the projected greater benefit in order to avoid incurring the projected lesser but ethically unacceptable harm.

Lessons

The two mini-cases show that some young and burgeoning subfields of bioengineering raise contentious ethical issues, here involving research activity and publication. These issues appear to be quite different in nature than the ones explored earlier that centered on whether engineering activity directly causes harms or creates unreasonable risks of harm to individual affected parties. The issue of whether it is ethically responsible for a bioengineering research team to publish results with the potential to yield both important medical benefits and individual and societal harms, and if so, under what conditions, and the issue of whether it is ethically responsible for neural engineers to participate in brain-chip-based cognitive enhancement endeavors, and if so, under what conditions, seem worlds removed from the safety-related issues discussed in the Citicorp, Bhopal, Pinto, and Hyatt Regency cases.

However, it should be noted that in both bioengineering mini-cases, (projected) harms *are* involved, although in less immediate and less direct ways. Moreover, some of the projected downstream, indirect harms in question are *societal* harms, ones that some believe could or would result, in societies such as the United States, from untimely publication of research about the biosynthesis of potent addictive drugs, or from allowing access to brain-chip-based cognitive enhancement to hinge on ability to pay the (substantial) going market price. This situation recalls points made in Chapters 2 and 3: that it is important that engineers have a robust notion of harm, not a narrow one limited to physical and financial harm to individuals, and that engineers should be attentive not just to harms related to their work that are tangible, directly caused, and that manifest themselves immediately but also to those that are intangible, indirectly caused, and only manifest themselves downstream.

Whether openly publishing the results of research on how to biosynthesize opioids or other classes of potent drugs is ethically acceptable and responsible arguably depends not just on the magnitudes of the projected benefits and potential harms and on whether they are equitably distributed but also on contingent characteristics of the society in which the publication would take place and in which efforts would be made, by researchers and public officials, to prevent illicit production and distribution of the drugs in question by unauthorized individuals.

Implanting microchips in carefully chosen locations of human brains with the goal of achieving electrical stimulation–induced enhancement of specific human cognitive capabilities or skills is a possibility that could materialize within the next half century. Under what conditions, if any, would it be ethically permissible or responsible for a bioengineer to be actively involved in R&D work aimed at actualizing that possibility? That is a difficult question to answer at this point. An important reason is that it is not immediately clear whether permitting brain-chip cognitive enhancement interventions in a society with something like a market economy would actually exacerbate social inequality by setting back equal opportunity for economic advancement. It is one thing to invoke that disturbing scenario as something that *could* possibly occur but quite another to make a compelling case that it definitely *will* occur. However, other things being equal, the possibility that allowing brain-chip-based cognitive enhancement to be made available to those and only those who can afford to pay the going market price would actually intensify socioeconomic inequality is one that seems more credible given the central roles of information and cognitive skills in advanced economies.

Some will reply that, as with the introduction of numerous technological innovations in societies with free-market economies, over time neural enhancement will become much more affordable, and hence more widely accessible. However, it cannot be reasonably assumed without further ado that the standard evolving access pattern exemplified in the diffusion of mass-produced consumer electronic innovations, such as the DVD player, smartphone, and HD television—from initial access limited to the affluent few, to eventual widespread access to the nonaffluent many—will be repeated in the case of the bioengineering innovation of cognitive enhancement. Granted, economies of scale will likely lead to substantial reductions in the cost of customized brain chips, but such economies are likely to prove elusive as regards major reductions in the costs of chip implantation, calibration, and maintenance.

Judgments about whether it would be ethically responsible for neural engineers to participate in cognitive enhancement R&D should not hinge solely on the fact that "enhancement" rather than "therapy" is involved. Those judgments should be based on, among other factors, the kinds of technology-based enhancement interventions being considered, *prior* ethically scrupulous research into their effects on the brains and psyches of those undergoing them, and realistic assessment of the extent to which allowing such endeavors to proceed in a society with something approximating an information-centric market economy will actually contribute to causing incremental societal harm by undermining the core ethical values of liberty and equal opportunity.[612]

Discussion Questions

1. Does researchers' long-recognized ethical right to communicate to the research community the methods and results of their inquiries show that publishing a paper with dual-use potential, that is, one

[612]The ethical analyses of the two mini-cases have a common element. In both, it was suggested that a "realistic assessment" of some sort should play an important role in making judgments about the ethical responsibilities of bioengineers. In one case, what is needed is realistic assessment of the likelihood that a governance scheme can be devised that will actually preclude or make extremely difficult the illicit bioproduction of dangerous opioids. In the other, what is needed is realistic assessment of the likelihood that allowing access to brain-chip-based cognitive enhancement to hinge on ability to pay the going market price will exacerbate existing socioeconomic inequality.

containing findings that promise benefits but also could be turned to harmful account in society, is always the ethically responsible thing to do? Why or why not?

2. Under what circumstances, if any, might a bioengineer have a derivative ethical responsibility to refrain temporarily from publishing or to substantially redact a paper reporting the results of her or his research in an accessible scholarly journal? Give an example in which the circumstances are such that the bioengineer would arguably have an ethical responsibility to not publish or to substantially redact her or his paper, at least temporarily. If you do not think there are any such circumstances, indicate why you think that is the case. Do any FEREs play a role in your answer? Explain.

3. Suppose you were a member of the Smolke research team in early 2015, and the team was considering submitting a paper describing its complete biosynthesis of opioids in yeast for publication in *Science*. What, if anything, do you believe you and/or the team should have done to ensure that publication of the paper in *Science* later that year was an ethically acceptable and responsible thing for the team to do? Justify your view. Do any FEREs have a bearing on your answer? Explain.

4. Are T/S-based interventions in human brains for "therapeutic purposes" always ethically acceptable? Why or why not?

5. Are T/S-based interventions in human brains for cognitive "enhancement" purposes always ethically irresponsible and unacceptable? Why or why not?

6. Do bioengineers specializing in neural engineering have an ethical responsibility to avoid involvement in R&D work on T/S-based interventions for the purpose of realizing human cognitive enhancement? Why or why not?

CHAPTER 5

Key Case Ideas and Lessons

This chapter will recapitulate the key ideas used in exploring the ethical dimension of the cases and underscore some noteworthy lessons learned therefrom.

THE LEADING PRECEPT OF MOST CURRENT CODES OF ENGINEERING ETHICS

The cardinal precept in most current codes of engineering ethics is that "Engineers, in the fulfillment of their professional duties, shall: (1) hold paramount the safety, health, and welfare of the public."[1] However, in the first half of the twentieth century, the leading precept of most such codes was that engineers should be loyal to their employer or client. The post-WWII elevation of the "hold paramount" precept to the top spot in most current codes of engineering ethics probably reflects wider recognition of the powerful effects of engineering on contemporary society. Contemporary codes with "hold paramount" precepts effectively urge today's engineers to regard society as a whole as their omnipresent background client. Society at large enables much if not most contemporary engineering work, and its interests must take priority when they conflict with private interests of the engineer or the private firm or client for which he or she works. The notion that the foremost ethical responsibility of engineers in their professional work is to protect the safety, health, and welfare of the public is critically important. Engineering students and practicing engineers should internalize this idea and shape their professional conduct accordingly.

[1] See, for example, National Society of Professional Engineers (2007), Fundamental Canon #1.

A key issue in coming decades will be what, besides public safety and health should be designated as components of public welfare. Given the increasing importance of and pressure on environmental resources in contemporary society, *public access to salubrious environmental resources* and *progress toward a sustainable societal way of life* could well emerge as widely recognized components of public welfare over the next few decades. Similarly, given the pervasiveness and increasing criticality of information and information technology in contemporary life, *a social order characterized by user-friendly computer interfaces, a reasonable level of individual informational privacy,* and *universal access to basic Internet service* could also become widely recognized elements of public welfare. If so, they would achieve the status of being important social goods that engineers would have an ethical responsibility to not harm or jeopardize through their work

THE FERES

In elaborating the "ethical responsibilities approach" to evaluating engineering conduct from an ethics perspective, I articulated a foundational idea of this book: that the most fundamental ethical responsibility of engineers is to "combat harm." That notion was then unpacked into four fundamental ethical responsibilities of engineers, or FEREs. The first three—FERE1, FERE2, and FERE3—articulate general ethical responsibilities of engineers to combat harm in three ways: by *not causing* harm or creating an unreasonable risk of harm, by *trying to prevent* harm or an unreasonable risk of harm, and by *trying to alert and inform* about impending harm or an unreasonable risk of harm. The basic idea is that all engineers have these three fundamental ethical responsibilities vis-à-vis their work and work about which they are technically knowledgeable, and they have those responsibilities toward all those affected or likely to be affected by that work.

FERE4, also linkable to harm, can be viewed as a kind of implicit quid pro quo contract: employed engineers have an ethical responsibility to work to the best of their abilities to serve the interests of their respective employers or clients,[2] but only when the employer's or client's interests

---

[2] In working to the best of their abilities to serve those legitimate interests, engineers are presumably trying to benefit and avoid harming employers who treat them fairly rather than harming or exploiting them.

are "legitimate" and only as long as the engineers are treated fairly and reasonably by their employer/client.[3]

Another key idea of this book is that the applicable FEREs can and should be brought to bear upon concrete engineering situations to determine specific "derivative ethical responsibilities" of engineers in those situations. Examples of FERE-based derivative ethical responsibilities include adhering to the spirit as well as the letter of regulations or building codes in one's engineering practice; trying to persuade a client, under appropriate circumstances, that pursuing a course of action different from one initially stipulated by the client may be in its best interest; designing products that embody ethical and local cultural compatibility; and, when working with a radical or paradigm-departing product design, being extra vigilant and taking extra precautions so as to avoid causing harm or an unreasonable risk of harm by unwittingly treating the radical design as if it were a conventional, paradigm-conforming one.

In short, in concrete engineering contexts, the FERES are translatable into specific ethical responsibilities incumbent on the engineer(s) in those situations. The ability to recognize which FEREs apply to a particular engineering situation and being able to apply those that do to derive specific ethical responsibilities relevant to that situation are important skills. Leaving the ethical responsibilities of the engineer at the general level of the FEREs invites her or him to pay them mere lip service or to express mere ritual approval of the FEREs while remaining unaware that there are specific ethical responsibilities incumbent on her or him in the concrete situation in question.

ETHICS AND THE SOCIOLOGY OF CONTEMPORARY ENGINEERING

The sociological situation of the engineer in the early twenty-first century is an ethically challenging one. A healthy majority of engineers in the contemporary West work for medium- or large-sized private firms

[3] If an engineer is not an employee of a firm and not a consultant to a client but rather is working independently on a project, FERE4 might still apply. The engineer in question can be viewed as an "employee" of society at large. Given that society has in many ways made it possible for the engineer to practice, as long as the engineer is treated fairly by society in general—for example, as long as society does not arbitrarily restrict a person's attempts to earn a livelihood as an engineer or arbitrarily appropriate his or her intellectual property—the engineer has an ethical responsibility to do his or her best to serve society's legitimate interests.

whose top priorities include realizing a healthy return on their investments and maximizing relatively short-term profit. The resources available through such firms make much contemporary engineering work possible. However, engineers in such firms are at significant risk of being faced with conflicts of interest. Moreover, they rarely receive support from their employers, professional associations, governments, or (in most cases) the law if they resist or oppose employer demands or expectations that they act in ways the engineers deem ethically irresponsible.[4] Contemporary engineers working in small-scale IT start-up firms may not be subject to the same ethics-related pressures as their counterparts in established, large-scale firms. They are, however, subject to other ethical challenges, including ones related to raising capital, recruiting and retaining talent, protecting intellectual property, and designing software that does not violate noteworthy user or societal interests. It is vital that the education of contemporary engineers equip them to meet such challenges in a responsible way, rather than accepting uncritical engineer-employee adaptation to prevailing, possibly ethically problematic, employer practices.

AN ETHICALLY PROBLEMATIC PATTERN OF ENGINEERING PRACTICE

As seen in the Newton MessagePad, Pinto, and Westway cases, an ethically problematic pattern of practice recurs in contemporary engineering. It is sometimes difficult for engineers to bring themselves to oppose, resist, or reject tight time constraints imposed on them, ones that they know or suspect are overly compressed and may jeopardize safety or preclude rational decision-making. The problem is that upstream acquiescence in such a constrained schedule, while convenient, often leads to unfortunate downstream outcomes. This pattern of engineering practice might be called the *upstream acquiescence, downstream misfortune* pattern. The "misfortune" may take the form of misconduct and/or the occurrence of some significant harmful outcome. In the cases discussed, acquiescence in upstream compression may have contributed to the occurrence of unfortunate downstream phenomena, such as inadequate new-product-safety

---

[4] The kinds of support that professional engineering associations and the law provide to engineers who wish to engage in ethically responsible engineering practice are explored in Chapter 6.

testing, negative health effects on some engineers and/or some of those close to them, and use of the results of a truncated research study by a party who realized it was incapable of providing the evidence needed to support the conclusion the party defended.

WHISTLEBLOWING AND ETHICAL RESPONSIBILITY

FERE2 states that engineers have a fundamental ethical responsibility to try to prevent harm or an unreasonable risk of harm to parties affected by their work, work with which they are involved, or work of others about which engineers re technically knowledgeable. But when does this general ethical responsibility require that an engineer publically blow the whistle on her or his employer or client? For an engineer to have an ethical responsibility to publicly blow the whistle to try to prevent harm or an unreasonable risk of harm, it was argued that five conditions must be satisfied. First, the projected harm linked to the engineering work in question must be serious in nature and substantial in magnitude. Second, the engineer must have brought the harm in question to the attention of her or his immediate supervisor and that supervisor's immediate superior. Third, the engineer must not have received a response from either of those parties or their superiors that promised to lead to the prevention, mitigation, or prevention of the recurrence of the harm in question. Fourth, the engineer must have good grounds for thinking that fair-minded people literate about the area of engineering in question would take her or his claims about the harm in question seriously if they were disclosed to them. Fifth, the engineer must believe that there is at least a decent chance that, together with her or his evidence, insider knowledge, and technical authority/reputation, going public with her or his concerns will prevent or significantly lessen the harm (or unreasonable risk of harm) in question, or at least its recurrence.

The cases explored in Chapter 4 show that in engineering, whistle-blowing can occur in at least three types of circumstances, depending on the temporal relationship between the whistleblowing action and the harm in question. In **remote-anticipatory whistleblowing**, the would-be harm-preventing action is taken *long before* the projected harm is expected to materialize, even as early as the research phase (Parnas/SDI). In **imminent-anticipatory whistleblowing**, the harm-prevention attempt is made by the whistleblower *shortly before* the time the harm in question is at serious risk of occurring (LeMessurier/ Citicorp Center). In **post facto**

whistleblowing, the attempt at harm prevention occurs *after* the harm has occurred, either to bring to light the factors that actually contributed to causing it or to prevent repetition of that harm in the future (Boisjoly/ Challenger).

Whether an engineer has an FERE2-based ethical responsibility to blow the whistle publicly in a specific situation may also depend on whether his or her effort to *prevent* harm is likely to result in harm to the whistleblower or others as a by-product, and on the relationship between the magnitude, probability of occurrence, and distribution of the harm the engineer is trying to prevent, and the magnitude, probability of occurrence, and distribution of the harm that may result from blowing the whistle. On some occasions that relationship will be hard to pin down with confidence, making it difficult for the engineer to know with certainty whether he or she has an ethical responsibility to blow the whistle.

RISK AND THE IDEALIZATION OF TECHNOLOGY IN SOCIETY

When an engineer-evaluator evaluates the risk of a new technology or technological system, it is important that she or he avoid idealization. That is to say, she or he must not base her or his risk assessment on the idealized assumption that the new technology or system will function exactly as designed, or on the utopian belief that the social institutions and organizations expected to interact with the technology will function perfectly. It is naive (or disingenuous) for an engineer to contend that the risk of the new technology or system depends only on its built-in design features. The engineer-risk evaluator must also take into consideration a range of contingent, context-specific social factors related to the technology.[5] These include the training, experience, professionalism, and schedules of the workers involved with the technology; the track records, resources, and cultures of the companies that will construct, operate, and maintain the technology; and the track records, leadership, resources, and cultures of the government agencies that will regulate the technology. In evaluating risk, taking into consideration only the inherent design features of the technology or technical system in question is apt to lead the engineer to give the public an overly optimistic evaluation of the

[5] For an excellent critique of risk evaluation based on idealized notions of technology and society, see Beder and Shortland (1992).

new technology's risk, thereby unduly skewing public debate about its acceptability. The engineer has a derivative ethical responsibility to avoid such idealization and misrepresentation. She or he must ensure that her or his risk evaluations also reflect *the contingent social realities related to all aspects of the technology's life cycle.* Both risk evaluations, one based on built-in technical design features and the other based on contingent social realities, must be tolerably low for the risk of proceeding with the new technology or technological system to be "acceptable."

That the idealization of technology and the neglect or downplaying of contingent social realities can lead to an overoptimistic risk evaluation is clear from the Bhopal case. There, the technical design features of the MIC plant were arguably eventually overwhelmed by contingent social realities of the situation, including flawed plant operations, inadequate worker training, plant maintenance problems, ineffective zoning, infrastructural deficiencies, weak regulatory practices, inadequate emergency-planning practices, and serious budgetary constraints.

ETHICAL RESPONSIBILITY AND THE CULTURE OF THE ENGINEERING WORKPLACE

Determination to act in an ethically responsible manner may sometimes lead or require an engineer to violate a cultural value or behavioral norm in his or her field of engineering, a cultural norm in the engineer's firm, or a behavioral norm in a work group of which he or she is a member. In the SDI case, David Parnas went against the cultural grain that allegedly prevailed on the panel of experts of which he was a member. He refused to do what he believed to be an apparently common practice among the other members of his advisory committee, namely, accept or approve spending government money for research on work related to a project that the acceptor or approver believed could never be known to be trustworthy.

Having a solid grasp of what being an ethically responsible practitioner requires of an engineer can position him or her to selectively choose to diverge from his or her firm's cultural values, behavioral norms, and entrenched practices. If he or she has no idea of what his or her specific ethical responsibilities are in a given engineering situation, the likelihood is that the engineer will take the path of least resistance and adapt to the prevailing workplace culture, perhaps to his or her downstream legal or psychological regret.

AN OVERLOOKED ETHICAL
RESPONSIBILITY OF ENGINEERS

It is widely recognized that, under certain circumstances, engineers have ethical responsibilities to their employers, clients, and the public affected by their products and systems. It is less widely recognized that engineers can also have ethical responsibilities to their work groups, colleagues, and those who report to them. What seems to have escaped explicit recognition to date is that engineers can also have ethical responsibilities to significant others, partners, spouses, and other family members. Engineering managers' decisions, accepted without opposition or acquiesced in by engineering employees who report to them, can have a profound effect on the well-being of their family members. The cultures and practices of some engineering workplaces can be so demanding that FERE3's "try to alert and inform about impending harm and unreasonable risk of harm" provision implies that the engineers have an ethical responsibility to disclose upstream to significant others or family members onerous new demands on their time. They should disclose the threat to the engineer's health and well-being of a new, more arduous engineering work regime and possible negative psychosocial consequences of the new situation for family members. However, engineers are *not* entitled to assume that their spouses, significant others, or family members have an obligation to *adapt* to whatever changes in work regime the engineer undertakes or is forced to adopt regardless of its costs to those closest to her or him.

Martin and Schinzinger contend that new engineering projects can be viewed as social experiments.[6] What medical experiments do to medical practitioners, engineering "experiments" arguably do to engineering practitioners: namely, impose on the engineer an ethical responsibility to secure the informed consent of those likely to be "experimented upon." Martin and Schinzinger focus on innovative engineering projects as macrosocial experiments. Analogously, major changes in engineering workplace regimes can be viewed as *micro*social experiments. Since they may sometimes engender risks of harm to the engineer's family members, the would-be ethically responsible engineer must inform those individuals of risks in the offing and seek to obtain upstream their informed consent to the interpersonal and intrapersonal costs they may be at risk of bearing downstream.

---

[6]Martin and Schinzinger (2005).

AN ENGINEERING PROFESSIONAL

At times, calling someone an "engineering professional" means simply that the individual thus designated holds an engineering job for which he or she had appropriate technical school training and that he or she has had significant workplace experience in that kind of job. I call that being an engineering professional in the "weak" sense. But the expression can also be understood in a more robust, "strong" sense.

"Engineering professional" in the strong sense means someone who, in addition to holding an engineering job and having appropriate training and workplace experience, *integrates abiding primary concern for protecting and promoting societal well-being into her or his everyday engineering practice.* There is more to being an engineering professional in the strong sense—and to being an ethically responsible engineer—than being a experienced technical virtuoso or a technically competent "hired gun."

It is important to resist the reduction or dilution of the phrase "engineering professional" to the point that it means only "someone with appropriate training and experience who has worked competently in a certain field of engineering for a significant period of time." Effort should be made to preserve the strong sense of "engineering professional," meaning a qualified and experienced engineer who, in the performance of his or her professional work, consistently prioritizes protecting and promoting the safety, health, and welfare of the public.[7] A technically competent engineer who takes the FEREs seriously and tries to determine and fulfill the specific derivative ethical responsibilities that emerge from application of the FEREs to specific engineering situations qualifies as an "engineering professional" in the strong sense.

ENGINEERING DESIGN, PARADIGM DEPARTURE, AND THE ETHICS OF PRECAUTION

A critically important, overarching lesson emphasized in this book concerns the ethical responsibilities of the engineer doing radical (as opposed to normal) engineering design work. Such work involves significant departure from the dominant paradigm for the kind of product being designed in one or more dimensions, namely, the product's characteristic

[7]For valuable historical perspective on the terms "profession" and "professional," see Barker (1992).

configuration, key *operating principle*, constituent *material, method(s)* of analysis used in its design, *processes* of production and operation, and *use(s)* for that type of technological item. An engineer doing such work has an FERE1-based ethical responsibility to be attentive to whether the product or system being designed marks a substantial departure from the conventional paradigm, and if it does, to be extra vigilant and, as a precaution, check and double-check to see if it can be analyzed, produced, operated, or maintained in the ways that conventional products that conform to the dominant paradigm properly are, or whether appropriate changes must be made.

Uncritical or unwitting adherence to the reigning paradigm of practice vis-à-vis a particular kind of engineering product, especially when working on an innovative product, process, or system, is ethically irresponsible, for such adherence invites potentially harmful or unreasonably risky outcomes. Doing *radical engineering design work imposes significant incremental precautionary responsibilities on the engineer(s) involved.* Doing radical engineering design work with conventional or standard engineering design practices can heighten risk and therefore be ethically problematic.

The rewards engineers receive for "designing outside the box" carry with them an ethical burden, namely, the responsibility to check, double-check, and carefully confirm that paradigm-departing products are not being analyzed, constructed, operated, and maintained in normal, conventional ways and thereby creating significant, often unrecognized risks of harm.

One risk that lurks in much engineering practice stems from designing something that is significantly paradigm-departing and yet, under pressure, treating it in a conventional, business-as-usual manner. Such a course of action can engender unexpected lethal risk, as seen in the *Challenger* and Citicorp cases. The engineer has an FERE1-based derivative ethical responsibility to be always alert to and to avoid that form of negligence.

NORMALIZATION OF RISK AND ROUTINIZATION OF THE EXPERIMENTAL

Two ethically problematic kinds of deeds in engineering practice are "normalizing risk" and "routinizing the experimental." *Normalizing risk* means labeling the risk posed by a new technological product or system as "expected" and as not posing a serious threat to safety, and hence as an "acceptable" risk. Such labeling can affect how seriously engineers take a risk and how diligently they address it. Normalizing risk can dilute the determination urgently to find and eliminate the source(s) of that risk.

Routinizing the experimental refers to the decision, often taken under financial or temporal pressure, to prematurely designate an experimental technological product or system as proven, standard, or routine. This gives the impression that it no longer warrants the level of scrutiny and precaution it received when it was labeled and regarded as experimental and nonroutine.

Economic or political interests can fuel both moves. If a nonengineer, such as a politician or business person, promotes the premature routinizing of an experimental technology, the engineer qualified to judge whether that relabeling is premature may have an FERE2-derived ethical responsibility to speak up if she or he believes the announced change increases the risk of harm from operating the still-experimental technology in a routine fashion. Both phenomena were arguably present in the *Challenger* case.

TECHNOLOGY TRANSFER AND ETHICAL RESPONSIBILITY

Technology transfer from a more developed country (MDC) to a less developed country (LDC) is not simply a matter of moving hardware between them. It is a complex *cultural* process of transplanting something created in an MDC, and which embodies various MDC cultural elements, into an LDC likely to have a quite different culture.

If the transfer of a technology from an MDC to an LDC is to count as an ethically responsible process of international technology transfer (ERPITT), on the part of both the transferring company and/or country and the recipient company and/or country, then the following critical condition must be satisfied:

> *All* requirements—technical, political-economic, and cultural—for the safe and effective operation in the LDC of the technology being transferred from the MDC must be satisfied.

Until that condition is satisfied, engineers involved, whether functioning as managers or practicing engineers, have an ethical responsibility to try to assure the transfer either does not take place, or the technological system is not activated, made operational, or, if paused, resumed. This is so even if the recipient LDC seeks to accelerate the transfer process. In cases of international technology transfer from an MDC firm to an LDC firm in which the technology has the potential to cause death, injury, or disease if not operated properly, would-be ethically responsible engineers should strive to identify and indicate clearly, to the MDC firm and the LDC firm and government, all the requirements—technical, social, and sociotechnical—for the safe and effective operation of the transferred or to-be-transferred technology, especially those currently unsatisfied or fulfilled in a subpar way in the LDC.

It has been more than three decades since the horrific Bhopal disaster occurred. However, it is far from clear that this critical lesson—that it is essential that all requirements for the safe and effective operation of the transferred technology and technological process or system be satisfied in the recipient LDC before the transferred technology is launched or allowed to resume operation—has been learned and made operational by countries and firms involved in transferring and receiving risky technologies. Moving hardware from an MDC to an LDC does not in any way guarantee the "cultural resources for safety" that existed and may have been reasonably taken for granted in the MDC either already exist in, will automatically be transferred to the LDC along with the hardware, or can readily be created in the LDC. That pivotal fact must be indelibly etched in the consciousness of all engineers involved in MDC-LDC transfer of potent risky technologies. Technology transfer is a complex process in which the cultural requirements for the safe and effective operation of the transferred technology must be consistent with rather than in tension with the characteristics of the cultures of the recipient country and firm.

"TWO CULTURES" AND ETHICAL RESPONSIBILITY

Tension can exist in a complex technical project between two groups: engineers impatient to get on with the job at hand, and systematic, fastidious researchers whose priority is to first understand the reality of a situation in all its richness and complexity, even if doing so takes a long time and expends substantial financial resources. The increasing role of research and researchers in engineering activity has engendered a new kind of "two cultures problem": tension between the quite different cultures of these two groups. Monetary incentives and pressures on engineers to finish jobs on schedule can exacerbate that cultural tension and press some engineers to engage in ethically problematic conduct to overcome the bottleneck. The aborted Westway project was arguably a case in point.

DECONTEXTUALIZATION

In several of the cases, a key contributory causal factor in the unfortunate outcomes that occurred was that the engineering design or production work was in some sense or other decontextualized or undercontextualized. The people who carried it out may have had an overly narrow context in mind when doing their design work (DeVille/Seville). Such situations can lead engineers to overlook problematic consequences of their work that stem from interactions with elements left outside that narrow system context. Alternatively, either deliberately or unwittingly, engineers involved may fail to take into consideration how their products will be made and diffused, whether they will be carefully regulated, for what purposes they will be used and to what foreseeable effects, in the actual social contexts of production and use. Such failure can have serious harmful consequences (Bhopal, Topf & Sons) or create significant risks of harm (biosynthesis of opioids, neural enhancement).

On a brighter note, there is evidence among younger nanotechnology practitioners that the long-entrenched researcher belief that they do not have any ethical responsibilities outside the limited context of the research laboratory is changing. Many nanotech researchers now believe that under certain conditions researchers also have ethical responsibilities that pertain to the larger social context, for example, to provide early warnings if they have reason to think their research is likely to be applied downstream in ways that pose an unreasonable risk of harm to society

The most important take-away lesson for engineering students and practicing engineers to absorb about "context" from these cases as a group is *to avoid decontextualizing and undercontextualizing engineering work.* Put differently, engineers should beware of doing their design or production work without paying attention to factors outside the technology itself, or taking into consideration factors drawn only from within a narrow context or system, or disregarding or paying only fleeting attention to factors in the larger contexts in which the fruits of their work will be deployed, used, and take effect. The Parnas/SDI, Bhopal, and Topf & Sons cases illustrate the dangers of decontextualization. In contrast,

Martin Fisher's product design work in Kenya illustrates the benefits of paying careful attention to robust social context and acting accordingly.

THE POLITICIZATION AND ECONOMIZATION OF ENGINEERING DECISION-MAKING

An important theme in several of the cases is the politicization and economization of engineering decision-making. Sometimes an engineer may give political and/or economic concerns priority over technical considerations in engineering decision-making. On other occasions, an engineer may acquiesce in someone else's, for example, his or her manager or company executive, doing the same. The affected decision-making may concern which technology to select for a project; which manufacturer to choose; which specifications to choose or features to include in or withhold from an engineering product or system design; or how a technology is to be operated, regulated, or maintained. Cases in which the politicization or economization of engineering judgment seem to have played a role include *Challenger*, Pinto, and Bhopal.

What makes this phenomenon ethically problematic is that making engineering decisions primarily on the basis of political or economic concerns instead of on technical considerations is apt to yield suboptimal technological outcomes that carry heightened risks of harm to users or other affected parties. Engineers who initiate, acquiesce in, or agree to implement such decision-making are effectively endorsing and facilitating dubious and sometimes unjust decision outcomes. Consequently, those with potent political-economic interests typically carry the day and benefit, while the harms or heightened risks of harm that may result from such prioritization are typically borne by users and less powerful parties who have not been involved in the decision-making process.

NEGLIGENCE

In reviewing the case studies, it quickly becomes clear that several involved ethical misconduct in the form of acts of *commission*, while in others, the ethical misconduct took the form of acts of *omission*—failures to do something, or failures to do something with the care properly expected of someone in the agent's position. In the Citicorp case study, organizationally speaking, LeMessurier seems not to have established and maintained cultural norms in his firm that would prevent safety-related structural changes in a building from being approved by any engineering employee without their first being run by and approved by the person most technically qualified to assess them, namely, LeMessurier himself. In addition, it appears that he did not check and recheck carefully to see whether the standard approach used to analyze a conventional high-rise building's structural support system, one that applied to conventional buildings with their columns at the four corners, also applied to the radical new structural design he had devised, with its supportive columns located at the midpoint of each of the four sides.

In the Bhopal case, some managers and engineers seem to have been remiss in allowing the MIC plant technology to be transferred and the new Bhopal MIC plant to commence and continue operations without having ensured that all conditions—technical, political-economic, and cultural—for its safe and effective operation were and remained satisfied.

Finally, the *Challenger* case appears to encompass at least two instances of imprudent acts of omission. First, MTI engineer-managers failed to provide adequate enabling resources and to give a high priority to addressing Boisjoly's urgent recommendation that a serious effort be launched to redesign the solid rocket booster's field and nozzle joints. Second, with respect to the launch recommendation process, some mid-level NASA officials were apparently negligent in not communicating to top NASA managers the serious objections to launching the next morning that had been voiced by MTI rocket-booster-specialist engineers on the eve of launch.

Negligent acts of omission that contribute to causing harm or help create an unreasonable risk of harm can be as much at odds with FERE1 and ethical responsibilities derived therefrom as any deliberate harm-causing act of commission.

WORKPLACE CULTURE AND THE ETHICALLY RESPONSIBLE ENGINEER

As shown in the Citicorp, Bhopal, and nanotechnology cases, the *culture* of a laboratory, design office, firm, or the wider society can exercise important influence on whether an engineer fulfills her/his ethical responsibilities. When engineering actions are being explained and justified, adequate attention is usually given to technical considerations, foreground economic and political influences, and the intellectual skills and character traits of the engineers whose conduct is being examined. However, cultural influences on engineering conduct are often given short shrift. Serious attention is rarely given in explaining engineering conduct to the cultures of the social units in which engineers operate and/or with which they interact, probably because these units operate in the background and silently set the stage for engineering action. Accounts of engineering action need to take into consideration the characteristic *mentalities* (values, core beliefs, norms, world views, assumptions, etc.); characteristic *social forms* (e.g., institutions, organizations, groups, roles, status system, rituals); characteristic *material artifacts and ways of making them*; and characteristic *personality traits and behavior patterns* (including professional practices) that prevail in the social unit(s) in which engineers are active or with which they interact in doing their work.

A social unit's *cultural system* comprises these four kinds of interacting background cultural elements, and the cultural system presses and sometimes shapes engineering conduct, for good and/or ill, sometimes exerting a decisive influence on engineering conduct and outcomes. The Bhopal episode is one case in point; the safety cultures of quite different degrees of strength that existed in the 13 nanotechnology research labs the author studied in 2005–06 is another; Google's and/or Apple's engineering workplace cultures may also be examples. More attention needs to be paid to creating and sustaining cultural systems in engineering workplaces that support and foster responsible engineering conduct instead of overlooking or tolerating ones that impede, deter, or penalize it.

CONFLICTS OF INTEREST

As seen in several of the cases, contemporary engineers (and other technical professionals) can find themselves in situations in which they are faced with significant conflicts of interest (COIs). Two factors make engineers' COIs an extremely important and recurrent ethical phenomenon in contemporary Western societies: the potential of much engineering activity to affect the public interest in major ways, and the heavy pressures that a firm's managers or executives can put on its engineers to prioritize the firm's private economic interests because of the large financial stakes the firm often has in winning or retaining lucrative contracts for engineering services and/or in completing contracted engineering work on time.

When an engineer working as a consultant to a client or as an employee of a firm is faced with a conflict of interest, he or she is likely to be pulled in opposing directions and usually must make a decision that prioritizes one or the other of the conflicting interests. Most often, COIs in engineering involve an engineer deciding whether to serve a private interest, for example, his or her own financial interest and/or that of his or her employer or client, or some facet of the public interest, such as the interest in protecting public safety, public health, and national security, and in not degrading or wasting public environmental or financial resources.

Although many, perhaps most, conflicts of interest involve tension between a private financial interest and the public interest, engineering-related COIs occasionally involve tension between competing *private* interests. For example, an engineer could find herself or himself in a situation where she or he has to choose between enhancing her or his private financial interest by devoting much more of her or his time to engaging in remunerative consulting activity, and adequately serving the private educational interest of her or his graduate students as their advisor by making herself or himself sufficiently available (for mentoring and help with their research), something that might be possible only by limiting the amount of consulting work that the advisor does.

Similarly, although *economic* interests may come to mind first when one thinks about COIs, a COI in engineering need not involve *any* economic interests. For example, an engineer might be torn between his or her private interest in devoting more time to his or her research for purely intellectual and/or status reasons, and adequately serving the aforementioned interest of his advisees in having reasonable access to their engineering advisor and mentor.

An engineer may occasionally find herself or himself in a situation in which it looks like she or he has a bona fide or real conflict of interest, but in fact, for one reason or another, the perceived COI may be only *apparent*. For example, an engineer may be a member of a public decision-making body, for example, a city council, that must choose which of several engineering firms, including the engineer's, will be awarded a major contract for engineering services. Clearly, the potential for a conflict of interest exists in such a situation. If, however, unbeknown to the general public, the engineer has already recused herself or himself from the decision-making process, the seeming conflict of interest would be apparent, not real, and the engineer would not be at risk of acting ethically irresponsibly unless she or he tried to influence a voting colleague.

Which FEREs pertain to engineering COIs? As noted, when an engineer is pulled in opposite directions by a COI, tension between one of his or her private interests and the public interest is often involved. In such cases, if the engineer decides to serve one or more of his or her private interests, especially if they are financial in nature, such that the public interest suffers, then he or she is not acting in accordance with FERE4. As argued in Chapter 3, because society enables much if not most contemporary engineering work and is the engineer's ultimate omnipresent background client, when private interests conflict with the public interest, the latter deserves to be given priority over the former.

Suppose an engineer faces the following conflict of interest: she or he has an intellectual and professional interest in spending more time pursuing her or his own research or development work, and her or his graduate student advisees have an educational interest in getting reasonable access time to their advisor and/or mentor. If the engineer elects to prioritize the former interest at the expense of the latter, then the engineer's action may violate FERE1, because if the engineer has accepted certain students as her or his graduate students, then they are entitled to have significant time with their advisor. The engineer's failure to be reasonably accessible to her or his students arguably violates their protectable interest in having such access. This failure can be viewed as inflicting a harm or an unreasonable risk of harm on the students, for it will likely retard their progress in their graduate work. Hence, the engineer's choice to further her or his own private interest, and consequently fail to make herself or himself reasonably available to her or his students, would indirectly harm or create an unreasonable risk of harming those students, which would be incompatible with FERE1.

It is not difficult to imagine circumstances in which a research engineer's decision to prioritize or to continue to prioritize pursuit of his or her own work at the expense of being reasonably available to his or her own graduate students could amount to failure to try to prevent harm or the unreasonable risk of harm to those students. For example, the engineer might attach such a high priority to intensive pursuit of his or her own research work that he or she might not make time to intervene to try get his or her students back on track in their graduate studies when they have been set back in their academic pursuits by receiving insufficient attention from their engineer-advisor. Failure to intervene to try and prevent further harm under such circumstances would arguably be inconsistent with FERE2.

"DESIGN FOR . . ."

Besides designing products and systems that meet performance specifications, respect cost restrictions, and are safe to use and operate, the design engineer has derivative ethical responsibilities to pay close attention to at least three other kinds of consideration: *manufacturability*, *ethical compatibility*, and *local cultural compatibility*.

One contributory cause of the Hyatt Regency Walkways accident (Case 14) was that the lead structural engineer arguably did not design for manufacturability (DFM). That is to say, he did not design the six original continuous hanger rods intended to connect the second and fourth floor walkways to the ceiling above the hotel atrium in a way that was straightforward and cost-effective to fabricate. This seems to have led the fabricator to propose and/or make a change from the original continuous one-rod hanger design to a two-rod hanger design, with all the problems that resulted therefrom. In short, apart from the additional cost that failure to DFM can impose on a company or client, failure to DFM can engender a significant risk of harm to those interacting with an altered final product or system. This suggests that engineers have an FERE1-based derivative ethical responsibility to DFM, as long as doing so does not pose a significant incremental risk to safety.

In Case 5, a client engaged software engineer Diane to design a database management system (DBMS). Because of the low level of privacy protection the client elected to have built into the system, Diane believed implementing the client's selection would yield a DBMS that would make sensitive employee data vulnerable to access by unauthorized parties. This could result in some employees' being harmed by parties who were able to hack the system. In effect, Diane's client asked her to design a DBMS that when implemented would facilitate the violation of an important ethical value—(informational) privacy. Engineers have an FERE1-based derivative ethical responsibility to design for ethical compatibility (DFEC) That is to say, they have a derivative ethical responsibility to not create or acquiesce in the creation of a product design with properties that invite or facilitate the product's being used in ways that would violate one or more important ethical values and contribute to bringing about harms.

One of the most admirable features of engineer Martin Fisher's design work at KickStart International (Case 16) was that he and his colleagues strived to endow their products with properties allowing them to fit harmoniously into the local cultural system of the society for which they were intended. Failure to design for local cultural compatibility (DFLCC) would have indirectly harmed and/or put at significant or unreasonable risk of harm those who purchased those products for agricultural use, where the harms in question ranged from financial loss to ostracism and physical injury. Therefore, engineers have an FERE1-based derivative ethical responsibility to DFLCC in their design work. This ethical responsibility is especially important in contexts of international technology transfer, when a product being designed in a more developed country (MDC) is intended for use by residents of or workers in a less developed country (LDC), one likely to have a significantly different local culture and perhaps inadequate cultural resources for safety.

Not only did Fisher and KickStart design their products to be compatible with the local culture, they also made sure their products made good use of local environmental resources rather than ones necessitating import. Thus the block press uses local earth, and the sunflower press uses seeds from locally grown sunflower plants, not imported materials. Thus, one could say that Fisher's products exemplify a philosophy of design for local cultural and environmental compatibility.

There may well be additional "design for . . ." ethics constraints on engineering design work besides the three just discussed. For example, in the discussion of the Cadillac Seville/Deville case—Case 1—it was suggested the engineers involved might have had an FERE1-based derivative ethical responsibility to design for (harm-free) mass producibility. By this is meant that in doing her or his design work on a product, including endowing it with various physical and performance properties, the engineer should keep in mind that the product may be mass-produced. If it is, its negative impact on the environment may be aggregative, the result of the effects of thousands if not hundreds of thousands or millions of individually relatively innocuous mass-produced items embodying the original design properties. Another emerging ethics-related "design for" constraint on engineering design work, one that did not arise in the cases of Chapter 4, is *design for sustainability* or, more specifically, *design for recyclability*.

ETHICAL ISSUES RELATED TO ENGINEERING RESEARCH

As discussed in Chapter 2, research has become an increasingly prominent feature of engineering work in recent decades. Several of the cases involved ethical issues related to research. A useful way to group some of those issues is to distinguish those that pertain to the actual *practice* (i.e., conduct or performance) of engineering research from those that pertain to the *publication* of such research. In each of those categories, ethical issues can be raised by both "how" and "whether" certain conduct is carried out.

As regards the *practice* of engineering research, ethical issues can be raised by *how* (i.e., the manner in which) such research is conducted or carried out. For example, as noted in the discussion of the Schön case (Case 3), there is consensus in the research community that falsifying or fabricating data in any phase of the research process, including during the "execution" phase, is deeply ethically irresponsible (under FERE1) and constitutes research misconduct. Similarly, in research involving human subjects, including some nanotechnology and neuroengineering research, "how" also matters ethically. Procedurally, human subjects researchers have an upstream ethical responsibility to accurately inform the humans involved about all risks the research involves and to obtain their informed consent to voluntarily participate in it.[8] In the nanotechnology case, another "how" phenomenon was noted: researchers taking shortcuts that violate lab safety rules, a practice arguably incompatible with FERE1. Researchers have FERE1- and FERE2-based derivative ethical responsibilities to avoid such problematic practices.

As for "whether"-related ethical issues related to the practice of engineering research, the brain-chip-based cognitive enhancement case raised the difficult ethical issues of *whether* it would be ethically permissible for neural engineers to participate in that line of R&D work, and if so, under what conditions, and *whether* there were conditions under which neural engineers could have an FERE1-based derivative ethical responsibility to *not* engage in such work. Besides possibly being a violation of some core human subjects research principle or guideline, such as those elaborated in the Belmont Report, one condition that could engender an ethical responsibility not to participate in such enhancement work would be if involvement in it would foreseeably contribute to bringing about significantly worsened social inequality, arguably a societal harm. Given the increasingly information-centric nature of many contemporary societies with market economies, the possibility that such an outcome might result cannot be cavalierly dismissed.

---

[8] One of the most seminal works to date about the ethical principles and values to be respected in research involving human subjects is *The Belmont Report*. It argues that three "Basic Ethical Principles"—"respect for persons," "beneficence," and "justice"—must be respected in all such research, and elaborates three "Applications" of the three principles to the conduct of human subjects research, namely, "informed consent," "risk/benefit assessment," and the (just) "selection of research subjects." See National Commission for the Protection of Human Subjects of Biomedical and Behavioral Research (1979). With *The Belmont Report* as background, in 1991 the U.S. federal government drew up a "Federal Policy for the Protection of Human Subjects," often called "The Common Rule." Eventually 16 departments of the U.S. federal department adopted the Common Rule. Revisions were made in the Common Rule over the years and "The Final Rule" was published in the *Federal Register* on January 19, 2017. See https://www.gpo.gov/fdsys/pkg/FR-2017-01-19/pdf/2017-01058.pdf}.

As for the *publication* of research, ethical issues are sometimes raised by *how* (i.e., the manner in which) the engineering research publication process is handled, particularly when the research in question is carried out collaboratively and the resultant papers are coauthored. As noted in discussion of the Schön case, questions arose about a number of papers on which he was listed as a coauthor. In particular, questions arose about the ethical acceptability of how the status of "coauthor" was handled by certain listed authors, and about how the coauthors of research papers review their manuscript.

In contrast to ethical issues raised by "how" aspects of publication, sometimes, ethical issues can arise about "whether" aspects. For example, if a researcher or team of researchers is considering publishing an account of the results of "dual-use research of concern," the issue can arise as to *whether* it is ethically acceptable and responsible for a researcher or team to publish its research results and methods at a certain point in time in a particular society, and if so, under what conditions. A related ethical issue is *whether* a researcher has a derivative ethical responsibility to refrain from publishing his or her unredacted report of research results and methods, either for the time being or permanently, and if so, under what conditions. Those issues presented themselves and were explored in the opioid biosynthesis case.

FACTORS CONDUCIVE TO ENGINEERING MISCONDUCT

Review of the cases reveals that various sorts of factors played roles in inducing or pressing certain engineers to engage in engineering misconduct. Some of the factors involved personal characteristics of engineers. Others were technical in nature and involved features of engineering products, processes, and systems. Yet others were social-contextual in nature and involved features of multilayered engineering situations, for example, characteristics of work groups, organizations, and macrosocial institutions and policies. What follows is a partial list of factors, distilled from the cases, that can induce, press, or be conducive to engineering misconduct.

- pressure being put on engineers—at the workplace, organizational, or macrosocial level—to compromise their design, research, development, testing, production, or maintenance work for temporal or financial reasons;

- engineers having a sense of work-related duty that impels them to prioritize serving something other than the health, safety, and welfare of the public, for example, the orders of supervisors, employers, clients, or national governments;
- engineers being excessively proud of or psychologically invested in their product designs, to the point of being unable or unwilling to recognize, acknowledge, or remedy their shortcomings;
- engineers lacking a strong ethical value system that could enable them to resist external or internal pressures to make or acquiesce in ethically problematic product or system designs or to engage or acquiesce in ethically questionable engineering practices;
- engineers outsourcing critical engineering tasks to less qualified technical personnel and then not overseeing and conscientiously verifying the validity of their work;
- an organizational culture that requires, tolerates, or does not significantly penalize misconduct by its engineers;
- organizational complacency in addressing matters of engineering risk brought to management attention by front-line engineer-employees;
- an organizational reward system that incentivizes or tolerates ethically questionable funding, research, or publication practices;
- conflict between the meticulous, "proceed slowly and systematically" culture of a group of researchers, and the "damn the torpedoes, full speed ahead!" culture of a group of development or construction engineers, when both are involved with the same complex, costly, and risky project;
- facile optimism or negligence regarding the importance of cultural resources for safety to responsible engineering practice on the part of engineers and engineer-managers involved in risky processes of technology transfer from highly industrialized countries to relatively newly industrialized countries; and
- the proposal or prescription of a paradigm-departing engineering product or system design.

The reader is encouraged to extract from the cases additional noteworthy general factors, whether individual, social (including organizational and cultural), or technical in nature, that can induce, press, or be conducive to engineering misconduct.

CHAPTER 6

Resources and Options for Ethically Responsible Engineers

Whether someone becomes an ethically responsible engineer clearly depends partly on her or his personal values, ideals, and resolve. However, society can facilitate engineers' being ethically responsible in their professional practice by introducing appropriate incentives, social innovations, and legislation to enable, benefit, and protect those who want to fulfill their ethical responsibilities. In the contemporary United States, the social infrastructure supportive of ethically responsible engineering practice is modest. Nevertheless, it is in the interest of would-be ethically responsible engineers to be aware of the resources and options available to them—and of their limitations. Discussion follows of three kinds of relevant resources and options, two social and one individual in nature.[1]

ORGANIZATIONAL RESOURCES

The phrase "organizational resources" refers both to organizations that are presumptively supportive of ethically responsible engineering practice, and to innovations within organizations that could promote such practice.

1. *Organizations*. When thinking about support for ethically responsible engineers, the first organizations likely to come to mind are the *professional engineering societies*. These organizations presumably stand ready to support engineer-members who are penalized or at risk of being

[1]The case studies in Chapter 4 can be viewed as *intellectual* resources for would-be ethically responsible engineers. They exhibit a range of behaviors in ethically challenging engineering situations, some exemplifying and others incompatible with being an ethically responsible engineer.

penalized by their employers for opposing or disclosing employer actions or practices the members believe imperil public safety, health, or welfare.

An example in which a professional engineering society provided such support occurred in the early 1970s. The IEEE supported three engineers: systems engineer Holger Hjortsvang, electrical engineer Robert Bruder, and programmer analyst Max Blankenzee,[2] who had been working for San Francisco's Bay Area Rapid Transit (BART) while the system was being designed, built, and tested. In January 1972, after not getting constructive responses to concerns they had expressed at lower organizational levels about the safety of Westinghouse Computer Systems' innovative automatic train-control (ATC) system, the three engineers took their concerns to a member of the BART board of directors. He, in turn, brought them to the attention of the full board and the top management of BART.

After a hearing, the board voted to reject the three engineers' criticisms and supported BART management, which contended that the concerns had no merit. The latter then gave the three engineers a choice: resign or be fired. When they refused to resign, they were fired for insubordination, lying about anonymous memos about safety they had authored, and violating organizational procedures by contacting the board of directors. On October 2, 1972, shortly after the initiation of partial revenue service, a BART train overran the system's Fremont station "as a result of an ATC failure."[3] Several passengers were injured. Later, a study by "a special panel of distinguished engineers" validated the three BART engineers' concerns.[4] The three sued BART for "breach of contract [of employment] and deprivation of constitutional rights."[5]

In late 1974, IEEE submitted an amicus curiae brief to the court in support of the engineers' suit. The brief urged the court to rule

> that an engineer's contract of employment includes as a matter of law an implied term that such engineer will protect the public safety, and that discharge of an engineer solely or in substantial part because he [sic] acted to protect the public safety constitutes a breach of such implied term.[6]

---

[2] Hjortsvang was a senior member of IEEE. See Stephen Unger (1994), p. 21. In a letter of July 20, 1973, Bruder wrote Unger that he was formerly a member of IEEE and had "switched to the NSPE [National Society of Professional Engineers] when he went into construction work. Max Blankenzee was not an IEEE member." (Stephen Unger email to the author, August 26, 2014).

[3] Unger (1994), p. 24.

[4] Ibid., p. 23.

[5] Ibid., p. 24. See also Friedlander (1974).

[6] Unger (1994), p. 260.

Shortly after the brief was filed, the engineers reached an out-of-court settlement with BART. In 1978, IEEE's Committee on the Social Implications of Technology (CSIT) awarded the three engineers the first IEEE CSIT Award for Outstanding Service in the Public Interest.[7]

This intervention notwithstanding, on the whole the support given by U.S. professional engineering societies to members at risk of being penalized for striving to engage in ethically responsible engineering practice has been tepid. What might account for this kind of response? One hypothesis is that professional engineering societies are concerned about possibly being held legally liable if they support members with ethics concerns, for example, by giving them advice about what to do about ethical responsibility issues involving their companies. A professional engineering society might worry that it would be liable to a charge of being in receipt of proprietary corporate information from the engineer(s) it is supporting. However, this fear-of-liability hypothesis is rendered suspect by the fact that company lawsuits against professional engineering societies have been extremely rare.[8] Moreover, professional engineering societies could protect themselves against that possibility by purchasing liability insurance.[9]

A second explanatory hypothesis is that professional engineering societies fear possible losses of income and members. To the extent that companies pay for or reimburse their engineer employees for their professional engineering society membership dues, professional engineering societies might fear that if they support members embroiled in ethics disputes with their employers, the latter might decide to prohibit their employees from being society members and/or stop paying or reimbursing their membership dues. Professional engineering societies with significant numbers of engineer-members employed by large companies could be especially reluctant to support an engineer with an ethical responsibility concern employed by such a company. The potential income loss to the society could be substantial. However, this fear-of-loss-of-members-and-dues hypothesis is also unconvincing. To the best of my knowledge,

---

[7]Ibid., p. 27.

Another episode in which IEEE committees were helpful to an engineer attempting to fulfill her or his ethical responsibilities was the 1977–78 case of systems analyst Virginia Edgerton. In 1979, Edgerton received the second IEEE CSIT Award for Outstanding Service in the Public Interest. For details and discussion of this case, see Unger (1994), pp. 27–30 and 263–280.

[8]One of those rare examples is American Society of Mechanical Engineering v. Hydrolevel Corp., 456 U.S. 556 (1982), http://caselaw.findlaw.com/us-supreme-court/456/556.

[9]I owe this point to Stephen Unger.

company payment for or reimbursement of its engineer-employees' engineering society membership dues is not a widespread practice.

Stephen Unger, professor emeritus of computer science and electrical engineering at Columbia University, who chaired the IEEE Ethics Committee in 1997 and 1998, has offered a more plausible, conflict-of-interest-based account of why professional engineering society support of ethically responsible engineering practice has been modest:

> My explanation for the lack of [professional] society support for ethical engineers is that the ranks of engineer society officers are largely filled with engineering managers (some at very high levels), and professors. The former, very naturally, are company and management oriented, while the latter are often in the position of appealing to companies for research contracts, summer jobs, sabbatical leave jobs, jobs for students, etc.[10]

A different kind of organization pertinent to the cause of ethically responsible practice by technical professionals is the *government oversight unit*. The Office of Research Integrity (ORI), a unit of the U.S. Department of Health and Human Services, probes charges of researcher misconduct when government money is involved.[11] A technical professional can contact ORI if he or she is concerned about possible research misconduct by colleagues or about being personally pressured to engage in such misconduct. However, to date, ORI has dealt exclusively with cases of misconduct in *science* research, not in *engineering* research.[12] The author knows of no U.S. government office that plays a role vis-à-vis engineering research and development comparable to that of ORI vis-à-vis scientific research, especially in the biomedical area.

A third kind of organization occasionally helpful to engineers concerned with ethically responsible engineering practice is the *nonprofit, nonpartisan public interest organization with the mission of aiding whistleblowers*. An example of this kind of organization is the Government Accountability Project (GAP)[13]. GAP's mission is "to ensure government and corporate accountability by advancing occupational free speech,

---

[10]Stephen Unger, email to the author, August 26, 2014. Unger also pointed out to me weaknesses in the two explanatory hypotheses discussed above.

[11]https://ori.hhs.gov/case_summary.

[12]The ORI website—http://ori.dhhs.gov/case_summary—provides "Case Summaries" for all administrative actions that "were imposed due to findings of research misconduct." It lists 37 case summaries of incidents resolved between "2008 and Older" and mid-2017. Of the 37 cases, none involves engineering research. Only one, involving bioinformatics, is even remotely related to engineering.

[13]https://www.whistleblower.org.

defending whistleblowers and presenting their verified concerns to appropriate officials, groups, or journalists."[14] GAP is also active in efforts to strengthen whistleblower protection laws. Founded in 1977, it claims to have helped "well over 6,000 whistleblowers."[15] GAP assists whistleblowers in a wide range of professional fields, including engineering.

2. *Intra-organizational innovations.* One intra-organizational innovation has emerged in recent decades that could be useful to engineers concerned about the effects of their firms' actions on the health, safety, or welfare of the public: *the ombuds office.* The ombuds office, often called "the ombudsman's office," is a neutral, sometimes independent unit of a private or public organization. Individual employees, including engineers, can approach the ombuds office confidentially to voice concerns or complaints about matters that bear on public health, safety, or welfare, or unfair treatment of employees. The ombuds office is usually authorized by the firm to investigate complaints filed with it and highlight the misconduct involved, make recommendations, or mediate fair settlements of issues between an employee and her or his employer, or between employees.

A handful of university ombuds offices have been created to provide services to engineers and/or engineering students. However, there appear to be very few such offices in private industry.[16]

In 2014, General Motors Corporation faced multiple lawsuits arising from the fact that, between 2001 and 2014, millions of GM cars were sold with ignition switches some GM employees, including some engineers, knew for years were defective.[17] By August 2015, GM's Ignition Compensation Claims Administrator had accepted that defective switches were implicated in at least 124 deaths and 274 injuries, with GM paying out hundreds of millions of dollars in claims.[18] Reflecting

[14] https://www.whistleblower.org/our-history-and-mission.

[15] Ibid.

[16] Universities or engineering schools with ombuds offices that can relate to the interests of engineering students or practicing engineers include Purdue, Cornell, Texas A&M University College of Engineering, Clemson, Pennsylvania State University College of Engineering, Georgia Institute of Technology, University of Pennsylvania, and University of Michigan. Certain government organizations, such as NASA and the Argonne National Laboratory, have established engineering-related ombuds offices. Most ombuds offices are found in public-sector organizations. Very few private companies have bona fide ombudsman's offices. A rare exception is United Technologies Corporation. See http://www.utc.com /How-We-Work/Ethics-And-Compliance/Pages/Ombudsman-Program.aspx.

[17] Ivory (2014).

[18] Shepardson (2015).

on this situation, consumer advocate Ralph Nader made a suggestion to Mary T. Barra, GM CEO and chairman since January 15, 2014:

> I would suggest that she establish an independent ombudsman, where conscientious engineers who have been muzzled by their cost-concerned bosses can go anonymously to the ombudsman, who has a direct line to the CEO and president of GM.[19]

Three other intra-organizational innovations deserve brief mention. First, the National Institute of Engineering Ethics (NIEE),[20] the IEEE, and the Online Ethics Center for Engineering and Science (OECES)[21] once had *ethics hotlines* or *help lines* that engineers could use anonymously to report ethics concerns or seek information and advice about ethics issues they were facing at work. Those services seem to no longer exist.[22]

Second, one professional engineering society considered establishing an *ethics support fund* from voluntary professional society member donations. The idea was to use the collected money to assist engineers facing engineering-related ethical issues.[23]

Third, for years the American Association of University Professors (AAUP) has maintained a *censure list*, that is, a list of institutions whose administrations AAUP has censured for actions or policies it believes have curtailed or threatened academic freedom and/or otherwise violated the rights of faculty members at those institutions.[24] One could imagine professional engineering societies establishing their own censure lists. Each engineering society could publicize a list of employers and clients it had censured for allegedly firing or penalizing one or more member employees for actions they took to protect public safety, health, or welfare.

[19] Coscarelli (2014).

[20] The NIEE has been housed in the Murdoch Center for Engineering Professionalism at Texas Tech University since 2001. See http://www.depts.ttu.edu/murdoughcenter/center/index.php and http://www.depts.ttu.edu/murdoughcenter/center/niee/index.php.

[21] The OECES is managed by the Center for Engineering, Ethics, and Society, a unit of the National Academy of Engineering.

[22] IEEE had an "ethics hotline" from August 1996 until February 1998. The June 2014 edition of the *IEEE Constitution and Bylaws* states the professional society's current position on giving advice to practicing engineers: "Neither the Ethics and Member Conduct Committee nor any of its members shall solicit or otherwise invite complaints, nor shall they provide advice to individuals." (See http://www.ieee.org/documents/ieee_constitution_and_bylaws.pdf, p. 300.11.

[23] For discussion of the defunct IEEE ethics hotline and the proposed-but-rejected IEEE ethics support fund, see Unger (1999), pp. 36–40.

[24] See http://www.aaup.org/our-programs/academic-freedom/censure-list.

Were censure lists to be established, companies might think twice before firing or penalizing an engineer-employee for opposing company action or practice in the name of protecting public safety, health, or welfare. Being on such a list might seriously hinder such companies' efforts to recruit or retain critical engineering talent. To the author's knowledge, no professional engineering society has ever established a censure list.

In short, while intra-organizational innovations have significant potential value, to date their adoption and influence have been limited.

LEGAL RESOURCES AND OPTIONS

Several U.S. federal laws offer modest levels of support to engineers who wish to practice their profession in an ethically responsible manner. Four of the most noteworthy are the False Claims Act (FCA) of 1863;[25] the National Environmental Policy Act (NEPA) of 1970;[26] the Foreign Corrupt Practices Act (FCPA) of 1977;[27] and the Whistleblower Protection Act (WPA) of 1989.[28]

The False Claims Act of 1863[29] allows an individual who believes that a person or company, usually a federal contractor, has knowingly defrauded the government, to sue that party on behalf of the government in hopes of recovering all or part of the money in question. Such a suit is known as a *qui tam* action. If such a suit succeeds, the plaintiff is entitled to up to 30% of the money recovered, depending on whether the U.S. Department of Justice joined the plaintiff's suit against the defendant, for example, a company for which the plaintiff had previously worked.[30]

Such suits have occasionally been successful.[31] However, the FCA also has serious limitations. Recall that Nira Schwartz's *qui tam* action against TRW and Boeing N. America was quashed. That happened when defendant TRW threatened to subpoena and introduce in court a number

---

[25] http://www.justice.gov/civil/docs_forms/C-FRAUDS_FCA_Primer.pdf.

[26] http://www.epa.gov/compliance/basics/nepa.html.

[27] http://www.justice.gov/criminal/fraud/fcpa/.

[28] https://www.gpo.gov/fdsys/pkg/STATUTE-103/pdf/STATUTE-103-Pg16.pdf. See also http://en.wikipedia.org/wiki/Whistleblower_Protection_Act.

[29] 31 U.S.C. §§ 3729–3733. The False Claims Act was revised in 1986 and has been amended several times since.

[30] U.S. Department of Justice (2011), p. 3.

[31] For a chronological list of whistleblowing cases, some of which involved *qui tam* actions brought under the False Claims Act, see http://en.wikipedia.org/wiki/List_of_whistleblowers.

of confidential government technical documents about U.S. national missile defense system specifications the Department of Defense did not want to become public. Because of this, and perhaps also to avoid embarrassment, the federal government invoked "the state secrets privilege," at which point the presiding judge quashed Schwartz's suit before she had a chance to make her case in court.[32]

The National Environmental Policy Act (NEPA) of 1970[33] requires that any federal agency in the executive branch that is planning a federal action likely to significantly affect the environment must prepare a detailed document, called an "environmental impact statement" (EIS), and submit it to the Environmental Protection Agency (EPA). The EIS must assess the environmental impact of the proposed action and of alternatives to it.[34] The public, other federal agencies, and individual private parties, such as environmental protection organizations, can provide input into the EIS, comment on it, and challenge its adequacy or accuracy in court. If the review of a submitted EIS reveals or can be shown to imply that proceeding with the project will violate a provision of some federal law, for example, the Endangered Species Act or the Clean Water Act, the project in question may be delayed, revised, or rejected. Thus, engineers concerned about the effects of an engineering project on the environment may be able to use NEPA to challenge the undertaking. If it is shown that the EIS for a public project was distorted or otherwise defective, perhaps to secure a permit to proceed, a judge can delay or effectively terminate the project, as happened in the Westway case.

The Foreign Corrupt Practices Act (FCPA) of 1977[35] prohibits American individuals and companies from making payments, directly or indirectly, with money or anything else of value, to foreign governmental officials for the purpose of obtaining or retaining business with foreign firms and governments. Some technology companies that sell their products internationally, such as aerospace, telecommunication, pharmaceutical, and medical device firms, have in the past been charged with violating the FCPA.[36] The FCPA could conceivably help engineers refuse employer orders to engage in ethically problematic actions or practices illegal under the FCPA.

[32] See Chapter 4, Case 13.
[33] 42 U.S.C. 4321 *et seq.*, specifically Title I, Sec. 102.
[34] One alternative that must be considered is that of doing nothing.
[35] 15 U.S.C. §§ 78dd-1. The FCPA was amended in 1998.
[36] For a year-by-year list of the Securities and Exchange Commission enforcement actions under the FCPA, see https://www.sec.gov/spotlight/fcpa/fcpa-cases.shtml.

The Whistleblower Protection Act (WPA) of 1989[37] was intended to protect covered *federal employees* who make a protected disclosure of information they reasonably believe evidences illegal or improper activities by *government agencies*. The law forbids government agencies from engaging in a wide range of "prohibited personnel practices," such as firing and demotion, against protected federal employee-whistleblowers, including engineers. The WPA was intended "to help eliminate wrongdoing within the Government"[38] by protecting government employees who disclose such wrongdoing. The WPA prohibits a federal agency on whom a federal worker has blown the whistle from retaliating against that employee because she or he disclosed information she or he believed to be true about an alleged agency "violation of any law, rule, [or] government regulation"; "gross mismanagement"; "gross waste of funds"; "abuse of authority"; or any "substantial and specific danger to public health or safety"[39] caused by agency action.

However, as implemented, several limitations of the WPA became clear. For example, in *Garcetti v. Ceballos*[40] the U.S. Supreme Court held that

> when public employees make statements pursuant to their official duties, the employees are not speaking as citizens for First Amendment purposes, and the Constitution does not insulate their communications from employer discipline.[41]

This lack of protection is a significant disincentive to federal employees who are considering whether to publicly blow the whistle on their government employer.

The government entities most heavily involved in investigating and adjudicating whistleblowing claims by federal employees are the Office of Special Counsel (OSC), which investigates whistleblower complaints; the Merit Systems Protection Board (MSPB), which makes judicial decisions on whistleblower complaints; and the Federal Circuit Court of Appeals (FCCA), which hears appeals of MSPB decisions.

For various reasons, in recent decades these entities have rarely decided in favor of the whistleblowers. "On average each month, more than

---

[37] Public Law 101–12. The WPA of 1989 was strengthened by the Whistleblower Protection Enhancement (WPEA) Act of 2012, Public Law 112–199.

[38] Whitaker (2007).

[39] "Whistleblower Protection Act," http://en.wikipedia.org/wiki/Whistleblower_Protection_Act.

[40] Garcetti v. Ceballos, 547 U.S. 410 (2006), https://www.law.cornell.edu/supct/html/04–473.ZS.html.

[41] Stout (2006).

15 whistleblowers would lose initial decisions from administrative hearings at the Merit Systems Protection Board, while fewer than one would prevail."[42] As for the FCCA, of the 227 federal whistleblower cases it heard between 1994 and June 2012, only 3 were decided in favor of whistleblowers.[43] A 2010 MSPB study of federal employee whistleblowing elicited a 58% response rate from the 71,970 federal employees in 24 surveyed federal departments and agencies. When those who had blown the whistle on their departments or agencies were asked, "Within the last 12 months, have you personally experienced some type of reprisal or threat of reprisal by management for having reported an activity?" more than a third [36.2%] answered in the affirmative.[44] Such realities have made it difficult in practice for federal employee whistleblowers to prevail and, equally important, have probably deterred many would-be whistleblowers from stepping forward to disclose government agency misconduct.

The Whistleblower Protection Enhancement Act (WPEA) of 2012, passed unanimously in the U.S. Senate and the House of representatives, was intended to strengthen federal employee whistleblower protection.[45] It broadened the scope of protected disclosures that *Garcetti v. Ceballos* and other federal court decisions had narrowed, extended whistleblower protections to government scientists who disclosed government agency censorship or made disclosures related to the integrity of the scientific process, expanded the costs that whistleblowers subjected to retaliation for their disclosures could recover, and expanded the domain of "prohibited personnel practices" to include the enforcement of nondisclosure agreements that do not also notify employees of their whistleblower rights and protections.[46] Whether WPEA will improve the incentives for federal employee whistleblowing and enhance whistleblower protections in practice remains to be seen.[47]

[42] Government Accountability Project (2015), "Whistleblower Protection Enhancement Act," http://www.whistleblower.org/whistleblower-protection-enhancement-act-wpea.
[43] Ibid.
[44] U.S. Merit Systems Protection Board (2011), p. 11.
[45] For the text of the WPEA, see http://www.gpo.gov/fdsys/pkg/BILLS-112s743enr/pdf/BILLS-112s743enr.pdf. For discussion of the most important changes made by the WPEA, see http://www.whistleblower.org/node/664.
[46] American Bar Association (2012).
[47] Regarding protection for whistleblowing employees of companies in the *private* sector, the 2013 defense authorization bill revised U.S. federal statutes so that employees working for contractors to the U.S. Department of Defense or, under a four-year pilot program, for a company contracted by other federal agencies (with the exception of intelligence agencies) were granted new protections. For covered federal contracts signed on or after July 1,

* * *

These four laws provide modest levels of (different kinds of) support to engineers who seek to engage in ethically responsible engineering practice. The FCA financially incentivizes engineers who believe fraud has been committed against the government to sue to recover damage to the public purse by companies for which they worked. The NEPA provides a way an engineer might be able to block a project that he or she believes will significantly damage or degrade a natural or human-made environment. The FCPA establishes major penalties for companies that engage in corrupt business practices abroad, such as bribery and kickbacks on inflated government contracts. Finally, the WPA prohibits retaliation in the form of prohibited personnel practices against covered federal whistleblower employees who disclose specified forms of government agency misconduct.

Besides filing lawsuits based on these pieces of legislation, one other potential legal remedy open to some whistleblowing engineers deserves mention, namely, a type of lawsuit that can be brought against an employer who fires an engineer for making public concerns that her or his employer's actions are harming or pose a threat of harm to the public interest. Under certain conditions, a person employed in the private sector—hence not covered by the Whistleblower Protection Act—has the right to bring a *wrongful termination of employment suit* against her or his former employer. Some engineers have succeeded with wrongful termination suits they filed for loss of employment they claimed was in retaliation for their being whistleblowers, that is, for speaking out publicly about employer activities they contended were illegal, harmful to the public, or posed a risk of harm to the public.

Consider the case of Walter Tamosaitis. He earned a PhD degree in systems engineering and engineering management, is a registered professional engineer, had decades of experience in chemical and nuclear plants, and expertise in chemical mixing technology.[48] In 2013, he filed a wrongful termination suit against the San Francisco–based engineering, design,

---

2013, contractor and subcontractor employees are protected against reprisals for blowing the whistle over alleged contractor or subcontractor waste, fraud, or abuse. See http://docs .house.gov/billsthisweek/20121217/CRPT-112HRPT-705.pdf, secs. 827–828, pp. 496–518.

Certain state governments arguably provide more encouragement of and protection for whistleblowing employees than does the U.S. federal government. One example is California, which has its own California Whistleblower Protection Act. For the protections this act affords, see https://www.dir.ca.gov/dlse/WhistleblowersNotice.pdf.

[48] See State of Washington Superior Court for Benton County (2010).

and construction firm for which he had worked for 44 years and which had fired him earlier that year.

In 2011, Tamosaitis had warned publicly about what he held were serious safety problems with a waste treatment plant being built by his employer, at the Department of Energy's Hanford Nuclear Site, in Washington. The purpose of the plant is to turn highly toxic, radioactive sludge, a by-product of making plutonium for nuclear weapons, into solid glass supposedly capable of being safely buried for centuries.

According to Tamosaitis, who was leading a team of 100 scientists and engineers working on the vitrification effort, after going public with concerns he had about the future safety of the vitrification plant being constructed, he was removed from his managerial position, banned from the waste treatment plant, demoted, and, two years later, fired. His former employer disputed the engineer's claims. However, the company that acquired Tamosaitis's former employer settled the lawsuit with him in August 2015. The engineer received $4.1 million.[49] After the settlement, Tamosaitis stated, "Hopefully, I have sent a message to young engineers to keep their honesty, integrity and courage intact."[50]

EMPLOYMENT-RELATED OPTIONS

Whereas the organizational and legal categories of resources and options are societal in nature, a third category involves individual engineers taking independent initiatives on their own. Engineers interested in practicing engineering in an ethically responsible way have several employment-related options open to them.

Engineering students or young engineers in the job market can choose to exercise special care in selecting their employers. They can elect to base their decisions partly on whether a prospective employer supports the idea that engineers should be able to object to company actions and practices they believe jeopardize public safety and/or welfare without being dismissed, penalized, or ostracized at work for doing so.[51]

An engineer concerned about ethically responsible engineering practice at work can attempt to forge alliances or form united fronts with

[49] See Vartabedian (2015) and Cary (2015).
[50] Vartabedian (2015).
[51] Unfortunately, it is often difficult for a prospective engineer-employee to get a reliable reading on a prospective employer's attitude toward and policies regarding ethically responsible engineering practice by its engineer-employees.

like-minded engineers at his or her company, with a view to opposing alleged company misconduct *as a group* when it occurs. The efficacy of this strategy will depend on whether the company in question would be more reluctant to lose the talent, skill, and experience of those possessing a significant part of its intellectual capital than to lose the knowledge and skills of a solitary engineer.

The individual engineer who declines or is unable to forge an alliance with like-minded fellow engineer-employees at her or his firm can strive to be recognized as an excellent engineering practitioner. This would presumably make the company or client more reluctant to terminate her or his employment because she or he voiced or otherwise acted on an ethical responsibility concern. All things being equal, the greater the acknowledged professional excellence of an engineer who cares about ethically responsible engineering practice, the less easily intimidated into acquiescence such a person is likely to be at the prospect of being demoted or of losing her or his job for opposing company or client actions or practices. Highly competent individual engineers are likely to have viable employment options if they get fired, and that may embolden them to resist deeds, practices, and demands of their employers or clients they regard as ethically irresponsible.

An engineer determined to ensure that his or her professional practice is ethically responsible could decide to start his or her own company and make it one in which conscientious, ethically responsible engineering practice is a cornerstone cultural value and behavioral norm.

A currently employed engineer could decide to seek a new employer because of concerns about ethically responsible engineering practice at the existing employer. However, doing so could be rendered difficult or impossible in the short term if the engineer signed a legally binding restrictive confidentiality or nondisclosure agreement with her or his initial employer. The same might be the case if the initial employer threatens to blacklist an engineer thinking of departing because of concerns about ethically responsible engineering practice.

Some engineers might seek to avoid the potential conflicts of interest encountered in many engineering jobs in industry and hope to gain the freedom to act in an ethically responsible manner by working in academia. However, working in academia does not guarantee that the engineer will be spared ethical responsibility challenges. For example, an academic engineer's research could be funded by a corporation about which he or she comes to have a public safety, environmental protection, or other ethical responsibility concern. Such a situation would confront the engineer with significant conflict of interest. The private firm funding an academic engineer's research could, for its own commercial reasons, make provision or continuation of its financial support for that research contingent on the academic engineer's willingness to delay or restrict the content about the research in his or her publications as the company wishes. Alternatively, the firm could make provision of funding for the academic engineer's research hinge on his or her willingness to publish quickly, at the behest of the funding company, preliminary findings the researcher regards as too premature to be published.

* * *

The modest social infrastructure supportive of ethically responsible engineering practice currently in place in the contemporary United States suggests that in most cases an engineer-employee's decision to be an ethically responsible practitioner will not be fueled by the availability of societal protection against employer retribution, or by the prospect of economic reward for acting responsibly by disclosing malfeasance. An engineer's decision to engage in ethically responsible engineering practice is more likely to stem from belief that acting in such a manner is the ethically right thing for a professional engineer to do, from the fact that doing so will leave the engineer with a clear conscience about her or his professional activity, and/or from the influence on the engineer of compelling examples of ethically responsible engineering practice set by respected teachers, work colleagues, or other engineering role models.

CHAPTER 7

Conclusion

To conclude this book, I shall characterize my general approach to exploring ethical issues in engineering in a little more detail. I shall do so by discussing similarities and differences between my approach and that of the author of a recent critique of how ethical issues have been studied in U.S. engineering education.[1]

BUCCIARELLI'S CRITIQUE OF U.S. ENGINEERING-ETHICS EDUCATION

Louis Bucciarelli notes that ABET,[2] the organization that evaluates and accredits most postsecondary engineering programs in the U.S., "recommends the study of ethics so that students attain 'an understanding of professional and ethical responsibility.'"[3] However, he has "never felt comfortable with this directive and with the way the subject is generally taught."[4] Bucciarelli's discomfort with what he believes to be "the way the subject is generally taught" stems from his belief that the cases used are "artificial" and

> neglect the social nature of day-to-day engineering. In focusing solely on an *individual* agent's possible courses of action, they oversimplify; they are not a valid abstraction.[5]

---

[1] Bucciarelli (2008).
[2] For details on ABET, whose corporate name is "Accreditation Board for Engineering and Technology," see http://www.abet.org/History/.
[3] Bucciarelli (2008), p. 141.
[4] Ibid.
[5] Ibid.

He grants that "[e]thics ought not be neglected in engineering education[.]"[6] However, he claims that "to learn about the social, the organizational—even the political—complexities of [engineering] practice" is "more fundamental and prerequisite" for engineering students.[7] Bucciarelli also rues traditional U.S. engineering-ethics education because of what he sees as its tendency to

> see every task and challenge an engineer faces as a problem to be solved—by an individual alone.[8]

He illustrates his concerns about traditional engineering-ethics education by discussing two cases. The first involves a young "control systems engineer" assigned to solve "some problems [with] the autopilot software for commercial Airbus jets."[9] The engineer learns from pilots that under certain conditions the autopilot takes control actions that "can cause the plane to stall and potentially crash." She or he has been asked to "figure out what's going wrong," even as the planes in question continue to be flown using the problematic software. The case ends with the posing of a blunt ethics question to students who are asked to assume the role of the Airbus engineer: "What should you do?"[10]

The second case he discusses is the space shuttle *Challenger* disaster.[11] For Bucciarelli, the focus of this case is the off-line caucus of four Morton-Thiokol Inc. (MTI) senior managers—all engineers by training— with a group of MTI engineers with expertise in booster rockets. This caucus took place during a break the MTI managers requested from the

---

[6]Ibid.
[7]Ibid. While Bucciarelli regards learning about these "complexities of practice" as "more fundamental and prerequisite" than "ethics" in engineering education, it is unclear what he believes learning about these complexities of practice is "more fundamental and prerequisite" *for*. If the goal he has in mind is to produce engineers with a robust sense of and commitment to ethically responsible engineering practice, it is not clear that learning about social "complexities of practice" is "more fundamental and prerequisite" to realizing that goal than is studying engineering ethics. If that is in fact the goal Bucciarelli has in mind, perhaps he regards learning about social complexities of engineering practice as more fundamental and prerequisite to realizing it because he equates ethics education for engineers with the narrow, distorted form of it he believes has long dominated in the United States. Be that as it may, in this book, attention to social "complexities of practice" is integral to, not an alternative to, ethical analysis of several of the cases explored.
[8]Bucciarelli (2008), p. 141.
[9]Ibid., p. 142. See also Meckl (2003).
[10]Bucciarelli (2008), p. 142.
[11]Bucciarelli references the write-up of this case by faculty from Texas A&M University. See Texas A&M University (1993b).

launch-recommendation teleconference between MTI and NASA officials the evening before the shuttle's fateful launch.

As discussed in Case 8, several of the MTI engineers present warned the senior MTI engineer-managers about what they regarded as the serious risk of O-ring failure in cold weather like that expected the next morning at launch time. They urged them to maintain MTI's initial recommendation against launching *Challenger* if the ambient temperature was less than 53°F. However, after Jerald Mason, V.P. of Wasatch Operations, told Robert Lund, V.P. of Engineering, "Take off your engineering hat and put on your management hat,"[12] the MTI senior engineer-managers present voted unanimously to reverse the company's earlier recommendation and give NASA what it sought: MTI's recommendation to launch the next morning.

What troubles Bucciarelli about these two examples?

> It is the lack of any substantive treatment, none the less recognition, of the social context of the situations with which we confront our students in these scenarios that is the source of my unease. The focus on the *individual* agent, hanging there, alone, confronting some identifiable ethical question strikes me as too simple a representation of what an engineer faces in practice.[13]

Reflecting this concern about neglecting social context, Bucciarelli notes that the narrative of the Airbus software case does not indicate whether the control systems engineer was part of a team or acting alone. He apparently believes this detail could make a difference in what the engineer should do. Bucciarelli probably regards posing the "What should you do?" question as premature, since the engineer's work situation was not adequately specified.

What disturbs Bucciarelli about the *Challenger* case? Three of his comments provide clues. First, he wants to "move beyond myth-making about whistleblowing."[14] Second, he states that

> instrumental, rational analysis never suffices in practice. . . . What appeared to Boisjoly as hard evidence for failure [of the seals in the solid rocket booster joints], appeared to others as ambiguous, tentative, unconvincing.[15]

[12] Bucciarelli (2008), p. 142.
[13] Ibid., p. 143.
[14] Ibid., p. 144.
[15] Ibid., pp. 144–145. Rather than asserting that the erosion-temperature evidence "appeared to" Boisjoly as "hard evidence for failure," I suggest that in light of the available erosion-temperature evidence, Boisjoly believed the rocket booster O-rings were at serious

Third, he holds that

> in the design, operation, and maintenance of complex engineering systems, responsibility is shared among participants in a project. Seeking a root cause and/or fingering a single individual to blame (or praise) for failure (or success) . . . is likely to be a superficial rendering of events and, as such, does a disservice to the intelligence of our students.[16]

These comments reflect a central Bucciarellian tenet: that engineering work is a *social* process, not the sum of actions of solitary individual engineers.

His first comment expresses his concern that the Boisjoly whistleblowing episode could encourage "myth-making," something he believes fosters simplistic thinking about engineering by highlighting the deeds of individual "heroes" and "villains."

His second comment suggests that Bucciarelli believes the disagreement in the caucus over what to recommend to NASA about launching the next morning was *not* the result of MTI engineers Roger Boisjoly and Arne Thompson[17] taking a rational position after looking at the erosion-temperature evidence, and four "others"—the four MTI senior managers—taking an irrational position after looking at the same evidence. Rather, according to Bucciarelli, the disagreement resulted from the fact that the erosion-temperature evidence "appeared to" Boisjoly as "hard evidence for failure," whereas it "appeared to" "others"—presumably the senior engineer-managers—as "ambiguous, tentative, [and] unconvincing."

Bucciarelli's third comment expresses his belief that looking for "root causes" of unfortunate engineering outcomes, and blaming or praising "single individual[s]" (instead of groups) for failures or successes, is apt to lead to superficial, misleading accounts of engineering activity.

In short, Bucciarelli seeks to replace what he sees as the dominant pedagogical focus on the ethical responsibilities of *individual engineers* with emphasis on "the responsibility of individuals as part of a collective, a shared responsibility."[18]

I am ambivalent about Bucciarelli's concerns. I agree with him that "myth-making about whistleblowing" is regrettable; however, resorting

---

risk of failing if the shuttle was launched under the conditions expected at launch time the next day.

[16] Ibid., p. 145.

[17] Thompson was supervisor of rocket motor cases at MTI. See Rogers (1986), p. 104.

[18] Bucciarelli (2008), p. 144.

to "myth-making" is not inevitable. Several accounts of Boisjoly's conduct in the teleconference and caucus, and in the postdisaster inquiry
that culminated in his revelations to the presidential commission, avoided
myth-making about whistleblowing.[19]

As for the disagreement in the off-line caucus, I suspect it stemmed
not from the erosion evidence's having *appeared* differently to different
parties but from the different *postures* the engineers adopted toward the
same available evidence: a precautionary posture in Boisjoly's case, and an
innocent-until-proven-guilty posture in the case of the engineer-managers.

In contrast to Bucciarelli, given how upset several NASA managers
were with MTI's initial recommendation to not launch, I suspect it was
the MTI senior engineer-managers' sudden concern about the economic
ramifications[20] of MTI's possibly losing its lucrative status of being the
sole provider of solid rocket boosters to NASA that explains their sudden
switch from the usual "guilty-until-proven-innocent" to an "innocent-
until-proven-guilty" posture regarding the shuttle launch decision.[21] The
claim that the O-ring erosion evidence "appeared" differently to the engineers than it did to the engineer-managers is not needed to explain the
disagreement in the off-line caucus that led to reversal of the initial recommendation. The opposing positions taken on whether to recommend
launch were shaped by the different postures adopted toward the same
evidence, postures that were affected by individual factors, cultural factors, and political-economic interests.

Regarding seeking "root causes" and singling out individuals for
blame or praise for engineering failures or successes, I agree that doing
so may sometimes yield superficial accounts of engineering episodes.[22]

---

[19] Bell and Esch (1987); Boisjoly (1987); Rogers (1986), ch. 5; and Texas A&M University (1993b).

[20] "[I]t was common internal company knowledge that MTI was in the process of nego-
tiating a lower price [for the boosters] prior to the *Challenger* disaster to protect their [sic]
single-source contract status." Boisjoly (1987), p. 12. But the economic ramifications were
actually far greater. At the time, MTI was planning to compete for a new NASA contract
to "build the next generation of solid-fuel rocket boosters for the space shuttle." When the
contract was awarded in April 1989, to Lockheed, NASA indicated that "the contract . . .
should be worth about $1.2 billion." Leary (1989).

[21] It is important to recall Boisjoly's words: "I must emphasize that MTI Management
fully supported the original decision to not launch below 53°F (12°C) prior to the caucus."
Boisjoly (1987), p. 8. After their about-face, triggered by NASA's angry response to the
original recommendation, the engineer-managers took the position that since the line engineers had not "proven" it was unsafe to launch, it was okay to launch. Unfortunately, the
absence of proof that something is unsafe is not proof that it is safe.

[22] Union Carbide's explanation of the Bhopal disaster comes to mind in this connection.

However, such accounts are no more an inevitable consequence of ethical evaluation of the actions of individual engineers than are myth-making accounts of engineer-whistleblowers. Attentiveness to an engineering episode's context in its full complexity, and to the wide range of factors that often contribute to causing an engineering outcome, from precipitating and stimulating to enabling and facilitating factors, can help avoid simplistic accounts.

A FOUNDATIONAL-CONTEXTUAL ETHICAL RESPONSIBILITIES APPROACH

How does the general approach to exploring ethical issues in engineering reflected in this book differ from Bucciarelli's preferred approach?

I agree with Bucciarelli that U.S. engineering-ethics education has leaned too heavily on engineering society codes of ethics, has used too many contrived hypothetical and thin real cases, has neglected the social nature of engineering work, and has sought tidy "solutions" to ethical problems. However, in reacting against these shortcomings, Bucciarelli has shelved or consigned to limbo the task of making ethical judgments about the conduct of individual engineers. He appears to be much more interested in exposing engineering students to "the social complexities of engineering practice"[23] and probing "the social responsibility of the [engineering] profession as a whole"[24] than in making ethical judgments about the conduct of individual engineers.

I also agree with Bucciarelli that much if not most contemporary engineering work is carried out in teams and other social groups. Such work units often have their own cultures in addition to operating in organizations whose overarching cultures also sometimes shape that work.

However, recognizing the "social nature of day-to-day engineering"[25] and viewing "engineering as a social process"[26] do not require abandoning the task of making ethical judgments about the conduct of individual engineers. One can criticize rule-and-case reliance on codes of engineering ethics without indicting all approaches to passing ethical judgment on the conduct of individual engineers. Bucciarelli appears to want to substitute studies that illuminate "the social, the organizational—even the

[23] Ibid., pp. 144 and 145.
[24] Ibid., p. 146.
[25] Ibid., p. 141.
[26] Ibid., p. 145.

political—complexities of [engineering] practice"[27] for code-of-ethics-based studies that yield ethical judgments about the conduct of individual engineers. I prefer to *expand* the range of factors taken into account in making *non-code-based* ethical judgments about the conduct of individual engineers. Bucciarelli appears to want to *replace* traditional code-of-ethics-based engineering-ethical education with study of social-scientific accounts of engineering practice, such as ethnographies and social constructivist analyses of engineering work. I favor developing a *non-code-based* form of ethical assessment of the conduct of individual engineers that is *enriched* by concepts, perspectives, and findings of social-scientific studies of engineering practice.

Determining the ethical responsibilities of individual engineers should remain an important focus of engineering-ethical inquiry. What needs to change is how we ascertain those responsibilities, especially the specific ethical responsibilities of engineers in concrete situations. In determining the latter, attention *should* be paid to the social nature of contemporary engineering work. But careful attention should also be given to the fact that engineering work unfolds in *multileveled, multidimensional contexts.* A distinguishing feature of my approach is that it requires attention to be paid to both the social processes and the sociocultural contexts of engineering work and its outcomes. That can be done without abandoning the task of determining the context-specific ethical responsibilities of individual engineers.

In a nutshell, the specific ethical responsibilities of individual engineers in concrete engineering situations should be determined after considering five categories of factors:

1. The **fundamental ethical responsibilities of engineers** (the FEREs).
2. The **noteworthy sociotechnical features**[28] **of the engineering work** in question (Is the work an example of normal/conventional or radical/paradigm-departing engineering design? Is it done by a solitary engineer or a team of engineers? Is it upstream basic research, detailed design, or downstream development or maintenance work? Does it utilize physical models or computer simulation?).

---

[27] Ibid., p. 141.
[28] Bucciarelli urges that careful attention be paid to the *social* features of engineering work. I prefer to speak of the *sociotechnical* features of specific examples of engineering work. In some such features, the social element will dominate; in others, the technical element will dominate. In yet others, the social and technical elements may be inextricably intertwined and be more or less equally important and influential.

3. The **noteworthy features of the social-organizational context** in which the engineering work in question takes place (Does the work occur in a company currently under heavy economic pressure? Does it occur in a company whose culture presses engineers to fulfill their marching orders uncritically or in one that affords its engineers decisional latitude rooted in concern for ethically responsible engineering practice? Is the engineering work in question done by employees or consultants? Does it take place in a large long-established firm or a small start-up? Does it unfold in a single organizational unit, or is it a collaboration involving multiple organizational units, perhaps with different cultures?).

4. The **noteworthy features of the macrosocietal context(s)** in which the fruits of the engineering activity in question are diffused and used (Will the resultant engineering product be used in loosely and/or tightly regulated industrial, governmental, or consumer environments? Will it be used in a less developed society and/or a more developed society? Will it be diffused into a society with a rapidly expanding, stagnant, or contracting economy?).

5. The **likely noteworthy harm-and-well-being-related consequences,** for all affected and likely-to-be-affected parties, of the application and use of the fruits of the engineering work in question (Whose well-being is likely to be affected by the products or systems in question? Which parties among those affected are likely to realize benefits that enhance their well-being, and which are likely to bear harm-related costs and risks that may dilute or diminish it? What kinds of benefits, harms, and risks of what magnitudes are involved, and how are they likely to be distributed? To what extents are the likely distributions of harms, risks, and benefits consistent with Rawls's Difference Principle?).

In Chapter 3, I referred to taking "an ethical responsibilities approach" to making ethical judgments about engineering conduct. In light of the just-listed five categories of factors, the general approach taken in this book to making ethical judgments about engineering conduct would perhaps be better termed **a foundational-contextual ethical responsibilities approach**. It requires identification of the applicable FEREs, scrutiny of the sociotechnical processes and organizational and social contexts of the specific engineering work and its products, and estimation of the effects of the socially situated engineering work and its products on the well-being of all affected and likely-to-be-affected parties.

TWO QUOTATIONS

Two quotations capture the spirit of the approach taken in this book to exploring ethical issues in engineering:

> [W]hile [ethical] integrity without [technical] knowledge is weak and use-less, [technical] knowledge without [ethical] integrity is dangerous and dreadful.
>   —Samuel Johnson[29]

> It is not enough that you should understand about applied science in order that your work may increase man's blessings. Concern for the man himself and his fate must always form the chief interest of all technical endeavors; concern for the great unsolved problems of the organization of labor and the distribution of goods, in order that the creations of our mind shall be a blessing and not a curse to mankind. Never forget this in the midst of your diagrams and equations.
>   —Albert Einstein[30]

Technical expertise per se is not enough to make someone an excellent engineering professional in contemporary society. Nor does making engineering students aware of the social complexities of engineering practice, while desirable, suffice to produce such an engineer. Not even a combination of those two factors will do the trick. The Johnson quotation helps us realize that as the fruits of engineering endeavor become more potent and pervasive in contemporary society, it is more critical than ever that engineers inform their work and shape its products with the readings of a sensitive ethical compass.

Given the pressures to which they are subject, the temptation is strong for contemporary engineers to immerse themselves in and become pre-occupied with the daunting technical details and intellectual and organizational challenges of their work. The Einstein quotation reminds us it is vital that they not become cognitively myopic and decontextualize or undercontextualize that work. Attentiveness to the multiple *levels* of context within which engineering work unfolds—at the work group, departmental, divisional, and macro-organizational levels; and the local, regional, national, and global societal levels—and to the multiple *dimensions* of each of these contexts—political-economic, sociological, cultural, technical, environmental, and agental[31] is essential to under-

---

[29] Samuel Johnson (1759). The words in brackets are the present author's.
[30] "Einstein Sees Lack in Applying Science," *New York Times*, February 17, 1931, p. 6.
[31] The "agental" dimension of context refers to the engineering and other actors present in a particular engineering context, and their individual and group characteristics.

standing engineering-in-society interactions and to making thoughtful ethical judgments about engineering conduct. Having a better grasp of the social nature and sociotechnical features of most engineering work may be "prerequisite"[32] for making such judgments. However, it is not a *substitute* for making such judgments. Situating engineering work in its full multileveled, multidimensional context is essential to grasping its causes and consequences, and hence to making defensible ethical judgments about the conduct of individual engineers.

Relying on untutored moral intuition, religious beliefs, or the letter of a code of engineering ethics does not suffice to assure ethically responsible conduct in engineering. In the twenty-first century, engineers must be mindful of the fundamental ethical responsibilities of engineering professionals; develop the ability to make context-sensitive ethical judgments that take into consideration the full range of likely well-being-related consequences of their actions, practices, products, and systems; and make their professional practice reflect that mindfulness and that ability.

Ethically responsible engineers in contemporary society combine state-of-the-art technical expertise with mindfulness of the fundamental ethical responsibilities of engineers. They recognize the key socio-technical features of the engineering activity in which they are involved, and embrace a robust and evolving notion of harm. They are attentive to the multiple multidimensional social contexts in which their work unfolds and its end products are used, and take into account how their work and its fruit bear, directly and indirectly, on the well-being of all affected parties. Educating students to become ethically responsible engineers is neither easy nor widespread. It is, however, extremely important. Done properly, it will increase the likelihood that engineers' work will continue to be, in Einstein's words, "a blessing, and not a curse" to humankind.

---

[32]Bucciarelli (2008), p. 141.

BIBLIOGRAPHY

Abate, Tom. (2015). "Stanford Researchers Genetically Engineer Yeast to Produce Opioids," *Stanford News*, August 15, http://news.stanford.edu/2015/08/13/opioids-yeast-smolke-081315/.

Accreditation Board for Engineering and Technology. (2014). http://www.abet.org.

Agre, Philip E., and Mailloux, Christine A. (1997). "Social Choice about Privacy: Intelligent Vehicle-Highway Systems in the United States," in *Human Values and the Design of Computer Technology*, ed. Batya Friedman (Center for the Study of Language and Information: Stanford), pp. 289–310.

American Bar Association. (2012). "Congress Strengthens Whistleblower Protections for Federal Employees," November-December 2012 (Special Feature), http://www.americanbar.org/content/newsletter/groups/labor_law/ll_flash/1212_abalel_flash/lel_flash12_2012spec.html.

American Institute of Chemical Engineers (2015). "Code of Ethics," https://www.aiche.org/about/code-ethics.

American Society of Civil Engineers. (2007). "The Hyatt Regency Walkway Collapse," http://asce.org/question-of-ethics-articles/jan-2007/.

———. (2009). "ASCE Code of Ethics," October, http://www.asce.org/uploadedFiles/About_ASCE/Ethics/Content_Pieces/CodeofEthics2006.pdf}.

American Society of Mechanical Engineers. (2012). "Code of Ethics of Engineers," https://www.asme.org/getmedia/9eb36017-fa98-477e-8a73-77b04b36d410/p157_ethics.aspx}.

Anderson, Ronald E., Johnson, Deborah G., Otterbein, Donald, and Parole, Judith. (1993). "Using the New ACM Code of Ethics in Decision Making," *Communications of the ACM*, Vol. 36, No. 2, February, pp. 98–106; http://www.acm.org/about/p98-anderson.pdf.

Association for Computing Machinery. (1992). "ACM Code of Ethics and Professional Conduct," October, http://www.acm.org/about/code-of-ethics.

Augustine, Norman. (2002). "Ethics and the Second Law of Thermodynamics," *The Bridge*, Vol. 32, No.3, pp. 4–7, http://www.nae.edu/Publications/Bridge/EngineeringEthics7377/EthicsandtheSecondLawofThermodynamics.aspx.

Auschwitz-Birkenau Memorial and Museum. (2014). "The Number of Victims," http://en.auschwitz.org/h/index.php?option=com_content&task=view&id=14&Itemid=13.

Barker, Stephen F. (1992). "What Is a Profession?" *Professional Ethics*, Vol. 1, Nos. 1–2, pp. 73–99.

Baura, Gail. (2006). *Engineering Ethics: An Industrial Perspective* (Elsevier: Burlington, MA).

"Bay Area Rapid Transit." https://en.wikipedia.org/wiki/Bay_Area_Rapid_Transit.

Beasley, Malcolm, Datta, Supriyo, Kogelnik, Herwig, Kroemer, Herbert, and Monroe, Don. (2002). *Report of the Investigation Committee on the Possibility of Scientific Misconduct in the Work of Hendrik Schön and Coauthors,* http://publish.aps.org/reports/lucentrep.pdf, September, pp. 1–19 plus Appendices A–H.

Beauchamp, Tom. (2013) "The Principle of Beneficence in Applied Ethics," *The Stanford Encyclopedia of Philosophy* (Winter 2016 ed.), ed. Edward N. Zalta, https://plato.stanford.edu/archives/win2016/entries/principle-beneficence/.

Beder, Sharon, and Shortland, Michael. (1992). "Siting a Hazardous Waste Facility: The Tangled Web of Risk Communication," *Public Understanding of Science*, Vol. 1, pp. 139–160.

Bell, Trudy, and Esch, Karl. (1987). "The Fatal Flaw in Flight 51-L," *IEEE Spectrum*, Vol. 24, No. 2 (February), pp. 36–51.

Berkes, Howard. (2012). "Remembering Roger Boisjoly: He Tried to Stop Shuttle Challenger Launch," February 6, http://www.npr.org/blogs/thetwo-way/2012/02/06/146490064/remembering-roger-boisjoly-he-tried-to-stop-shuttle-challenger-launch.

"Bhopal Disaster." http://en.wikipedia.org/wiki/Bhopal_disaster.

Birsch, Douglas, and Fielder, John H., eds. (1994). *The Ford Pinto Case: A Study in Applied Ethics, Business, and Technology* (State University of New York Press: Albany).

Boisjoly, Roger. (1987). "Ethical Decisions—Morton Thiokol and the Space Shuttle *Challenger* Disaster," *ASME*, 87-WA/TS-4, pp. 1–13.

Brey, Philip. (2009). "Biomedical Engineering Ethics," in *A Companion to the Philosophy of Technology*, ed. Jan Kyrre Berg Olsen, Stig Andur Pedersen, and Vincent F. Hendricks (Blackwell, Oxford).

Broughton, Edward. (2005). "The Bhopal Disaster and Its Aftermath: A Review," *Environmental Health*, Vol. 4, No. 6, https://www.ncbi.nlm.nih.gov/pmc/articles/PMC1142333/.

Boyd, Gerald M. (1986). "White House Finds No Pressure to Launch," *New York Times*, April 4.

British Military Court. (1946). Hamburg, Germany, "The Zyklon B Case, Trial of Bruno Tesch and Two Others," March 1–8, in United Nations War Crimes Commission, *Law Reports of Trials of War Criminals*, Vol. 1, pp. 93–103, http://www.loc.gov/rr/frd/Military_Law/pdf/Law-Reports_Vol-1.pdf.

Broad, William J. (2000a). "Missile Contractor Doctored Tests, Ex-Employee Charges," *New York Times*, March 7, http://www.nytimes.com/2000/03/07/us/missile-contractor-doctored-tests-ex-employee-charges.html.

———. (2000b). "Antimissile Testing Is Rigged to Hide a Flaw, Critics Say," *New York Times*, June 9, http://www.nytimes.com/2000/06/09/us/antimissile-testing-is-rigged-to-hide-a-flaw-critics-say.html.

———. (2002). "U.S. Is Tightening Rules on Keeping Scientific Secrets," *New York Times*, February 17, http://www.nytimes.com/2002/02/17/us/nation-challenged-domestic-security-us-tightening-rules-keeping-scientific.html.

———. (2003a). "U.S. Seeks Dismissal of Suit by Critic of Missile Defense," *New York Times*, February 3, http://www.nytimes.com/2003/02/03/politics/03MISS.html.

———. (2003b). "Missile-Defense Critic's Suit Is Dismissed," *New York Times*, March 8, www.nytimes.com/2003/03/08/us/missile-defense-critic-s-suit-is -dismissed.html.

Bucciarelli, Louis L. (2008). "Ethics and Engineering Education," *European Journal of Engineering Education*, Vol. 33, No. 2 (May), pp. 141–149.

Carey, Benedict. (2016). "Chip, Implanted in Brain, Helps Paralyzed Man Regain Control of Hand," *New York Times*, April 13, https://www.nytimes.com/2016 /04/14/health/paralysis-limb-reanimation-brain-chip.html.

Carlton, Jim. (1997). *Apple: The Inside Story of Intrigue, Egomania, and Business Blunders* (Times Business/Random House: New York).

Cary, Annette. (2015). "Hanford Whistleblower Wins; Tamosaitis to Receive $4.1 Million Settlement," *Tri-City Herald*, August 12, http://www.tri -cityherald.com/news/local/hanford/article32240640.html.

Center for Constitutional Rights. (2007). "FAQs: What Are State Secrets," October 17, https://ccrjustice.org/home/get-involved/tools-resources/fact-sheets -and-faqs/faqs-what-are-state-secrets.

Chandler Jr., Alfred D. (1977). *The Visible Hand: The Managerial Revolution in American Business* (Harvard University Press: Cambridge).

Chouhan, T. C. (2005). "The Unfolding of Bhopal Disaster," *Journal of Loss Prevention in the Process Industries*, Vol. 18, pp. 205–208, http://apps.engr.utexas .edu/ethics/standards/bhopal/Article%2002%20-%20The%20unfolding %20of . . . pdf.

Cohen, Jon. (2002). "Designer Bugs," *The Atlantic*, July/August, http://www .theatlantic.com/magazine/archive/2002/07/designer-bugs/378484/.

Committee on Science, Engineering, and Public Policy of the National Academies. (2009). *On Being a Scientist: A Guide to Responsible Conduct in Research*, 3rd ed. (National Academies Press: Washington, DC).

Congressional Budget Office. (1984). "Analysis of the Costs of the Administration's Strategic Defense Initiative, 1985–1989," May 1, Table 2, https://www .cbo.gov/publication/15155.

Constant II, Edward. (1980). *The Origins of the Turbojet Revolution* (Johns Hopkins University Press: Baltimore).

Cook, Richard C. (2006). Challenger *Revealed: An Insider's Account of How the Reagan Administration Caused the Greatest Tragedy of the Space Age* (Thunder's Mouth Press: New York).

Corn, Joseph J. (2011). *User Unfriendly: Consumer Struggles with Personal Technologies, from Clocks and Sewing Machines to Cars and Computers* (Johns Hopkins University: Baltimore).

Coscarelli, Joe. (2014). "Ralph Nader on the General Motors Disaster and How to End 'Cover-Your-Ass' Corporate Culture," *New York Magazine*, April 3, http://nymag.com/daily/intelligencer/2014/04/ralph-nader-interview-on-general -motors.html.

Crichton, Michael. (2002). *Prey* (Harper: New York).

Cullinan, Paul. (2004). "Case Study of the Bhopal Incident," *Encyclopedia of Life Support Systems*, Vol. I: *Environmental Toxicology and Human Health*, http:// www.eolss.net/sample-chapters/c09/e4-12-02-04.pdf.

Cushman, John H. (1995). "Half-Million Cadillacs Recalled in Federal Pollution Settlement," *New York Times*, December 1, http://www.nytimes

.com/1995/12/01/us/half-million-cadillacs-recalled-in-federal-pollution -settlement.html.

DeGeorge, Richard T. (1981). "Ethical Responsibilities of Engineers in Large Organizations: The Pinto Case," *Business and Professional Ethics Journal*, Vol. 1, No. 1, pp. 1–14, http://www.jstor.org/stable/27799725.

Diamond, Stuart. (1985a). "The Bhopal Disaster: How It Happened," *New York Times*, January 28, http://www.nytimes.com/1985/01/28/world/the-bhopal -disaster-how-it-happened.html.

———. (1985b). "The Disaster in Bhopal: Workers Recall the Horror," *New York Times*, January 30, http://www.nytimes.com/1985/01/30/world/the-disaster-in -bhopal-workers-recall-horror.html.

———. (1985c). "Disaster in Bhopal: Lessons for the Future," *New York Times*, February 3, http://www.nytimes.com/1985/02/03/world/the-disaster-in-bhopal -lessons-for-the-future.html.

Dockery, Colleen A., Hueckel-Weng, Ruth, Birbaumer, Neils, and Plewnia, Christian. (2009). "Enhancement of Planning Ability by Transcranial Direct Current Stimulation," *Journal of Neuroscience*, Vol. 29, No. 22 (June 3), pp. 7271–7277, https://www.researchgate.net/profile/Christian_Plewnia/publication /26264555_Enhancement_of_Planning_Ability_by_Transcranial_Direct _Current_Stimulation/links/53fe010e0cf2364ccc0a1d51/Enhancement-of -Planning-Ability-by-Transcranial-Direct-Current-Stimulation.pdf.

Dore, Ronald. (1989). "Technology in a World of National Frontiers," *World Development*, Vol. 17, No. 11, pp. 1665–1675.

Dowd, Maureen. (1984). "The Westway's Man in the Middle," *New York Times*, June 27, http://www.nytimes.com/1984/06/27/nyregion/man-in-the-news-the -westway-s-man-in-the-middle.html.

Dowie, Mark. (1977). "Pinto Madness," *Mother Jones*, Vol. 18, September/October, http://www.motherjones.com/politics/1977/09/pinto-madness?page=1.

Dutta, Sanjib. (2002). "The Bhopal Gas Tragedy," ICFAI Center for Management Research (ICMR), Hyderabad, India, http://www.econ.upf.edu/~lemenestrel /IMG/pdf/bhopal_gas_tragedy_dutta.pdf.

Dvorsky, George. (2016). "Brain Implant Enables Quadriplegic Man to Play *Guitar Hero* with His Hands," *Gizmodo*, April 13, http://gizmodo.com/brain -implant-enables-quadriplegic-man-to-play-guitar-h-1770566874.

Eckholm, Erik. (1999). "China Shifts on How to Resettle Million People for Giant Dam," *New York Times*, May 25, http://www.nytimes.com/1999/05/25 /world/china-shifts-on-how-to-resettle-million-people-for-giant-dam.html.

Edwards, Tim. (2002). "How Many Died in Bhopal?" http://www.bhopal.net/old _bhopal_net/death-toll.html.

Ehrenberg, Rachel. (2015). "Engineered Yeast Paves Way for Home-Brew Heroin," *Nature*, Vol. 521, No. 7552, May 18, http://www.nature.com/news /engineered-yeast-paves-way-for-home-brew-heroin-1.17566.

"Einsatzgruppen." http://en.wikipedia.org/wiki/Einsatzgruppen.

Endy, Drew, Galanie, Stephanie, and Smolke, Christina D. (2015). "Complete Absence of Thebaine Biosynthesis under Home-Brew Fermentation Conditions," August 13, http://www.biorxiv.org/content/early/2015/08/13/024299 (preprint).

"Engine Control Unit." http://en.wikipedia.org/wiki/Engine_control_unit#Control_of_Air.2FFuel_ratio.

Farah, Martha J. (2011). "Neuroscience and Neuroethics in the 21st Century," in *Oxford Handbook of Neuroethics*, ed. Judy Illes and Barbara J. Sahakian (Oxford University Press: Oxford), pp. 761–782.

Ferris, Susan, and Sandoval, Ricardo, (2004). "The Death of the Short-Handled Hoe," http://www.pbs.org/itvs/fightfields/book1.html.

Fisher, Madeline. (2008). "Study Uncovers How Ritalin Works in Brain to Boost Cognition, Focus Attention," June 24, http://news.wisc.edu/study-uncovers-how-ritalin-works-in-brain-to-boost-cognition-focus-attention/.

Fisher, Martin. (2006). "Income Is Development," *Innovation*, No. 1 (Winter), pp. 9–30, http://www.policyinnovations.org/ideas/policy_library/data/01375/_res/id=sa_File1/INNOV0101_KickStart.pdf.

Fleming, Gerald. (1993). "Engineers of Death," *New York Times*, July 18, Op-Ed.

Floel, A., Garraux, G., Xu, B., Breitenstein, C., Knecht, S., Herscovitch, P., and Cohen, L. G. (2008). "Levodopa Increases Memory Encoding and Dopamine Release in the Striatum in the Elderly," *Neurobiology of Aging*, Vol 29, pp. 267–279, https://www.ncbi.nlm.nih.gov/pmc/articles/PMC2323457/.

Fregni, Felipe, Boggio, Paulo S., Santos, Marcelo C., Lima, Moises, Vieira, Adriana L., Rigonatti, Sergio P., Silva, M. Teresa A., Barbosa, Egberto R., Nitsche, Michael A., and Pascual-Leone, Alvaro. (2006). "Noninvasive Cortical Stimulation With Transcranial Direct Current Stimulation in Parkinson's Disease," *Movement Disorder*, Vol. 21, No. 10, pp. 1693–1702, https://www.researchgate.net/profile/Alvaro_Pascual-Leone/publication/6968020_Noninvasive_cortical_stimulation_with_transcranial_direct_current_stimulation_in_Parkinson%27s_disease/links/55b74f3308ae9289a08bdf3c/Noninvasive-cortical-stimulation-with-transcranial-direct-current-stimulation-in-Parkinsons-disease.pdf.

Friedlander, Gordon D. (1974). "The Case of the Three Engineers vs. BART," *IEEE Spectrum*, October, pp. 69–76.

Funk, Cary, Kennedy, Brian, and Podrebarac Sciupac, Elizabeth. (2016a). "U.S. Public Opinion on the Future Use of Gene Editing," July 26, http://www.pewinternet.org/2016/07/26/u-s-public-opinion-on-the-future-use-of-gene-editing/.

———. (2016b). "U.S. Public Opinion on the Future Use of Synthetic Blood Substitutes," July 26, http://www.pewinternet.org/2016/07/26/the-publics-views-on-the-future-use-of-synthetic-blood-substitutes/.

———. (2016c). "Public Opinion on the Future Use of Brain Implants," July 26, http://www.pewinternet.org/2016/07/26/public-opinion-on-the-future-use-of-brain-implants/.

———. (2016d). "U.S. Public Wary of Biomedical Technologies to 'Enhance' Human Abilities," July 26, http://www.pewinternet.org/2016/07/26/u-s-public-wary-of-biomedical-technologies-to-enhance-human-abilities/.

Galanie, Stephanie, Thodey, Kate, Trenchard, Iris J., Interrante, Maria Filsingerm, and Smolke, Christina D. (2015). "Complete Biosynthesis of Opioids in Yeast," *Science*, Vol. 349, Issue 6252 (September 4), pp. 1095–1100, http://science.sciencemag.org/content/349/6252/1095.

Gates, Guilbert, Ewing, Jack, Russell, Karl, and Watkins, Derek. (2017). "How Volkswagen's 'Defeat Devices' Worked," *New York Times*, updated March 16, https://www.nytimes.com/interactive/2015/business/international/vw-diesel -emissions-scandal-explained.html.

Gawande, Atul. (2017). "It's Time to Adopt Electronic Prescriptions for Opioids," *Annals of Surgery*, Vol. 265, No. 4 (April), pp. 693–694, http://journals.lww .com/annalsofsurgery/Fulltext/2017/04000/It_s_Time_to_Adopt_Electronic _Prescriptions_for.12.aspx.

Gillum, Jack D. (2000). "The Engineer of Record and Design Responsibility," *Journal of Performance of Constructed Facilities*, Vol. 14, No. 2 (May), pp. 67–70.

Goldberger, Paul. (1982). "Design Change Cited in Hyatt Disaster," *New York Times*, February 26, http://www.nytimes.com/1982/02/26/us/design-change -cited-in-hyatt-disaster.html.

Graham, John D. (1991). "Product Liability and Motor Vehicle Safety," ch. 4 in *The Liability Maze: The Impact of Liability Law on Safety and Innovation*, ed. Peter W. Huber and Robert E. Litany (Brookings Institution Press: Washington, DC), pp. 120–190.

Guynn, Jessica. (2013). "Google Cuts $7-Million Settlement with States over Street View," *Los Angeles Times*, March 12, http://articles.latimes.com/2013 /mar/12/business/la-fi-tn-google-cuts-7million-settlement-with-states-over -street-view-20130312.

Harmon, Amy. (2017). "Human Gene Editing Receives Science Panel's Support," *New York Times*, February 14, https://www.nytimes.com/2017/02/14/health /human-gene-editing-panel.html.

Harris, Charles E., Pritchard, Michael S., Robins, Michael J., James, Ray, and Englehardt, Elaine. (2014). *Engineering Ethics: Concepts and Cases*, 5th ed. (Cengage Learning: Independence, KY).

Haskins, Paul J. (1983). "Collapse of Hotel's 'Skywalks in 1981 Is Still Reverberating in Kansas City," *New York Times*, March 29, http://www.nytimes.com /1983/03/29/us/collapse-of-hotel-s-skywalks-in-1981-is-still-reverberating-in -kansas-city.html?pagewanted=all.

Hayes, Dennis. (1989). *Behind the Silicon Curtain: The Seductions of Work in a Lonely Era* (South End Press: New York).

Hazarika, Sanjoy. (1989). "Bhopal Payments by Union Carbide Set at $470 Million," *New York Times*, February 15, http://www.nytimes.com/1989/02/15 /business/bhopal-payments-by-union-carbide-set-at-470-million.html.

Helft, Miguel. (2007). "Google Zooms In Too Close for Some," *New York Times*, June 1, http://www.nytimes.com/2007/06/01/technology/01private.html.

Hoffman, W. Michael. (1984). "The Ford Pinto," in *Business Ethics*, ed. W. Michael Hoffman and Jennifer Moore (New York: McGraw-Hill), pp. 412–420.

Hormby, Tom. (2010). "The Story behind Apple's Newton," *Gizmodo*, January 19, http://gizmodo.com/5452193/the-story-behind-apples-newton.

Huber, Peter W., and Litan, Robert E., eds. (1991). *The Liability Maze: The Impact of Liability Law on Safety and Innovation* (Brookings Institution Press: Washington, DC).

Hunt, Terence. (1986). "NASA Suggested Reagan Hail *Challenger* Mission in State of the Union," *AP News Archive*, March 12, http://www.apnewsarchive

.com/1986/NASA-Suggested-Reagan-Hail-Challenger-Mission-In-State-of
-Union/id-00a395472559b3afcd22de473da2e65f.

Ibrahim, Youssef M. (1989). "Successors Ready: U.S. Oilmen Bow Out of Their
Saudi Empire," *New York Times*, April 1, http://www.nytimes.com/1989/04
/01/world/successors-ready-us-oilmen-bow-out-of-their-saudi-empire.html
?pagewanted=all&src=pm.

International Committee of Medical Journal Editors. (2013). "Recommendations
for the Conduct, Reporting, Editing, and Publication of Scholarly Work in
Medical Journals," http://www.icmje.org/icmje-recommendations.pdf.

Ivory, Danielle. (2014). "G.M. Reveals It Was Told of Ignition Defect in '01,"
New York Times, March 12, http://www.nytimes.com/2014/03/13/business
/gm-reveals-it-was-told-of-ignition-defect-in-01.html?module=Search&
mabReward=relbias%3Ar%2C%7B%221%22%3A%22RI%3A6%22%7D.

Jackson, Ronald J., Ramsay, Alistair J., Christensen, Carina D., Beaton, Sandra,
Hall, Diane F., and Ramshaw, Ian A. (2001). "Expression of Mouse Interleu-
kin-4 by a Recombinant Ectromelia Virus Suppresses Cytolytic Lymphocyte
Responses and Overcomes Genetic Resistance to Mousepox," *Journal of Virology*, Vol. 75, No. 3 (February), pp. 1205–1210.

Jasanoff, Sheila. (2007). "Bhopal's Trials of Knowledge and Ignorance," *Isis*, Vol.
98, pp. 344–350.

Johnson, Deborah G. (1991). *Ethical Issues in Engineering* (Prentice-Hall: Englewood Cliffs, NJ).

———. (1993). *Computer Ethics*, 2nd ed. (Prentice Hall: Englewood Cliffs, NJ).

Johnson, Samuel. (1759). *The History of Rasselas, Prince of Abissinia*, Ch. 41,
http://andromeda.rutgers.edu/~jlynch/Texts/rasselas.html.

Joy, Bill. (2000). "Why the Future Doesn't Need Us," *Wired*, Vol. 8, No. 4 (April).

Kaiser, Jocelyn. (2017). "U.S. Panel Gives Green Light to Human Embryo Editing," *Science*, February 14, http://www.sciencemag.org/news/2017/02/us-panel
-gives-yellow-light-human-embryo-editing.

Kaplan, David, and Manners, Robert A. (1972). *Culture Theory* (Prentice-Hall:
Englewood Cliffs, NJ).

Kerzner, Harold. (2013). *Project Management: Case Studies*, 4th ed. (Wiley:
Hoboken).

Ketcham, Brian T. (1977). "The Auto and New York City's Decline," *New
York Times*, Letter to the editor, October 11, www.nytimes.com/1977/10/11
/archives/letter-on-westway.html.

Knigge, Volkhard. (2005). "Innocent Ovens," in the Exhibition Catalogue, "Engineers of the 'Final Solution': Topf and Sons—Builders of the Ovens at Auschwitz," (Weimar), http://www.topfundsoehne.de/download/medienmappe_ef
_en.pdf.

Kuhn, Thomas S. (1970). *The Structure of Scientific Revolutions*, 2nd ed. (University of Chicago Press: Chicago).

Landivar, Liana Christin. (2013). "Disparities in STEM Employment by Sex, Race,
and Hispanic Origin," September, http://www.census.gov/prod/2013pubs/acs
-24.pdf.

Lapierre, Dominique, and Moro, Javier. (2002). *Five Past Midnight in Bhopal*
(Warner Books: New York).

Leary, Warren E. (1989). "NASA Picks Lockheed and Aerojet," *New York Times*, April 22, http://www.nytimes.com/1989/04/22/business/nasa-picks-lockheed -and-aerojet.html.

Lederman, Leon M. (1999). "The Responsibility of the Scientist," *New York Times*, July 24.

LePage, Michael. (2015). "Home-Brew Heroin: Soon Anyone Will Be Able to Make Illegal Drugs," *New Scientist*, May 18, https://www.newscientist.com/article /dn27546-home-brew-heroin-soon-anyone-will-be-able-to-make-illegal-drugs/.

Leugenbiehl, Heinz C., and Davis, Michael. (1992). "Engineering Codes of Ethics: Analysis and Applications," http://ethics.iit.edu/publication/CODE—Exxon %20Module.pdf.

Levine, Richard. (1989). "West Side Highway's Demise: Nightmare for New York City Drivers," *New York Times*, January 8.

Lewis, Sandford. (2004). "The Bhopal Chemical Disaster: 20 Years without Justice," https://www.youtube.com/watch?v=0csW97x8d24#t=12.

Lohr, Steve, and Streitfeld, David. (2012). "Data Engineer in Google Case Is Identified," *New York Times*, April 30. http://www.nytimes.com/2012/05/01 /technology/engineer-in-googles-street-view-is-identified.html.

Luban, David. (1988). *Lawyers and Justice: An Ethical Study* (Princeton University Press: Princeton, NJ).

Lubasch, Arnold H. (1983). "Possible Conflict of Interest Cited On the Westway," *New York Times*, March 18, pp. B1, B5.

———. (1985a). "Westway Project Is Blocked Again by Federal Judge," *New York Times*, August 8.

———. (1985b). "U.S. Appeals Court Upholds Decision to Halt Westway," *New York Times*, September 12.

Markoff, John. (1993). "Marketer's Dream, Engineer's Nightmare," *New York Times*, December 12.

Marshall, R. D., Pfrang, E. O., Leyendecker, E. V., Woodward, K. A., Reed, R. P., Kasen, M. B., and Shives, T. R. (1982). *Investigation of the Kansas City Hyatt Regency Walkways Collapse.* (Washington, DC: UNT Digital Library), http:// ws680.nist.gov/publication/get_pdf.cfm?pub_id=908286.

Martin, Mike, and Schinzinger, Roland. (2005). "Engineering as Social Experimentation," *Ethics in Engineering*, 4th ed. (McGraw Hill: New York), pp. 88–116.

Masci, David. (2016). "Human Enhancement: The Scientific and Ethical Dimensions of Striving for Perfection," Pew Research Center, July 26, http:// www.pewinternet.org/essay/human-enhancement-the-scientific-and-ethical -dimensions-of-striving-for-perfection/.

Mazurka, Sanjoy. (1989). "Bhopal Payments by Union Carbide Set at $470 Million," *New York Times*, February 15.

McCracken, Harry. (2012). "Newton, Reconsidered," *Time*, June 1, http:// techland.time.com/2012/06/01/newton-reconsidered/.

McGinn, Robert (1990). *Science, Technology, and Society* (Prentice-Hall: Englewood Cliffs, NJ).

———. (1995). "The Engineer's Moral Right to Reputational Fairness," *Science and Engineering Ethics*, Vol. 1, No. 3, pp. 217–230.

————. (1997). "Optimization, Option Disclosure, and Problem Redefinition: Derivative Moral Obligations of Engineers and the Case of the Composite-Material Bicycle," *Professional Ethics*, Vol. 6, No. 1, pp. 5–25.

————. (2003). "'Mind the Gaps': An Empirical Approach to Engineering Ethics, 1997–2001," *Science and Engineering Ethics*, Vol. 9, No. 3, pp. 517–542.

————. (2008). "Ethics and Nanotechnology: Views of Nanotechnology Researchers," *Nanoethics*, Vol. 2, No. 2 (July), pp. 101–131.

————. (2010). "Ethical Responsibilities of Nanotechnology Researchers: A Short Guide," *Nanoethics*, Vol. 4, No. 1 (April), pp. 1–12.

————. (2013). "Discernment and Denial: Nanotechnology Researchers' Recognition of Ethical Responsibilities Related to Their Work," *Nanoethics*, Vol. 7, No. 2 (August), pp. 93–105.

Meckl, Peter H. (2003). "Integrating Ethics into an Undergraduate Control Systems Course," http://ethics.iit.edu/eac/post_workshop/Ethics%20in%20Control.pdf, pp. 1–4.

Mehta, Suketu. (2009). "A Cloud Still Hangs over Bhopal," *New York Times*, December 3.

Miller, Seumas, and Selgelid, Michael J. (2007). "Ethical and Philosophical Consideration of the Dual-Use Dilemma in the Biological Sciences," *Science and Engineering Ethics*, Vol. 13, pp. 523–580.

Mohr, Charles. (1985). "Scientist Quits Antimissile Panel, Saying Task Is Impossible," *New York* Times, July 12, http://www.nytimes.com/1985/07/12/world/scientist-quits-antimissile-panel-saying-task-is-impossible.html.

Moncarz, Piotr D., and Taylor, Robert. (2000). "Engineering Process Failure—Hyatt Walkway Collapse," *Journal of Performance of Constructed Facilities*, Vol. 14, No. 2 (May), pp. 46–50.

Morgenstern, Joe. (1995). "The Fifty-Nine Story Crisis," *The New Yorker*, May 29, pp. 45–53. Reprinted in *Journal of Professional Issues in Engineering Education and Practice* (1997), Vol. 123, No. 1, pp. 23–29.

Murray, Douglas L. (1982). "The Abolition of *El Cortito*, the Short-Handled Hoe: A Case Study in Social Conflict and State Policy in California Agriculture," *Social Problems*, Vol. 30, No. 1 (October), pp. 26–39.

National Commission for the Protection of Human Subjects of Biomedical and Behavioral Research. (1979). *The Belmont Report: Ethical Principles and Guidelines for the Protection of Human Subjects of Research*, April 18, https://www.hhs.gov/ohrp/regulations-and-policy/belmont-report/.

National Nanotechnology Infrastructure Network. (2008). http://www.nnin.org/education-training/graduate-professionals/international-winter-school/prior-years/2008.

National Nanotechnology Initiative. (2016). http://www.nano.gov/nanotech-101/what/nano-size.

National Society of Professional Engineers. (2007). "NSPE Code of Ethics," July, http://www.nspe.org/Ethics/CodeofEthics/index.html.

Newcombe, Ken. (1981). "Technology Assessment and Policy: Examples from Papua New Guinea," *International Social Science Journal*, Vol. 33, No. 3, p. 501.

O'Brien, Jennifer. (2010). "Ritalin Boosts Learning by Increasing Brain Plasticity," *New York Times*, March 8, https://www.ucsf.edu/news/2010/03/4376/ritalin -boosts-learning-increasing-brain-plasticity.

O'Brien, Kevin. (2009). "Google Threatened with Sanctions over Photo Mapping Service in Germany," *New York Times*, May 19, http://www.nytimes.com/2009/05/20/technology/companies/20google.html.

———. (2010). "244,000 Germans Opt Out of Google Mapping Service," *New York Times*, October 21, http://www.nytimes.com/2010/10/21/technology /21google.html.

O'Brien, Kevin, and Streitfeld, David. (2012). "Swiss Court Orders Modifications to Google Street View," *New York Times*, June 8, http://www.nytimes.com /2012/06/09/technology/09iht-google09.html.

Oreskes, Michael. (1985). "Westway Funds Trade-In Wins Federal Approval," *New York Times*, October 1, http://www.nytimes.com/1985/10/01/nyregion /westway-funds-trade-in-wins-federal-approval.html.

Oye, Kenneth A., Lawson, J. Chappell H., and Bubela, Tania. (2015). "Drugs: Regulate 'Home-Brew' Opiates," *Nature*, Vol. 521, Issue 7552 (May 18), http://www.nature.com/news/drugs-regulate-home-brew-opiates-1.17563.

Parnas, David Lorge. (1985). "Software Aspects of Strategic Defense Systems," *Communications of the ACM*, Vol. 28, No. 12, pp. 1326–1335; originally printed in *American Scientist*, Vol. 73, No. 5, 432–440.

———. (1987). "SDI: A Violation of Professional Responsibility," *Abacus*, Vol. 4, No. 2, pp. 46–52.

Peplow, Mark. (2016). "A Conversation with Christina Smolke," *ACS Central Science*, Vol. 2, No. 2, pp. 57–58. http://pubs.acs.org/doi/full/10.1021/acscentsci .6b00029.

Perlman, David. (2015). "Stanford Scientists Raise Hope, Concern with Synthetic Narcotics," *San Francisco Chronicle*, September 7, http://www.sfchronicle.com /science/article/Stanford-scientists-brew-synthetic-opiates-from-6489576.php.

Perlmutter, Emanuel. (1973). "Indefinite Closing Is Set for West Side Highway," *New York Times*, December 17.

Peterson, M. J. (2009a). "Bhopal Plant Disaster: Situation Summary," International Dimensions of Ethics Education in Science and Engineering, Science, Technology, and Society Initiative, University of Massachusetts, Amherst, March 20, pp. 1–8, http://scholarworks.umass.edu/edethicsinscience/4/; click on Bhopal_Summary.pdf.

———. (2009b). "Bhopal Plant Disaster—Appendix A: Chronology," February 26, pp. 1–19, http://scholarworks.umass.edu/edethicsinscience/4/; click on Bhopal_AChrono.pdf.

Petroski, Henry. (1992). *To Engineer Is Human* (Vintage: New York).

Pierre-Pierre, Garry. (1996). "After Twenty Years, Work Begins on Far Less Ambitious Westway," *New York Times*, April 2.

Pinkus, Rosa Lynn B., Shuman, Larry J., Hummon, Norman P., and Wolfe, Harvey. (1997). *Engineering Ethics: Balancing Cost, Schedule, and Risk—Lessons Learned From the Space Shuttle* (Cambridge University Press: Cambridge).

Polanyi, Michael. (1962). *Personal Knowledge* (Chicago: University of Chicago Press).

Polgreen, Lydia, and Kumar, Hari. (2010), "8 Former Executives Guilty in '84 Bhopal Chemical Leak," *New York Times*, June 7, http://www.nytimes.com /2010/06/08/world/asia/08bhopal.html.

Postol, Theodore A. (2002). "Why Missile Defense Won't Work," *Technology Review*, April, https://www.technologyreview.com/s/401407/why-missile-defense -wont-work/.

Postrel, Virginia. (2005). "In Silicon Valley, Job-Hopping Contributes to Innovation," *New York Times*, December 1, http://www.nytimes.com/2005/12/01 /business/in-silicon-valley-job-hopping-contributes-to-innovation.html.

Prüfer, Karl. (1942). Topf & Sons in-house memo, September 8, http://www .holocaust-history.org/auschwitz/topf/.

Raikow, David. (1993). "SDIO Changes Its Letterhead to BMDO," *Arms Control Today*, Vol. 23, June, p. 31, http://www.bits.de/NRANEU/docs/SDIO-BMDO .htm.

Rawls, John. (1999). *A Theory of Justice*, rev. ed. (Harvard University Press: Cambridge, MA).

Reich, Eugenie Samuel. (2009). *Plastic Fantastic* (Palgrave Macmillan: New York).

Reinhold, Robert. (1985). "Disaster in Bhopal: Where Does the Blame Lie?" *New York Times*, January 31.

Resnick, David. (2014). "What Is Ethics in Research and Why Is It Important?" http://www.niehs.nih.gov/research/resources/bioethics/whatis/index.cfm.

Reynolds, Terry. (1991). "The Engineer in Twentieth-Century America," in *The Engineer in America*, ed. Terry Reynolds (U. of Chicago Press: Chicago), pp. 169–190.

Robbins, William. (1985). "Engineers Are Held at Fault in '81 Hotel Disaster," *New York Times*, November 16, http://www.nytimes.com/1985/11/16/us /engineers-are-held-at-fault-in-81-hotel-disaster.html.

Roberts, Sam. (1983). "Army's Engineers to Study Westway for 2 More Years," *New York Times*, September 14, http://www.nytimes.com/1983/09/14 /nyregion/army-s-engineers-to-study-westway-for-2-more-years.html.

———. (1984a). "Battle of the Westway: Bitter 10-Year Saga of a Vision on Hold," *New York Times*, June 4, http://www.nytimes.com/1984/06/04 /nyregion/battle-of-the-westway-bitter-10-year-saga-of-a-vision-on-hold .html?pagewanted=all.

———. (1984b). "For Stalled Westway, a Time of Decision," *New York Times*, June 5, http://www.nytimes.com/1984/06/05/nyregion/for-stalled-westway-a -time-of-decision.html?pagewanted=all.

———. (1984c). "Hearings Open on Whether Westway Is 'Imperative' or Is a 'White Elephant,'" *New York Times*, June 27, p. B1, http://www.nytimes.com /1984/06/27/nyregion/hearings-open-on-whether-westway-is-imperative-or-is -a-white-elephant.html?pagewanted=all.

———. (1985). "Westway Landfill Wins the Support of Army Engineer," *New York Times*, January 25, http://www.nytimes.com/1985/01/25/nyregion/westway -landfill-wins-the-support-of-army-engineer.html?pagewanted=all.

Rogers, William P. (1986). *Report of the Presidential Commission on the Space Shuttle Challenger Accident* (Government Printing Office: Washington, DC), http://history.nasa.gov/rogersrep/genindex.htm.

Romero, Simon. (2000). "How a Byte of Knowledge Can Be Dangerous, Too," *New York Times*, April 23, http://www.nytimes.com/2000/04/23/weekinreview/the-world-when-villages-go-global-how-a-byte-of-knowledge-can-be-dangerous-too.html.

Rosenbaum, Eli. (1993). "German Company Got Cremation Patent," *New York Times*, July 27, 1993.

Royal Society and the Royal Academy of Engineering. (2004). *Nanoscience and Nanotechnologies: Opportunities and Uncertainties*, RS Policy Document 19/04, July.

Schanberg, Sydney H. (1983). "A Wake for Westway," *New York Times*, March 22.

———. (1984). "Westway's Sleaze Factor," *New York Times*, October 9.

Schwartz, Gary T. (1991). "The Myth of the Ford Pinto Case," *Rutgers Law Review*, Vol. 43, pp. 1013–1068.

Selgelid, Michael, and Weit, Loma. (2010). "The Mousepox Experience: An Interview with Ronald Jackson and Ian Ramshaw on Dual-Use Research," *EMBO Reports*, Vol. 11, No. 1 (January), pp. 18–24.

Service, Robert F. (2015). "Powerful Painkillers Can Now Be Made by Genetically Modified Yeast—Are Illegal Drugs Next?" *sciencemag.org/news*, August 13, http://www.sciencemag.org/news/2015/08/powerful-painkillers-can-now-be-made-genetically-modified-yeast-are-illegal-drugs-next.

Severy, Derwyn M., Brink, Harrison M., and Baird, Jack D. (1968). "Vehicle Design for Passenger Protection from High-Speed Read-End Collision," *SAE Transactions*, Vol. 77-A, Technical Paper No. 680744, pp. 3069–3107.

Shallit, Jeffrey. (2005). "Science, Pseudoscience, and the Three Stages of Truth," https://cs.uwaterloo.ca/~shallit/Papers/stages.pdf.

Shepardson, David. (2015). "GM Ignition Fund Ends Review Approving 124 Death Claims," *Detroit News* (Washington Bureau), August 3. http://www.detroitnews.com/story/business/autos/general-motors/2015/08/03/gm-ignition-fund-ends-review-approving-death-claims/31051683/.

Shetty, Salil. (2014). "Thirty Years On from Bhopal Disaster: Still Fighting for Justice," *Amnesty International*, December 2, https://www.amnesty.org/en/latest/news/2014/12/thirty-years-bhopal-disaster-still-fighting-justice/.

Sinsheimer, Robert. (1978). "The Presumptions of Science," in *Daedalus*, Vol. 107, No. 2 (Spring), pp. 23–35, http://www.jstor.org/stable/20024542.

Singh, Vijaita. (2016). "India Says 'No' to Google Street View," *The Hindu*, June 9, updated October 18,. http://www.thehindu.com/sci-tech/technology/India-says-%E2%80%98no%E2%80%99-to-Google-Street-View/article14414575.ece.

Sparing, Roland, Dafotakis, Manuel, Meister, Ingo G., Thirugnanasambandam, Nivethida, and Fink, Gereon R. (2008). "Enhancing Language Performance with Non-Invasive Brain Stimulation—A Transcranial Direct Current Stimulation Study in Healthy Humans," *Neuropsychologia*, Vol. 46, pp. 261–268, https://www.researchgate.net/profile/Ingo_Meister/publication/6036489_Enhancing_language_performance_with_non-invasive_brain_stimulation_-_A_transcranial_direct_current_stimulation_study_in_healthy_humans/links/00463514ca1bdddb63000000.pdf.

State of Connecticut, Office of the Attorney General. (2013). "Attorney General Announces $7 Million Multistate Settlement with Google over Street View Collection of WiFi Data," http://www.ct.gov/ag/cwp/view.asp?Q=520518.

State of New York Commission of Investigation. (1984). *The Westway Environmental Approval Process: The Dilution of State Authority*, May.

State of New York Department of Transportation. (2011). "Lower Manhattan Redevelopment: 9A/West Street Promenade," https://www.dot.ny.gov/content/delivery/Main-Projects/projects/x75984-home/x75984-repository/9A%20-%20West%20Street%20Promenade%20Summer%202011%20Newsletter.pdf.

State of Washington Superior Court for Benton County. (2010). *Walter L. Tamosaitis, Ph.D., an individual, and Sandra B. Tamosaitis, Plaintiffs, vs. Bechtel National, Inc., a Nevada Corporation, URS Corporation, a Nevada Corporation, Frank Russo, an individual, Gregory Ashley, an individual, William Gay, an individual, Dennis Hayes, an individual, and Cami Krumm, an individual, Defendants*, Case No.: 10–2-02357–4, http://ehstoday.com/site-files/ehstoday.com/files/archive/ehstoday.com/images/Tamosaitis-lawsuit.pdf.

Stevens, Jane Ellen. (2002). "Martin Makes a Middle Class," *San Francisco Chronicle*, December 8, http://www.sfgate.com/magazine/article/Martin-Makes-a-Middle-Class-Stanford-grad-2747565.php.

Stix, Gary. (1989). "Bhopal: A Tragedy in Waiting," *IEEE Spectrum*, Vol. 26 (June), pp. 47–50.

Stout, David. (2006). "Justices Set Limits on Public Employees' Speech Rights," *New York Times*, May 30, http://www.nytimes.com/2006/05/30/washington/30cnd-scotus.html.

Streitfeld, David. (2013). "Google Concedes That Drive-By Prying Violated Privacy," *New York Times*, March 12, http://www.nytimes.com/2013/03/13/technology/google-pays-fine-over-street-view-privacy-breach.html?pagewanted=all&_r=0.

Strobel, Lee Patrick. (1979). "Ford Ignored Pinto Fire Peril, Secret Memos Show," *Chicago Tribune*, October 13, http://archives.chicagotribune.com/1979/10/13/page/43 and http://archives.chicagotribune.com/1979/10/13/page/284.

———. (1980). *Reckless Homicide? Ford's Pinto Trial* (And Books: South Bend, IN).

———. (1994). "The Pinto Documents," Ch. 4 in *The Ford Pinto Case: A Study in Applied Ethics, Business, and Technology*, ed. Douglas Birsch and John H. Fielder (State University of New York Press: Albany), pp. 41–53.

Stuart, Reginald. (1980a). "Prosecutor in Ford Pinto Case Raises Possibility of Moving for a Mistrial," *New York Times*, February 7, http://query.nytimes.com/mem/archive/pdf?res=F60D1EFA345D11728DDDAE0894DA405B8084F1D3.

———. (1980b). "Ford Ends Pinto Defense; 17 Testified for Company," *New York Times*, March 4, http://timesmachine.nytimes.com/timesmachine/1980/03/04/issue.html.

Texas A&M University, Department of Philosophy and Department of Mechanical Engineering. (1993a). "The Kansas City Hyatt Regency Walkways Collapse," NSF Grant Number DIR-9012252, http://ethics.tamu.edu/Portals/3/Case%20Studies/HyattRegency.pdf.

———. (1993b). "The Space Shuttle *Challenger* Disaster," NSF Grant Number DIR-9012252, http://ethics.tamu.edu/Portals/3/Case%20Studies/Shuttle.pdf.

Twilley, Nicola. (2015). "Home-Brewed Heroin," *The New Yorker*, May 18, http://www.newyorker.com/tech/elements/home-brewed-heroin.

Tyabji, Nasir. (2012). "MIC at Bhopal and Virginia and the Indian Nuclear Liability Act," *Economic and Political Weekly*, Vol. 47, No. 41 (October 13), pp. 41–50), http://mpra.ub.uni-muenchen.de/49609/1/MPRA_paper_49609.pdf.

Unger, Stephen H. (1994). *Controlling Technology: Ethics and the Responsible Engineer*, 2nd ed. (John Wiley & Sons: New York).

———. (1999). "The Assault on IEEE Ethics Support," *IEEE Technology and Society Magazine* (Spring), pp. 36–40, http://www1.cs.columbia.edu/~unger/articles/assault.html.

Union of India, "Affidavit on Behalf of Union of India" (2006), re *Union Carbide Corporation, Petitioners, vs. Union of India and Others*, Supreme Court of India, Civil Appellate Jurisdiction, Civil Appeal No. 3187–3188 of 1998, October 26, http://storage.dow.com.edgesuite.net/dow.com/Bhopal/Aff%2026Oct06%20of%20UOI%20in%20SC%20in%20CA3187%20n%203188.pdf.

U.S. District Court, Central District of California. (2001). *United States of America ex rel Nira Schwartz, Plaintiff v. TRW, Inc., an Ohio Corporation, and Boeing N. America, a Delaware Corporation*, Defendants, Plaintiff's Fourth Amended Complaint, Case No. CV 96–3605 CM (RMCx), http://web.archive.org/web/20010613160512/http://fas.org/spp/starwars/program/news01/schwartz.pdf.

U.S. Environmental Protection Agency. (2012). "Basic Information," http://www.epa.gov/compliance/basics/nepa.html.

U.S. Federal Communications Commission. (2012). "In the Matter of Google Inc.: Notice of Apparent Liability for Forfeiture," April 13, https://www.scribd.com/fullscreen/91652398.

U.S. Merit Systems Protection Board. (2011). "Blowing the Whistle: Barriers to Federal Employees Making Disclosures," http://www.mspb.gov/netsearch/viewdocs.aspx?docnumber=662503&version=664475&application=ACROBAT.

Van Alstine, Mark. (1997). "The Cremation Process," http://www.nizkor.org/hweb/camps/auschwitz/crematoria/cremation-009.html.

Vandivier, Kermit. (1972). "Why Should My Conscience Bother Me?" in *In the Name of Profit*, ed. R. Heilbroner (Doubleday: New York), pp. 3–31.

Varma, Roli, and Varma, Daya R. (2005). "The Bhopal Disaster of 1984," *Bulletin of Science, Technology & Society*, Vol. 25, No. 1 (February), pp. 37–45.

Vartabedian, Ralph. (2015). "Hanford Nuclear Weapons Site Whistle-blower Wins $4.1-Million Settlement," *Los Angeles Times*, http://www.latimes.com/nation/la-na-hanford-whistleblower-settlement-20150813-story.html.

Vaughan, Diane. (1996). *The* Challenger *Launch Decision: Risky Technology, Culture and Deviance* (University of Chicago Press: Chicago).

Venice Sustainability Advisory Panel. (2009). "Final Report," (June), http://tenaya.ucsd.edu/~dettinge/VeniceReport.pdf.

Vincenti, Walter G. (1992). "Engineering Knowledge, Type of Design, and Level of Hierarchy: Further Thoughts about What Engineers Know," in *Technological Development and Science in the Industrial Age*, ed. P. Kroes and M. Bakker (Kluwer Academic: Dordrecht), pp. 19–21.

Viscusi, W. Kip (1991). *Reforming Products Liability* (Harvard University Press: Cambridge, MA).

Wahl, Volker. (2008). "Thuringia under American Occupation (April until July 1945)," http://www.lzt-thueringen.de/files/huringia_under_american _occupation.pdf.

Weisman, Steven R., with Hazarika, Sanjoy (1987). "Theory of Bhopal Sabotage Is Offered," *New York Times*, June 23, http://www.nytimes.com/1987/06/23 /world/theory-of-bhopal-sabotage-is-offered.html?pagewanted=all&src=pm.

Wenn, J. M., and A. Horsfield. (1990). "The Transfer of Technology Occasioned by the Senegal Village Water-Supply Project," *Water and Environment Journal*, Vol. 4, Issue 2 (April), pp. 148–153.

Whitaker, L. Paige. (2007). "The Whistleblower Protection Act: An Overview," *Congressional Research Service*, March 12, pp. 1–8. http://digitalcommons.ilr .cornell.edu/cgi/viewcontent.cgi?article=1031&context=crs.

Whitbeck, Caroline. (2011a). "Addendum: The Diane Hartley Case," *Online Ethics Center for Engineering* 1/14/2011 OEC Accessed Wednesday, March 15, 2017, www.onlineethics.org/Topics/ProfPractice/Exemplars/BehavingWell /lemesindex/DianeHartley.aspx.

———. (2011b). *Ethics in Engineering Practice and Research*, 2nd ed., (Cambridge University Press: New York).

Willey, Ronald. (2014). "Consider the Role of Safety Layers in the Bhopal Disaster," *CEP*, December, https://www.aiche.org/sites/default/files/cep/20141222_1 .pdf.

Walsh, Mary Williams. (1994). "Holocaust Haunts Heirs' Bid," *Los Angeles Times*, September 2, http://articles.latimes.com/1994-09-02/news/mn-33857 _1_east-germany.

Wuchty, Stefan, Jones, Benjamin F., and Uzzi, Brian. (2007). "The Increasing Dominance of Teams in Production of Knowledge," *Science*, Vol. 316, No. 5827 (May 18), pp. 1036–1039, http://www.kellogg.northwestern.edu/faculty /jones-ben/htm/Teams.ScienceExpress.pdf.

INDEX